CHAOS AND FRACTALS IN CHEMICAL ENGINEERING

Proceedings of the First National Conference

CHAOS AND FRACTALS IN CHEMICAL ENGINEERING

Rome, Italy 25 – 27 May 1994

Editors

G Biardi

M Giona

AR Giona

Università di Roma "La Sapienza"

World Scientific

Singapore • New Jersey • London • Hong Kong

Published by

World Scientific Publishing Co. Pte. Ltd.

P O Box 128, Farrer Road, Singapore 9128

USA office: Suite 1B, 1060 Main Street, River Edge, NJ 07661

UK office: 57 Shelton Street, London WC2H 9HE

CHAOS AND FRACTALS IN CHEMICAL ENGINEERING

ISBN: 981-02-1904-0

Printed in Singapore.

PREFACE

The InterUniversity Research Center on Disordered Systems and Fractals in Chemical Engineering was founded by the universities of Cagliari Genova and Roma and the Politechnic Institute of Milano to promote, develop and coordinate research on the dynamics and control of chaotic systems, on transport and kinetics in disordered and porous media, and on the application of fractal geometry to problems of interest in chemical engineering.

The first step taken by the Center was to organize a national confererence on Chaos and Fractals in Chemical Engineering (Roma, May 25-27 1994) with the aim of meeting all the Italian researchers involved in physical and applied problems in which the application of fractal models and chaotic dynamics represents a useful tool of investigation and analysis.

The interest aroused within the national community of chemical engineers by this conference and the quality of the articles presented prompted the Scientific Committee to publish the papers delivered after peer-reviewing.

This volume collects the accepted manuscripts, divided into the main areas developed in the conference.

My hope is that in the near future our Center will be able to gather together international researchers to organize an international conference aimed at stimulating a constructive exchange of ideas on fundamental problems of chemical engineering within the community. These topics involve the future cultural development of our discipline.

Finally, I thank the CNR (Consiglio Nazionale delle Ricerche, Comitato per la Chimica) for its contribution towards the publication of this volume.

Prof. Alessandro R. Giona
Chairman of the Center

CONTENTS

Chaos in Chemical Process Units

Time-Series Analysis and Neural Networks

Disordered Systems and Fractals

As noted by David Avnir (in the preface to the book *The Fractal Approach to Heterogeneous Chemistry*, J. Wiley, New York, 1989): "*The main contribution of fractal geometry to modern scientific thinking is not so much through the development of useful mathematical tools, as in helping many scientists to overcome the psycological barriers of treating problems which involve very complex geometries. Suddenly, non-trivial geometries could be approached in terms of defining questions for which probable routes to solutions could be identified and suggested. So, regardless of whether fractal geometry will eventually emerge as the preferred approach to problems of complex geometries or some other, more refined tools will be added, the main achievement of fractal geometry is already here: It has initiated a massive scientific effort of looking at problems, some very old, others very modern, of the interplay between natural and man-made shapes and structures, and the processes which form these structures or the processes which take place in the enviroments created by these structures.*"

The analysis of anomalous features of dynamic phenomena in fractal structures is certainly one of the most fascinating topics in the physical theory of fractals with relevant engineering applications.

In the field of diffusion, this research originated with the work of P.G. de Gennes (*La Recherche* 7, 1976, 919) related to random walks on percolation lattices at criticality (the famous problem of *the ant in the labyrinth*), and now is widely studied in applied sciences in connection with transport phenomena, porous media, catalyst structure and properties.

Of course the field is even wider than purely diffusive phenomena and encompasses all the branches of chemical physics and kinetics: rheology, thermodynamics, aggregation phenomena etc., as indicated by the selection of articles presented in this section.

A NUMERICAL LATTICE SIMULATOR OF TRANSPORT IN DISORDERED SYSTEMS

MASSIMILIANO GIONA
Dipartimento di Ingegneria Chimica, Universitá di Cagliari
Piazza d'Armi, 09123 Cagliari, ITALY

and

ALESSANDRA ADROVER
Dipartimento di Ingegneria Chimica, Universitá di Roma "La Sapienza
Via Eudossiana 18, 00184 Roma, ITALY

ABSTRACT

We analyze the various methods for the determination of transport parameters starting from lattice models of disordered structures. The exit-time equation and sorption-curve analysis are introduced to determine diffusion and dispersion parameters in percolation clusters and in mixed disordered systems. The problem of biased diffusion (estimation of permeability) is addressed by means of lattice Boltzmann simulations.

1. Introduction

The analysis of transport phenomena in disordered and fractal structures has mainly been developed starting from the equivalence between transport and probabilistic models[1,2] (random walk simulations). As an example, from the temporal behaviour of the mean square displacement $< r^2(t) >$ of a Brownian particle it is possible to obtain in regular diffusion an estimate of the diffusion coefficient D

$$< r^2(t) >= 2Dt \, , \tag{1}$$

or the value of the anomalous diffusion coefficient D_w

$$< r^2(t) >\sim t^{2/D_w} \, , \tag{2}$$

$D_w \geq 2$, if the random walk exhibits anomalies due to geometrical and topological disorder of the structure.

The application of similar methods to biased transport, i.e. in the presence of convective phenomena, has not furnished conclusive results[3,4,5]

We thus need practical and theoretical tools to analyze generic transport conditions in disordered media.

Before approaching this aspect, there is a fundamental preliminary question to be answered: in dealing with chemical engineering problems why do need to study transport and kinetics in disordered structures, instead of confining ourselves to application the of "good old" semiempirical correlations that our fathers in science have developed since the birth of chemical engineering as a autonomous discipline?

In answering this *philosophycal* problem it should be borne in mind that practically all the industrial processes making use of porous adsorbents and catalysts involve disordered porous structures or disordered distribution of active sites[6]. In the near future, the challenge in chemical engineering will lie not only in the development of new processes but also in the design of new microporous materials with specified transport and kinetic performances. We are at an important crossroad in chemical engineering: we can remain confined to our historical background of "tubes, pipes and stirred pots", as chemical engineering is usually depicted, or open ourselves to the broader scientific community of science (chemistry, physics, mathematics, biology) and apply our efforts to the design of microscopic structures similar to those developed in electronics with VLSI, waveguide and nanostructure technology, in which the detailed mathematical-physical modelling of the phenomenon (propagation of the electromagnetic field) finds its counterpart in a real technological progress and in practical spinn-offs for everyday life.

For these reasons, we believe that is important in an engineering perspective to improve our knowledge of transport mechanisms in disordered structures and to extend the range of investigation to every kind of transport, not confining ourselves to purely Brownian motion, as the physical literature on disorder and transport did in the recent past.

This article address the problem of numerical lattice simulators of transport parameters focusing on some numerical methods for solving partial differential schemes in disordered lattice models[7].

2. Numerical lattice simulation

A lucid project on the analysis of porous media has been developed by Adler[8,9,10] in the case of macroporous solids (sandstones). Starting from image analysis of the material, it is possible to reconstruct a lattice model having some statistical properties equivalent to those of the macroporous structure under consideration. In the analysis developed by Adler for isotropic homogeneous media the equivalence is limited to the porosity (void fraction) ε and to the correlation length Λ_c.

Of course, some fundamental questions arise as to whether these quantities characterize a porous medium completely or whether other statistical parameters should be included. For example, surface effects play an important role in the case of convective-transport, and further statistical parameters describing the surface structure should be taken into account. Other approaches to the characterization of porous structures have been presented by Dullien[11] and by Giona and Adrover[12].

A similar analysis can be extended to microporous structures starting from structural information about the chemical constitution and preparation of the solid (see e.g. Breck[13] in the case of zeolite adsorbents and Sulikowski[14] on the preparation of microporous adsorbents).

Given a lattice representation of a disordered porous structure, two problems arise: the numerical solution of transport equations and the definition of suitable methods for determining transport parameters. It should be observed that we confine ourselves to a *mesoscopic* level of investigation, which means that the balance equations are written in terms of mean field variables and not in terms of Hamiltonian or quantum description of the motion of single molecular entities (molecular dynamics).

A simple way of approaching the numerical study of transport in disordered lattice models is by means of finite difference methods[7,15] which are closely connected with discrete models of transport.

Let us consider a discrete lattice $\mathcal{G}$ in the case of purely Brownian diffusion. If $\mathbf{r}$ indicates the lattice coordinate and $c(\mathbf{r},t)$ the concentration of the transported entity, the balance equation can be written as

$$\frac{dc(\mathbf{r},t)}{dt} = \mathbf{H}[c(\mathbf{r},t)], \tag{3}$$

where the linear operator $\mathbf{H} = \{H_{\mathbf{r}\mathbf{r}'}\}$ is given by

$$H_{\mathbf{r}\mathbf{r}'} = \begin{cases} 0 & \mathbf{r}' \text{ is not connected to } \mathbf{r} \\ 1 & \mathbf{r}' \text{ is connected to } \mathbf{r} \\ -Z(\mathbf{r}) & \mathbf{r}' = \mathbf{r}\,, \end{cases} \tag{4}$$

where $Z(\mathbf{r})$ is the coordination number of site $\mathbf{r}$. Eq. (3) can be recast in the form of a linear system by considering the Laplace transform $\tilde{c}(\mathbf{r},s) = \mathcal{L}[c(\mathbf{r},t)]$,

$$\tilde{c}(\mathbf{r},s) = (s\mathbf{I} - \mathbf{H})^{-1}[c(\mathbf{r},0)]\,, \tag{5}$$

so that the propagator between two points on the lattice can be easily studied in the Laplace domain. This approach has been followed by Schwalm and Schwalm[16,17,18,19] in deriving a renormalization group approach to study the Schrodinger equation on fractal lattices. This method can be extended to an unlimited range of applications in the linear theory of transport in disordered structures, including the effects of biasing fields and adsorption. The only intrinsic limitation of this method is the linearity of the transport equations.

The discretized version of a transport equation can be approached in a similar way. Let us consider the case of a twodimensional diffusion equation (i.e. in the case of purely Brownian motion). If $\mathcal{M}$ is the pore structure (cluster of accessible sites permeable to transport) described by means of its characteristic function $\chi(i,j)$

$$\chi(i,j) = \begin{cases} 1, & (i,j) \in \mathcal{M} \\ 0, & \text{otherwise}\,, \end{cases} \tag{6}$$

the balance equation attains the form

$$c_{ij}^{(n+1)} = c_{ij}^{(n)} + \beta' \sum_{(h,h)\in I(i,j)} J_{hk\to ij}^{(n)}, \tag{7}$$

where $\beta' = \mathcal{D}\Delta t/\Delta x$ and $J^{(n)}_{hk\to ij}$ is the local flux at time n,

$$J^{(n)}_{hk\to ij} = \chi(h,k)(c^{(n)}_{hk} - c^{(n)}_{ij})/\Delta x \ . \tag{8}$$

Therefore, the simplest explicit algorithm for the analysis of transport in a disordered lattice structure reads

$$\begin{aligned} c^{(n+1)}_{ij} &= c^{(n)}_{ij} + \beta' \sum_{(h,k)\in I(i,j)} \chi(h,k) J^{(n)}_{hk\to ij} \\ &= c^{(n)}_{ij} + \beta \sum_{(h,k)\in I(i,j)} \chi(h,k)[c^{(n)}_{hk} - c^{(n)}_{ij}] \ . \end{aligned} \tag{9}$$

Eq. (9) can be generalized to a variety of problems including convective effects or nonlinearities in the diffusion coefficient[15].

It should be observed that in developing Eq. (9) we made use of an explicit algorithm. It is of course possible to formulate different numeric schemes by considering implicit time discretization. It is also worth noting the possibility of making use of finite-element algorithms in the study of transport in disordered models of porous structures[20,21].

3. Estimation of transport parameters

3.1. Exit-time equation and stochastic modelling

A simple way to estimate transport coefficients in disordered structures is to transform parabolic balance equations in the form of an elliptic equation in some characteristic variable related to the behaviour of transport parameters.

This can be done, as long as the transport equations are linear, by considering the volume-averaged mean exit time from a characteristic unit pore cell. Let $\mathcal{C}$ be the unit pore cell, whose boundary $\mathcal{B}$ can be decoupled into an exit boundary $\mathcal{B}_e$, from which particles can escape and an impermeable boundary $\mathcal{B}_i$ represented by the pore walls. Consequently, the probability density function $E(\mathcal{B}, \mathbf{x}, t)$ associated with the first transit time through $\mathcal{B}$ starting from a point $\mathbf{x} \in \mathcal{C}$ can be introduced[22,23,24].

Instead of considering $E(\mathcal{B}, \mathbf{x}, t)$, which satisfies the Kolmogorov backward equation, it is convenient to analyze the behaviour of its moments $\theta^{(n)}(\mathbf{x})$

$$\theta^{(n)}(\mathbf{x}) = \int_0^\infty t^n E(\mathcal{B}, \mathbf{x}, t) dt \ . \tag{10}$$

In the case of convection-diffusion transport, the moments $\theta^{(n)}(\mathbf{x})$ satisfy the equation

$$D\nabla^2\theta^{(n)} + \mathbf{v}(\mathbf{x}) \cdot \nabla\theta^{(n)} + n\theta^{(n-1)} = 0 \ , \tag{11}$$

with $\theta^{(0)}(\mathbf{x}) = 1$. In particular, the most interesting quantity in the moment hierarchy is the first-order moment $\theta = \theta^{(1)}$, which represents the mean exit time from the unit cell $\mathcal{C}$. According to Eq. (11), θ is the solution of the equation

$$D\nabla^2\theta + \mathbf{v} \cdot \nabla\theta + 1 = 0 \ , \tag{12}$$

to be solved with the boundary conditions

$$\theta(\mathbf{x}) = 0 \quad \mathbf{x} \in \mathcal{B}_e\,, \qquad \frac{\partial\theta(\mathbf{x})}{\partial n} = 0 \quad \mathbf{x} \in \mathcal{B}_i\,, \tag{13}$$

where $\partial/\partial n$ is the normal derivative.

In the case of purely Brownian motion, $\mathbf{v} = 0$,

$$<\theta>\sim \frac{L^2}{D_e}\,, \tag{14}$$

where $<\cdot>$ indicates the volume average, L the characteristic linear size of the complex unit cell and D_e the effective diffusion coefficient, which encompasses the hindrance effects due to the geometry of the cell. Eq. (14) can be used to estimate the accessibility factor G^0 defined as

$$D_e = DG^0 \tag{15}$$

from the knowledge of $<\theta^0>$ corresponding to a completely empty structure (void unit cell) and $<\theta>$ obtained by solving the exit-time equation by considering the complex geometry of the cell

$$\frac{<\theta^0>}{<\theta>} = G^0\,. \tag{16}$$

Eqs. (12)-(14) can also be applied if the velocity field derives from a purely deterministic potential distibution $\phi(\mathbf{x})$. In this case, by making use of the weak-force approximation[25]

$$\mathbf{v} = \mu_m(-\nabla\phi)\,, \tag{17}$$

where μ_m is particle mobility, and of the Einstein equation, $\mu_m/D = 1/k_BT$ (k_B being the Boltzmann constant) one arrives to the equation

$$\nabla^2\theta - Pe_E\nabla\phi^*\cdot\nabla\theta + \frac{1}{D} = 0\,, \tag{18}$$

where $\phi = E_o\phi^*$ (E_o being some characteristic energy $\phi^* \sim O(1)$) and the energy Peclet number $Pe_E = E_o/k_BT$. Eq. (18) is applied to predict the temperature behaviour of the diffusion coefficient $D_e = D_e(T)$ of noble gases in zeolitic materials[26].

3.2. Sorption curve analysis

Another interesting way of obtaining diffusion coefficients in a simple fashion is to consider the characteristic relaxation scaling from sorption curve experiments. This approach has its physical counterpart in the macroscopic study of diffusion through polymeric matrices but can be fruitfully applied to lattice simulations experiments.

Let us consider the diffusion equation in two dimensions $\{(x,y)|0 < x < L_x, 0 < y < L_y\}$

$$\frac{\partial c}{\partial t} = D\left(\frac{\partial^2 c}{\partial x^2} + \frac{\partial^2 c}{\partial y^2}\right) \tag{19}$$

and the corresponding sorption experiments from the lattice structure under consideration. Two different simple boundary-value problems can be defined: a) in the case in which the lattice structure is initially saturated at a given concentration c_o and release through one boundary ($x = L_x$)

$$\begin{aligned} c(x,y,0) &= c_o \\ \left.\frac{\partial c(x,y,t)}{\partial x}\right|_{x=0} &= 0\,, \qquad c(L_x,y,t) = 0 \\ \left.\frac{\partial c(x,y,t)}{\partial y}\right|_{y=0} &= 0\,, \qquad \left.\frac{\partial c(x,y,t)}{\partial y}\right|_{y=L_y} = 0 \end{aligned} \tag{20}$$

b) in the case of initial zero loading and of a forcing condition from one boundary ($x = 0$)

$$\begin{aligned} c(x,y,0) &= 0 \\ c(0,y,t) &= c_o\,, \qquad \left(\frac{\partial c(x,y,t)}{\partial t} + \frac{hD}{L_x}\frac{\partial c}{\partial x}\right)_{x=L_x} = 0 \\ \left.\frac{\partial c(x,y,t)}{\partial y}\right|_{y=0} &= 0\,, \qquad \left.\frac{\partial c(x,y,t)}{\partial y}\right|_{y=L_y} = 0 \end{aligned} \tag{21}$$

where $h = V_o/V$ is the ratio between the volume of the structure V_o and the recovery volume V.

By solving these schemes numerically, the fractional release M_t/M_∞ can be obtained

$$\frac{M_t}{M_\infty} = \int_0^t \int_{L_y} \left(-D\frac{\partial c}{\partial x}\right)_{x=L_x} dy\,dt \Big/ \int_0^\infty \int_{L_y} \left(-D\frac{\partial c}{\partial x}\right)_{x=L_x} dy\,dt\,. \tag{22}$$

The fractional release scales asymptotically as

$$1 - \frac{M_t}{M_\infty} \sim \exp(-\lambda_1^2 D_e t)\,, \tag{23}$$

λ_1 being the fundamental (lowest) eigenvalue associated with the boundary conditions: $\lambda_1 = \pi/(2L_x)$, in the case of Eq. (20), $\lambda_1 L_x \cos(\lambda_1 L_x) = h$, in the case of Eq. (21). Therefore, from the exponential relaxation of the fractional sorption curves it is possible to estimate D_e and consequently the accessibility factor G^0. It is convenient to define an accessibility factor G^ε expressed as

$$G^\varepsilon = G^0\varepsilon\,. \tag{24}$$

Figure 1 shows the values of G^ε obtained from the exponential relaxation measurements in two-dimensional site-percolation clusters far from criticality compared with

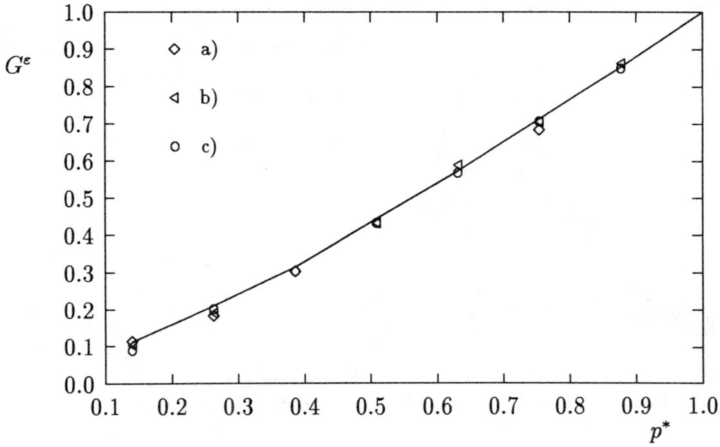

Figure 1: G^ε vs $p^* = (p - p_c)/(1 - p_c)$, $p = \varepsilon$ is the percolation probability, $p_c = 0.593$ the percolation threshold, in two-dimensional percolation clusters: a) Monte Carlo simulations; b) exit-time equation, Eq. (16), c) sorption-curve relaxation, with boundary condition (20).

the corresponding results deriving from the solution of the exit-time equation on the lattice and from Monte Carlo simulations[27]. It is important to observe that the two-boundary value schemes Eqs. (20), (21) can be equivalently used to estimate G^ε. This is peculiar of linear models. In the case of nonlinear transport schemes, the relaxation properties of the two boundary value schemes are completely different, as discussed in section 4.

Numerical evaluation of equivalent diffusivity in a generic model of a lattice structure can be easily performed by applying one of the approaches discussed in the last two sections. This analysis finds a direct application in all the separation processes in which molecular motion is hindered by purely geometric effects, as in the case of size-exclusion chromatography, in which a diffusional partition of solute molecules between the stationary and mobile phases is achieved and controlled by the dimenison of the solute particle compared with the average pore radius.

3.3. Diffusion coefficient in mixed systems

Sorption curve analysis can be fruitfully adopted in deriving transport properties in mixed disordered systems. A good introduction to the study of mixed disordered systems and random resistor networks is given by Roman[28]. Let us consider the classical diffusion equation in a structure composed of a random distribution of n distinct species A_i ($i = 1, .., n$). Let us further suppose that each of the $n \times n$ bonds A_i-A_j is characterized by a given diffusivity $D[i,j]$. Associated with $D[i,j]$,

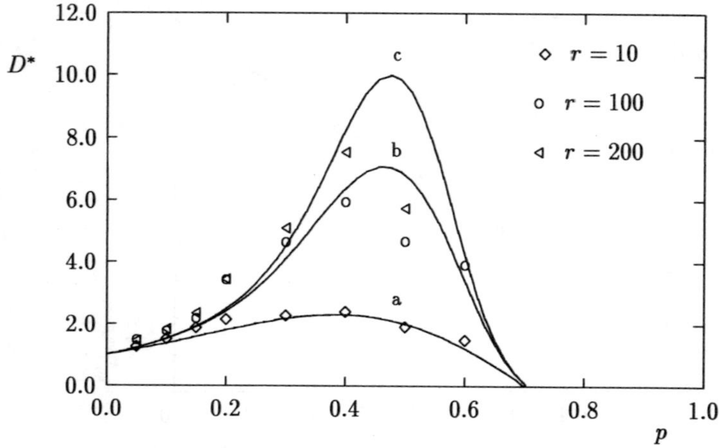

Figure 2: D^* vs p in a two-component random-resitor network as a function of the parameter $r = D_{12}/D_{11}$: $D_{11} = 1.0$ [a.u.], $D_{22} = 0.0$ [a.u.]. a) $r = 10$; b) $r = 100$; c) $r = 200$.

the hopping rates $w[i,j]$ can be defined through the relation[29]

$$D[i,j] = \frac{Za^2}{2d} w[i,j] , \tag{25}$$

where a is the lattice constant, d the dimension of the lattice, and Z the coordination number. Of course $D[i,j]$, $(w[i,j])$ is a symmetric matrix. Let $\mathbf{p} = (p_1, ..., p_n)$ be the probability vector associated with the random distribution of the species on the lattice. This model constitutes the classical random-resistor network (RRN) problem, extensively discussed by Roman et al.[29,30], especially in connection with dispersed ionic conductors. This model has some analogies with multicomponent percolation schemes[31]. The transport properties of RRN are developed by many authors[28,29,30,32,33]. Bunde et al.[32,33] discuss how to evaluate the transport coefficients by making use of Monte Carlo calculations (Boston termites), through the association of a characteristic time and a characteristic probability distribution of motion with each jump between nearest neighbouring sites. This transport problem can be successfully tackled in a simpler fashion by means of finite difference algorithms. In order to do so, we have to solve a linear diffusion equation, $\partial_t c - \nabla \cdot (D \nabla c) = 0$, by further considering that the spatial distribution of the diffusivities is highly heterogeneous. Following the same approach discussed previously, a simple finite-difference algorithm for solving the RRN problem in two dimensions (the extension to three dimensions is straighforward) read:

$$c_{i,j}^{(m+1)} = c_{i,j}^{(m)} + \beta \sum_{(h,k) \in I(i,j)} D[\chi(h,k), \chi(i,j)](c_{h,k}^{(m)} - c_{i,j}^{(m)}) , \tag{26}$$

where $c_{i,j}^{(m)}$ is the discretized concentration field at time m on the site (i,j); $\chi(h,k)$ is the generalized characteristic function of the lattice, $\chi(h,k) = q$ $(q = 1,...,n)$, if the site (h,k) is occupied by the species A_q; $I(i,j)$ is the set of the first nearest neighbours of the site (i,j) (four in square lattices); and $\beta = \Delta t/(\Delta x)^2$. Of course, since equation (26) represents an explicit algorithm, the following stability condition must hold true: $\beta D_{max} < 1/2$, where $D_{max} = \max_{i,j} D[i,j]$.

Relaxation properties of diffusion can be applied by considering the differential scheme (21). Figure 2 shows the resulting data for the effective diffusion coefficient $D^* = D_e/D_{11}$ obtained from the exponential scaling of $1 - M_t/M_\infty$. Simulations were performed on twodimensional percolation lattices of size 50×140 with $h = 0.1$. The error range is less than $0.05D_e$. In this figure, the solid lines corresponds to the predictions for D^* deriving from the Effective Medium Approximation (EMA) theory [34,29]. EMA theory furnishes a value of D^* which is the solution of the equation

$$p_{11}\frac{D^* - 1}{D^* + 1} + p_{12}\frac{D^* - r}{D^* + r} + p_{22} = 0 \; , \tag{27}$$

where $p_{11} = p_1^2$, $p_{22} = p_2^2$, $p_{12} = 1 - p_1^2 - p_2^2$. Comparison of the numerical simulation results with EMA theory indicates that for $p < p'_c$, $p'_c \simeq 0.407$ being the interface percolation threshold of the insulating phase (the second component), there is an acceptable level of agreement between EMA theory and simulations.

The same analysis can be applied to a generic distribution of species and to other problems involving transport in multicomponent systems. It is important to state that the proposed numerical approach is much simpler than the *Boston-termite simulation* (which is a particular random walk introduced to simulate random-resistor network problems) discussed by Bunde et al.[32] and is directly connected to the mean-field representation of the transport problem.

3.4. Dispersion in porous media

In section 3.1 we have considered the relation between the volume-averaged exit time and the diffusion coefficient in the case of purely Brownian motion or in the presence of a potential distribution on the assumption that the only source of stochasticity comes from thermal fluctuations.

In the study of dispersion phenomena in porous media in the presence of a convective velocity field, one should also consider the random fluctuations associated with the random distribution of the velocity inside the porous structure[35,36]. The exit-time equation can be used also in this case to predict dispersion parameters[23,24,37].

Starting from solute dynamics at microscopic scales written in terms of a stochastic differential equation

$$d\mathbf{x}(t) = \mathbf{v}(\mathbf{x}(t))dt + \sqrt{2D}d\xi(t) \; , \tag{28}$$

where $\mathbf{v}$ is the random velocity field, ξ a vector of Wiener processes and $\mathbf{x}$ the position vector of the solute particle, and by considering that thermal and velocity fluctuations are uncorrelated with each other, it is possible to derive for the axial dispersion

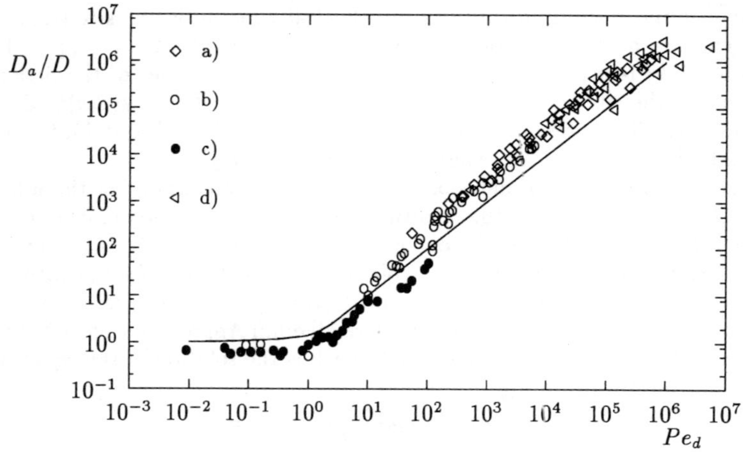

Figure 3: D_a/D vs particle Peclet number $Pe_p = d_p v/D$ (d_p being the particle diameter) in packed beds: comparison of the exit-time solution (line), Eq. (12) (by evaluating $< \theta >$ directly from a onedimensional approximation), with the experimental results: a) Ebach and White; b) Pfannkuch; c) Edwards and Richardson; d) Carberry and Bretton.

coefficient D_a the following expression

$$D_a = D + \alpha v^2 < \theta > , \tag{29}$$

where $\alpha \sim O(1)$ is a coefficient related to the correlation properties of the velocity fluctuations and $< \theta >$ the average exit-time obtained from the solution of Eq. (12). The simplest application of Eq. (29) is to consider a onedimensional approximation of the exit-time equation. Figure 3 shows the theoretical results obtained in this way[37] compared with literature data on dispersion in packed beds[38,39,40,41]. This analysis can be of course improved by considering more realistic lattice representations of the porous structure[42].

4. Nonlinear diffusion models

The finite-difference approach can be applied to the analysis of nonlinear transport in disordered systems. In this section we briefly analyze the dynamic properties of nonlinear diffusion models.

4.1. Statement of the problem

Let us consider the case of a diffusion process in the presence of a concentration-dependent diffusion coefficient

$$D = D_o \left(\frac{c}{c_o}\right)^q . \tag{30}$$

In this case, the diffusion equation becomes

$$\frac{\partial c}{\partial t} = \frac{D_o}{c_o^q (q+1)} \nabla^2 c^{q+1} . \tag{31}$$

4.2. Invariant function analysis

Let us analyze the infinite propagation in onedimension of the diffusion Eq. (31). In dimensionless terms, this becomes

$$\frac{\partial c}{\partial t} = \frac{\partial^2 c^{q+1}}{\partial x^2} . \tag{32}$$

The solution $c(x,t)$ can be assumed to be of the form

$$c(x,t) = t^{-\beta}\phi(xt^{-\beta}) = t^{-\beta}\phi(y) , \tag{33}$$

where $y = xt^{-\beta}$. By substituting Eq. (33) into Eq. (32) it follows that

$$-\beta t^{\beta(q+2)-1}(\phi + y\phi') = (q+1)[q(\phi')^2 + \phi^q \phi''] , \tag{34}$$

which implies that (ϕ depends exclusively on y)

$$\beta = \frac{1}{q+2} . \tag{35}$$

Consequently, the invariant function ϕ satisfies the equation

$$\phi^q \phi'' + q(\phi')^2 = -\frac{1}{(q+1)(q+2)}(\phi + y\phi) . \tag{36}$$

Eq. (36) can be solved with the initial condition

$$\phi(0) = 1 , \qquad \phi'(0) = 0 . \tag{37}$$

For $q = 0$, this reduces to the Gaussian behaviour $\phi(y) = \exp(-y^2/4)$. For q greater than one, there exists a finite value y^* of y for which $\phi(y^*) = 0$ and zero for $y > y^*$. This means that for each time t, the diffusion front will propagate up to a finite length $x_m(t) = y^* t^\beta$, and therefore diffusion is characterized by a finite propagation velocity, contrary to Brownian motion for which the propagation velocity is not finite[43].

Table 1: Scaling results for $D_w(q)$ obtained numerically compared with Eq. (40) in regular lattice. The disordered structure is a $2d$ percolation cluster above criticality, $p = 0.615$.

q	$2/D_w(q)$, Eq. (40)	$2/D_w(q)$, Simulations
0	1	0.75 ± 0.1
1	1/2	0.42 ± 0.05
2	1/3	0.32 ± 0.03
3	1/4	0.23 ± 0.03

4.3. Mean-square displacement

In the previous subsection we have implicitly calculated the scaling for the mean-square displacement for the nonlinear diffusion equation in one dimension from the scaling of the invariant function $R^2(t) =< r^2(t) >\sim t^{2\beta}$, with $\beta = 1/(q+2)$. In a similar way, in a d-dimensional space and in radial symmetric conditions,

$$c(r,t) = R^{-d}(t)\Phi(r/R(t)) \ , \tag{38}$$

from which it follows that

$$R^2(t) \sim t^{2/(2+qd)} = t^{2/D_w(q)} \tag{39}$$

For every q greater than zero, the mean-square displacement admits a power-law scaling with an exponent less than unity. This exponent is a consequence of the correlation properties in the motion deriving from a concentration dependent diffusion coefficient. In terms of the generalized diffusion exponent $D_w(q)$, Eq. (39) implies

$$D_w(q) = 2 + qd \ . \tag{40}$$

In the case of fractal structures, it follows from the same scaling arguments that

$$D_w(q) = D_w(0) + qD \ , \tag{41}$$

where $D_w(0)$ is the diffusion exponent for purely Brownian motion and D the fractal dimension. Table I reviews the scaling results for the mean-square displacement on 800×800 square percolation clusters slightly above criticality, $p = 0.615$. With the increasing of q it is difficult to ascertain from the data if disordered media exhibit anomalies in propagation, as in the case of Brownian motion. However, as discussed by Giona and Viola[15], nonlinear diffusion in loopless structures (e.g. DLA) presents some peculiar anomalous features which are not yet understood.

4.4. Diffusional relaxation in the nonlinear case

Relaxation behaviour in nonlinear diffusion depends closely on the boundary conditions. Let us consider Eq. (31) by assuming boundary conditions of the form (21). For $x = L_x$, the boundary condition should be modified to

$$(\partial c/\partial t + (hD_o/L_x)(c/c_o)^q \partial c/\partial x)|_{x=L_x} = 0.$$

Asymptotically, for large t it follows that

$$\frac{\partial c}{\partial t} \simeq D_o \frac{\partial^2 c}{\partial x^2} \tag{42}$$

and therefore the scaling of M_t/M_∞ follows Eq. (23), as in the linear case. On the other hand, in the case of boundary conditions of the form given by Eq. (20), the concentration c at $x = L_x$ tends asymptotically to zero. From the separation of variables (modal decomposition), $c(t,x) = \Theta(t)\Xi(x)$, it follows that

$$\frac{\Theta'(t)}{\Theta^q(t)} = \frac{D_o}{(q+1)c_o^q}\frac{(\Xi^{q+1}(x))''}{\Xi(x)} = \text{constant} \tag{43}$$

and therefore, for $q > 0$, $\Theta(t) \sim t^{-q}$, which implies

$$1 - \frac{M_t}{M_\infty} \sim \frac{1}{t^q} \, . \tag{44}$$

Consequently, exponential or power law relaxation can be observed depending on the boundary conditions. According to these results, sorption-curve analysis can also be applied to the nonlinear case. If one is interested in estimating of transport coefficients, the exponential scaling associated with Eq. (21) is preferable.

5. Biased diffusion and the LBE

In section 3.3, we have discussed an initial application of transport parameter estimation in the presence of convective fields. However, in many problems related with transport in porous media it is important to obtain a detailed picture of the velocity field inside a porous structure. For this purpose the Lattice Boltzmann Equation (LBE) proves to be a simple and powerful way of determining velocity distribution in porous structures (figure 4). The LBE has been applied by Rothman[44], Balasubramanian[45] et al. by considering cellular automata and by Succi et al.[46] by considering lattice Boltzmann simulations in terms of averaged populations.

Details on the LBE can be found elsewhere[47]. Here we focus on the estimation of permeability.

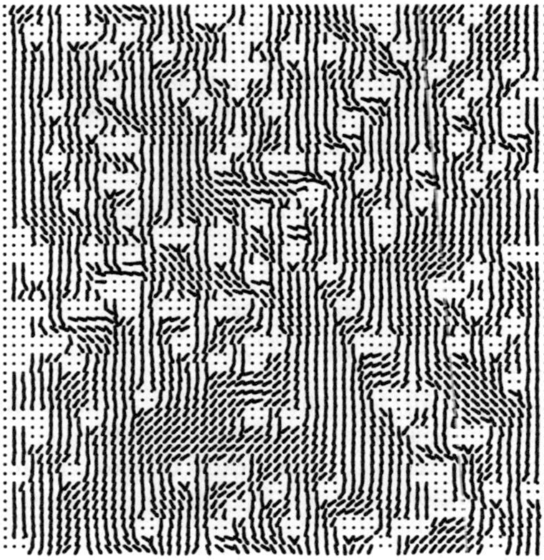

Figure 4: Velocity profile inside a two-dimensional percolation cluster, $p = 0.80$.

5.1. Convective transport and permeability

The permeability K of a porous medium is defined through the Darcy law

$$\mathbf{v} = \frac{K}{\mu}(-\nabla P)\,, \tag{45}$$

where μ is the viscosity of the fluid, ∇P the pressure gradient and $\mathbf{v}$ the seepage velocity. In LBE simulation, K can be estimated, by applying a given body force (which furnishes the desired pressure drop $\Delta P/L$) and by measuring the mass flow rate Q vs the pressure drop itself

$$Q = K\left(\frac{\Delta P}{L}\right)\frac{H}{\nu}\,, \tag{46}$$

H being the section of the duct and ν the kinematic viscosity. For an empty duct, the Poiseuille law furnishes $Q = (\Delta P/L)H^3/(12\nu)$, which corresponds to an equivalent permeability

$$K = \frac{H^2}{12}\,. \tag{47}$$

Simulation experiments were performed on twodimensional site-percolation clusters. The size of the lattice was 400×200, with a pressure drop directed towards the x-axis. The size of the lattice has been chosen to ensure a satisfactory averaging of the data. Figure 5 shows the relation between the dimensionless mass flow rate $Q^* =$

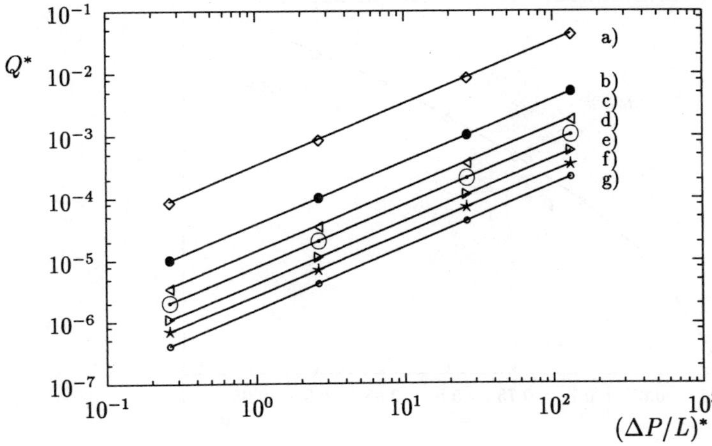

Figure 5: Q^* vs $(\Delta P/L)^*$ in twodimensional percolation clusters: a) $p = 0.99$, b) $p = 0.95$; c) $p = 0.90$; d) $p = 0.85$; e) $p = 0.80$; f) $p = 0.75$; g) $p = 0.70$.

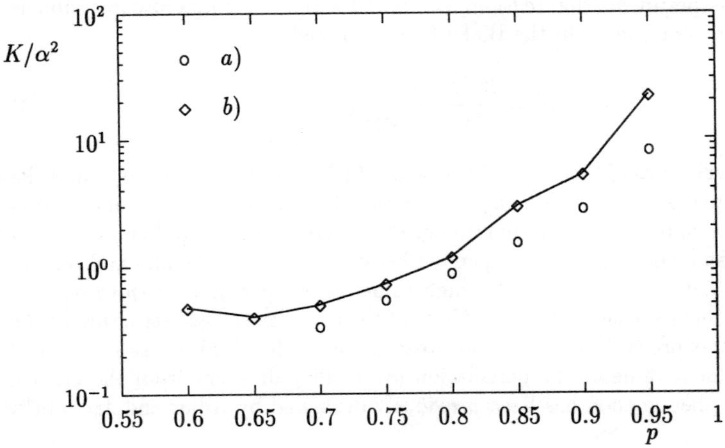

Figure 6: K/α^2 vs p in twodimensional percolation clusters. α is the size of the unit lattice site: a) numerical results obtained by applying the LBE; b) the Blake-Kozeny correlation.

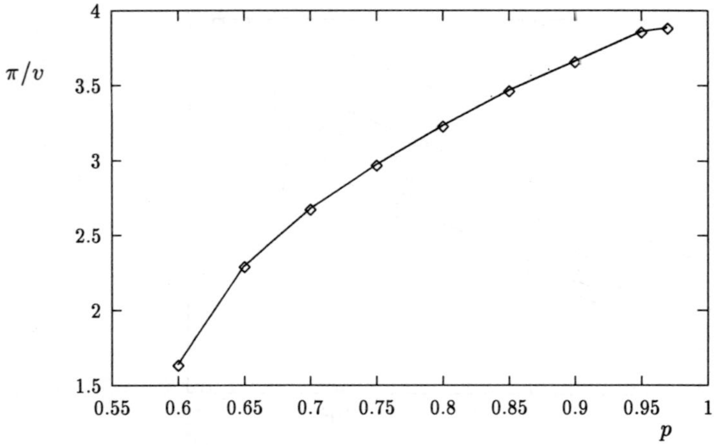

Figure 7: π/v vs p in two dimensional percolation clusters of size $L \times L$ with $L = 1000$.

$Q/(\rho H \nu)$ and the dimensionless pressure drop $(\Delta P/L)^* = (\Delta P/L)H^2/(\rho\nu^2)$, for different percolation clusters characterized by the value of the percolation probability p. The slope of these curves is equal to $K^* = K/H^2$. The numerical values obtained from simulation experiments (figure 6) are compared with simple macroscopic models (the Blake-Kozeny equation). In the Blake-Kozeny model,

$$K = \frac{36L_p^2}{150}\frac{\varepsilon^3}{(1-\varepsilon)^2}, \tag{48}$$

where L_p is the reciprocal of the ratio π/v between the local particle perimeter and the particle volume per unit site size α, $L_p = \alpha/(\pi/v)$. For twodimensional percolation clusters, the behaviour of π/v as a function of the percolation probability p is shown in figure 7. As can be observed, the agreement between simulation results and macroscopic models is not acceptable even for high values of the porosity $\varepsilon = p$ (moderate disorder). With the decreasing of p, the effects of the complex surface structure of the percolation clusters prevails and the Blake-Kozeny correlation exhibits an unphysical local minimum for a value of the percolation probability different from the critical value. A similar phenomenon has been previously discussed by Adler and Jacquin[48].

6. Concluding remarks

We have analyzed several approaches to the estimation of transport parameters in disordered lattice models. The application of these methods furnishes concrete answers in a variety of problems ranging from macroporous transport problems (dispersion and permeability in packed beds) to microporous transport. The methods employed are different - kinetic approach to transport (the LBE), stochastic modelling, direct numerical simulations - but the unifying concept in all the analyses is the need for an accurate statistical description of disorder in porous structures. In an engineering perspective, all the methods discussed can be integrated into a single simulator which, starting from a lattice model of a porous structure, is able to give velocity profiles inside the structure and to estimate diffusive and convective transport parameters.

Another interesting problem that is still open to further investigation is the estimation of transfer coefficients in turbulent condition in the presence of complex constraints (fractal boundary) induced by the geometry of the structure or by turbulent phenomena. A scaling analysis of the friction factor has been proposed by Punchocar and Drahos[49].

References

1. S. Havlin and D. Ben-Avraham, *Adv. Phys.* **37** (1987) 695.
2. A. Bunde and S. Havlin, eds., *Fractals and Disordered Systems* (Springer Verlag, Berlin, 1992).
3. D. Dhar, *J. Phys.* **A 17** (1984) L257.
4. R. B. Pandey *Phys. Rev.* **B 30** (1984) 489.
5. D. Stauffer, *J. Phys.* **A 18** (1985) 1827.
6. D. Avnir, ed., *The Fractal Approach to Heterogeneous Chemistry* (J. Wiley, New York, 1989).
7. M. Giona, A. Adrover and A. R. Giona, in *Fractals in the Natural and Applied Sciences*, M.M. Novak ed., (North Holland, Amsterdam, 1994), pp. 153-163.
8. P.M. Adler, *Porous Media - Geometry and Transports*, (Butterworth-Heinemann, Boston, 1993).
9. P. M. Adler, C. G. Jacquin and J. A. Quiblier, *Int. J. Multiphase Flow* **16** (1990) 691.
10. J. Quiblier, *J. Colloid Interface Sci.* **98** (1983) 84.
11. F. A. L. Dullien, *Transport in Porous Media* **6** (1991) 581.
12. M. Giona and A. Adrover, in *Characterization of Porous Solids III*, J. Rouquerol, F. Rodriguez-Renoso, K. S. W. Sing and K. K. Unger eds., (Elsevier, Amsterdam, 1994), pp. 197-206.
13. D. W. Breck, *Zeolite Molecular Sieves* (J. Wiley, New York, 1974).
14. B. Sulikowski, *J. Chem. Phys.* **97** (1993) 1420.
15. M. Giona and A. Viola, in *Fractals in the Natural and Applied Sciences*, M.M. Novak ed., (North Holland, Amsterdam, 1994), pp. 165-174.
16. W. Schwalm and M. Schwalm, *Phys. Rev.* **B 37** (1988) 5924.
17. W. Schwalm and M. Schwalm, *Phys. Rev.* **B 39** 81989) 12872.
18. W. Schwalm and M. Schwalm, *Phys. Rev.* **B 44** (1991) 382.
19. W. Schwalm and M. Schwalm, *Physica* **A 185** (1992) 195.
20. G. Chavent, G. Cohen and J. Jaffré, *Transport in Porous Media* **2** (1987) 465.

21. R. W. Lewis, P. J. Roberts and B. A. Schrefler, *Transport in Porous Media* **4** (1989) 319.
22. S. Karlin and H. M. Taylor, *A Second Course in Stochastic Processes* (Academic Press, New York, 1981).
23. R. N. Bhattacharya and V. K. Gupta, *Water Resour. Res.* **19** (1983) 938.
24. R. N. Bhattacharya and V. K. Gupta, in *Dynamics of Fluids in Hierarchical Porous Media*, J. H. Cushman ed., (Academic Press, London, 1990), pp. 61-96.
25. M. Doi and S. F. Edwards, *The Theory of Polymer Dynamics* (Oxford Science Publ., Oxford, 1989).
26. M. Giona and A. Adrover, *Chem. Engng. Sci.* **48** (1993) 1933.
27. M. Giona, A. Adrover and D. Spera, *Chem. Eng. Quart.* **7** (1993) 199.
28. H. E. Roman, *Static and Transport Properties of Disordered Systems: Random Fractal Structures and Composite Materials*, Habilitationschrift, Institut für Theoretische Physik, (Universität Hamburg , Hamburg, 1990).
29. A. G. Rojo and H. E. Roman, *Phys. Rev.* **B 36** (1988) 3696.
30. H. E. Roman and M. Youssouff, *Phys. Rev.* **B 36** (1987) 7285.
31. M. Giona and A. Adrover, *Phys. Rev.* **E 49** (1994) 5287.
32. A. Bunde, A. Coniglio, D. C. Hong and H. E. Stanley, *J. Phys.* **A 18** (1985) L137.
33. A. Bunde, W. Dieterich and H. E. Roman, *Phys. Rev. Lett.* **55** (1985) 5.
34. S. Kirkpatrick, *Rev. Mod. Phys.* **45** (1973) 574.
35. M. Sahimi, D. B. Hughes, L. E. Scriven and H. T. Davis, *Chem. Engng. Sci.* **41** (1986) 2103.
36. M. Sahimi, H. T. Heiba, H. T. Davis and L. E. Scriven, *Chem. Engng. Sci.* **41** (1986) 2123.
37. M. Giona and A. Adrover, *Environmetrics* **4** (1993) 255.
38. J.J. Carberry and R. H. Bretton, *AIChE J.* **3** (1958) 367.
39. E. A. Ebach and R.R. White, *AIChE J.* **4** (1958) 161.
40. M. F. Edwards and J. F. Richardson, *Chem. Engng. Sci.* **23** (1968) 109.
41. H. O. Pfannkuch, *Revue Inst. Francais du Petrole* **18** (1963) 215.
42. M. Giona, *Chaos Solitons and Fractals* **4** (1994) 461.
43. J. Crank, *Mathematics of Diffusion* (Claredon Press, Oxford, 1975).
44. D. H. Rothman, *Geophysics* **53** (1988) 509.
45. K. Balasubramanian, F. Hayot and W. F. Saam, *Phys. Rev.* **A 36** (1987) 2248.
46. S. Succi, E. Foti and F. Higuera, *Europhys. Lett.* **10** (1989) 433.
47. A. Adrover and M. Giustiniani, this book.
48. P. M. Adler and C. G. Jacquin, *Transport in Porous Media* **2** (1987) 553.
49. M. Punchocar and J. Drahos, in *Fractals in the Natural and Applied Sciences*, M. M. Novak ed., (North-Holland, Amsterdam, 1994) pp. 305-311.

LATTICE BOLTZMANN EQUATION AND NUMERICAL SIMULATION OF FLOW IN POROUS MEDIA

ALESSANDRA ADROVER and MANUELA GIUSTINIANI
Dipartimento di Ingegneria Chimica, Universitá di Roma "La Sapienza"
Via Eudossiana 18, 00184 Roma, ITALY

ABSTRACT

The Lattice-Boltzmann Equation (LBE) succesfully simulate incompressible Navier-Stokes equations, easily handling arbitrary complex boundary conditions. We discuss and analyze many different applications of the introduction of a body-force in the microdynamic LBE, highlighting the utility of this technique in the study of the fluid mechanical properties of disordered and fractal structures.

1. Introduction

In the last decade, Lattice Gas Automata (LGA) have been developed as an alternative method to direct simulations and spectral techniques for solving equations of fluid mechanics and transport[1].

LGA represent a simplified artificial microworld ruled by collision laws preserving mass and momentum. They can be usefully applied to describe macroscopic fluid dynamical phenomena, also providing microscopic details related to fluctuation[2] and correlation properties of the corresponding volume-averaged properties[3].

A significant improvement of LGA simulations can be obtained by considering the mean-field approximation of the particle-collision rules, called the Lattice-Boltzmann Equation (LBE). The LBE is essentially the kinetic equation resulting from ensemble-averaging of the discrete dynamics of LGA, on the assumption of molecular chaos[4,5,6].

A typical class of problems which can be addressed by means of the LBE is given by fluid dynamical phenomena in constrained geometry, for which the detailed structure of the porous medium is taken into account[3,7].

In this article we review the fundamental conceptual steps in the development of the LBE from Boolean up to mean-field formulation, also pointing out the advantages and disadvantages of the LBE formulation with respect to the LGA approach. We then discuss the modification of the collision-propagation microdynamic LBE by introducing a body-force[8,9] and analyze many different applications of the body-force formulation.

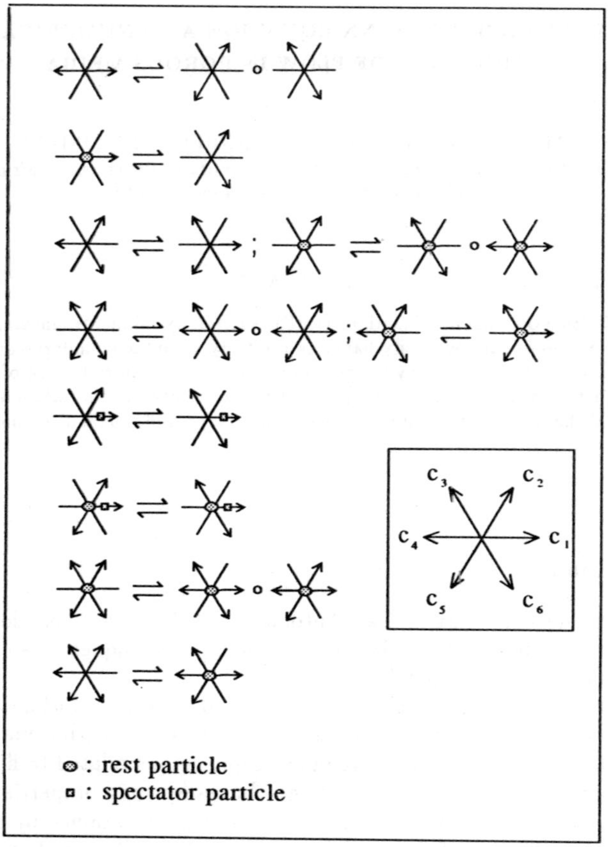

Figure 1: Collision rules of the 7-bit nondeterministic model B.

2. LGA: from microdynamic equation to Navier-Stokes equation

2.1. Microdynamic equation and collision rules

Mass and momentum conservation are the local rules governing the dynamics of LGA. These automata can be viewed as a set of pseudo-particles with unit mass and unit speed constrained to move along the links of a regular lattice[10]. The automaton state is entirely specified in terms of a set of Boolean variables $n_i(\mathbf{r}, t)$ taking the value one or zero according to whether the $\mathbf{r}$ site holds a particle moving along the i-th link or not. A set of collision rules governs momentum transfer between pseudo-particles as in a real fluid.

The simultaneous presence of two or more particles with the same speed at the same spatial location is forbidden (exclusion principle). Collisions at each lattice site and the subsequent propagation step define the spatial and temporal evolution of the

Boolean field $n_i(\mathbf{r},t)$, described by means of the microdynamic equation

$$n_i(\mathbf{r}+\mathbf{c}_i,t+1) = n_i(\mathbf{r},t)+\Omega_i(n)\,, \tag{1}$$
$$\Omega_i(n) = \sum_{ss'}(s'_i - s_i)\xi_{ss'}\prod_j n_j^{s_j}(1-n_j)^{(1-s_j)}\,, \tag{2}$$

where $\mathbf{c}_i$ (i=1,....,b) is the set of unit vectors connecting a given site of the lattice with its b nearest neighbours and Ω_i the collision operator, representing the change in the i-th occupation number due to collisional interaction. The nondeterministic collisions transforming the input state s into s' (s and s' being Boolean strings of b bits) are taken into account by introducing at each time and node and for any pair of states (s,s') of the Boolean variables $\xi_{ss'}$ such that

$$<\xi_{ss'}>= A(s\to s')\,,\forall(s,s')\,;\quad \sum_{s'}\xi_{ss'}=1\;\forall s\,, \tag{3}$$

where $A(s\to s')$ is the transition probability satisfying semi-detailed balance, which implies that $\sum_s A(s\to s')=1, \forall s'$. Because of mass and momentum conservation rules, the collision term fulfills the following relations:

$$\sum_{i=1}^{b}\Omega_i = 0\,,\quad \sum_{i=1}^{b}\Omega_i\mathbf{c}_i=\mathbf{0}\,. \tag{4}$$

Collisions conserve mass and momentum locally, whereas propagation conserves them globally. As a consequence of the one-speed nature of the automaton, total kinetic energy is automatically conserved, being indistinguishable from mass conservation.

In order to obtain the Navier-Stokes equation, as discussed in below, the lattice has to be symmetrical enough to ensure isotropy up to at least the fourth order tensor[10]. This is the case of the exagonal FHP lattice in two dimensions and of the face-centered hypercubic (FCHC) lattice in four dimensions (used for three-dimensional simulations).

Let us consider the FHP lattice, focusing on the set of collision rules capable of avoiding the presence of spurious invariants. In addition to total particle number, head-on collisions also conserve the difference of particle number in any pair of opposite directions, thus conserving four scalar quantities; the large scale dynamics of such a model differs drastically from real hydrodynamics. One way to remove spurious invariants is to introduce triple colisions. Figure 1 shows the set of collision rules with no spurious invariants called the FHPIII nondeterministic model B[11] (which will be used throughout this article). This is a seven-particle model including a rest particle, head-on collisions, triple collisions, head-on collision with a spectator particle (i.e. a particle unaffected in the collision), duals of head-on collisions and collisions involving up to five particles.

The spatial and temporal evolution of the automaton can be described by introducing the mean quantities per node N_i, the density ρ and mass current $\mathbf{j}$, thus defined

$$N_i(\mathbf{r},t) = \langle n_i(\mathbf{r},t) \rangle , \tag{5}$$

$$\rho(\mathbf{r},t) = \sum_i N_i(\mathbf{r},t) , \tag{6}$$

$$\mathbf{j}(\mathbf{r},t) = \sum_i \mathbf{c}_i N_i(\mathbf{r},t) = \rho(\mathbf{r},t)\mathbf{u}(\mathbf{r},t) , \tag{7}$$

$\mathbf{u}$ being the mean velocity and $< . >$ standing for ensemble average.

2.2. Low-speed equilibrium distribution

Non-deterministic LGA models satisfying semi-detailed balance and having no spurious invariants admit a universal equilibrium solution, completely factorized over all the nodes and all cells, with mean population N_i given by the Fermi-Dirac distribution (as can be expected from the exclusion principle), dependent only on the density and the mass current and independent of the transition probability[1].

For a model including b_m moving particles ($i \neq 0$) with unit mass ($k = 0$) and b_r rest particles ($i = 0$) with mass $m_k = 2^k, k \in (0, ..., b_r - 1)$, the equilibrium mean populations N_{ik}^{eq} read as:

$$N_{ik}^{eq} = \frac{1}{1 + \exp(m_k(h + \mathbf{q}\cdot\mathbf{c_i}))} , \tag{8}$$

where the Lagrange multipliers h and $\mathbf{q}$ can be calculated in terms of ρ and $\mathbf{j}$. For small u, the equilibrium mean population calculated perturbatively in power series of u up to the second order attains the form[12]

$$N_{i0}^{eq} = d\left(1 + \frac{D\rho}{c^2\rho_m}\mathbf{c}_i\cdot\mathbf{u} + G(\rho)(Q_{i\alpha\beta} + (c^2/D - c_s^2)\delta_{\alpha\beta})u_\alpha u_\beta\right) , \tag{9}$$

$$N_{0k}^{eq} = d_k\left(1 - m_k c_s^2 G(\rho)\frac{1-d_k}{1-d}\delta_{\alpha\beta}u_\alpha u_\beta\right) , \tag{10}$$

where

$$\rho_m = b_m d\, ; \quad \rho_r = \sum_{k=0}^{b_r-1} m_k d_k , \quad \rho = \rho_m + \rho_r , \tag{11}$$

$$G(\rho) = \frac{D^2\rho^2}{2c^4\rho_m^2}\frac{1-2d}{1-d} , \tag{12}$$

$$Q_{i\alpha\beta} = \begin{cases} c_{i\alpha}c_{i\beta} - (c^2/D)\delta_{\alpha\beta}, & i \neq 0 \\ 0, & i = 0 \end{cases} \tag{13}$$

$$c_s^2 = \frac{b_m d(1-d)c^2}{D(b_m d(1-d) + \sum_{k=0}^{b_r-1} m_k^2 d_k(1-d_k))} \tag{14}$$

c_s being the lattice speed of sound (see the next section). Throughout the article, the Greek subscript indicates the vector components referred to a Cartesian frame of reference.

2.3. Macrodynamical equations

On a very large lattice, space and time can be regarded as continuous by considering local equilibria varying slowly in space and time. Conservation relations (Eq. (4)), can be written in terms of mean population

$$\sum_{i,k} N_{ik}(\mathbf{r}+\mathbf{c}_i, t+1)m_k = \sum_{i,k} N_{ik}(\mathbf{r}, t)m_k \tag{15}$$

$$\sum_{i,k} N_{ik}(\mathbf{r}+\mathbf{c}_i, t+1)\mathbf{c}_i m_k = \sum_{i,k} N_{ik}(\mathbf{r}, t)\mathbf{c}_i m_k \tag{16}$$

Under the assumption of small Mach number, it is possible to use perturbative methods and multiscale technique to derive the macrodynamical equations, i.e. balance equations for the large scale and long-time behaviour of density and mass currents in the continuum limit. By considering up to second-order terms in velocity and gradients, we obtain[12]

$$\partial_t \rho + div(\rho \mathbf{u}) = 0 \tag{17}$$

$$\begin{aligned} \partial_t(\rho u_\alpha) + \partial_\beta(g(\rho)\rho u_\alpha u_\beta) &= -\partial_\alpha P(\rho, u) + \partial_\beta(\nu \partial_\beta(\rho u_\alpha)) \\ &+ \partial_\alpha((\frac{D-2}{D}\nu + \zeta) div(\rho \mathbf{u}) , \end{aligned} \tag{18}$$

where

$$g(\rho) = \frac{D}{D+2}\frac{\rho}{\rho_m}\frac{1-2d}{1-d} \tag{19}$$

$$P(\rho, u) = \frac{\rho_m c^2}{D} - \rho g(\rho)\frac{c_s^2}{c^2}\left(1 + \frac{D}{2} - \frac{c^2}{2c_s^2}\right)u^2 . \tag{20}$$

In Eq. (20) c_s is the speed of sound, since

$$\left(\frac{\partial P}{\partial \rho}\right)_{u=0} = \frac{b_m d(1-d)c^2}{D(b_m d(1-d) + \sum_{k=0}^{b_r-1} m_k^2 d_k(1-d_k))} = c_s^2 \tag{21}$$

and ν and ζ are the kinematic shear and bulk viscosities, given by

$$\nu = \nu_c + \nu_p = -\frac{b_m c^4}{D(D+2)}\psi(\rho) - \frac{c^2}{D(D+2)} ; \quad \zeta = \frac{b_m c^2}{D}X(\rho) - \frac{1}{2}\left(\frac{c^2}{D} - c_s^2\right) . \tag{22}$$

In Eqs. (17)-(20) $\partial_t = \partial/\partial t$, $\partial_\beta = \partial x_\beta, (\beta = 1, .., D)$ and the Einstein summation convention is adopted. The kinematic viscosity has two contributions of opposite signs: the *collision viscosity* ν_c depending on the collision rules adopted and the *propagation viscosity* which is a negative contribution due to the discrete nature of the lattice. In LG flow simulations it is useful to choose collision rules which minimize the viscosity and thereby maximize the Reynolds number; an increase in the number of collisions allowed determines a decrease in the mean free paths and consequently of the minimal viscosity characteristics of the model. It should be noted that the function $\psi(\rho)$ is still to be determined. It can be shown that, despite the presence of the factor $g(\rho)$, which is responsible of the non Galileian invariance of the microdynamical equation (18), the Navier Stokes equation for incompressible flow can be exactly recovered, freezing the density at a constant and uniform value ρ_o everywhere exept in the pressure term and rescaling time $t' \rightarrow t/g(\rho_o)$ and viscosity $\nu' \rightarrow \nu g(\rho_o)$. On the other hand simulation of compressible flow is still an open question. Different models have been proposed in order to eliminate the factor $g(\rho)$ (i.e. the non Galileian invariance) although the corresponding macrodynamical equation does not coincide exactly to the Navier Stokes equation for a viscous compressible fluid[13,14].

3. The Lattice Boltzmann Equation

In the Boltzmann approach, on the assumption that particles entering into collision have no prior correlation, the occupation numbers n_i in the microdynamical equations (1)-(2) are replaced by the corresponding mean population N_i[6,15]. By supplementing the average procedure with the statistical assumption of molecular chaos, one derives the Lattice Boltzamnn Equation (LBE), which in the linearized version (without rest particles for the sake of simplicity) reads as

$$N_i(\mathbf{r} + \mathbf{c_i}, t + 1) = N_i(\mathbf{r}, t) + \sum_{j=1} \Phi_{ij}(N_j(\mathbf{r}, t) - N_j^{eq}(\mathbf{r}, t)) \, , \tag{23}$$

$$\Phi_{ij} = \frac{\partial \Omega(N_i)}{\partial N_j} |_{N_i = d} \tag{24}$$

$$\Phi_{ij} = -\frac{1}{2} \sum_{ss'} (s_i - s'_i) A(s \rightarrow s') d^{p-1} (1-d)^{b-p-1} (s_j - s'_j) \, , \tag{25}$$

where $p = \sum_i s_i$ and Φ_{ij} is the linearized collision matrix (evaluated at the zero velocity equilibrium), which for of lattice symmetry reason is a circulant matrix. The advantage in using the linearized Boltzmann equation (which works with floating point variables) instead of the Boolean version (Eq. (1)) are: the suppression of the large amount of noise characteristic of LGA simulations; a straighforward implementation of 3D simulations; and a much more economical handling of hydrodynamic problems. On the other hand, Eq. (23) contains much less physical information than Eq. (1): all the physics related to particle correlations is lost and multispecies diffusion and reaction diffusion problems cannot be approached in a simple way[16,17].

There are, however, two important results following from the LBE: the irreversible approach to equilibrium and the explicit calculation of viscosity[1]. The functional form of $\psi(\rho)$ can be analytically determined and the collision viscosity attains the form

$$\nu_c = \frac{c^2}{D+2} \frac{\sum_{i\alpha\beta} Q_{i\alpha\beta}^2}{\sum_{ij\alpha\beta} Q_{i\alpha\beta} \Phi_{ij} Q_{j\alpha\beta}} \, . \tag{26}$$

The functional form of $X(\rho)$ for the bulk viscisity, can be determined in a similar fashion. In concise form, for the FHP lattice

$$\nu = -\frac{1}{4}\left(\frac{1}{2} + \frac{1}{\lambda}\right), \tag{27}$$

where λ is the eigenvalue of the linearized collision matrix having $Q_{i\alpha\beta}$ as eigenvector. Succi et al.[18] propose a new model, called LBE with enhanced collisions, based on the LBE in order to achieve maximum efficiency regardless of the collision rules. In LBE with enhanced collisions, the circulant matrix Φ_{ij} is determined exclusively on the basis of lattice symmetry and stability analysis, thus improving the efficiency of the method (in terms of the maximum Reynold number) and the algorithm implementation.

4. Body forces in LBE

The LBE successfully simulates Navier-Stokes equations in the case of incompressible flow and low Mach numbers. In this case, the pressure field P, Eq. (20) scales linearly with the density of the moving particle ρ_m. This represents an intrinsic constraint in the lattice gas dynamics, equivalent to an equation of state, which limits the possibility of forcing an arbitrary pressure drop without making use of an external body-force field. In the cellular model, this can be simply done by biasing collision rules in a given direction in order to produce a local net momentum source[1]. Equivalently, in LBE simulations pressure gradients and body forces can be included by simply adding a lattice body force[9,19,20] $f_i(\mathbf{r},t)$ to the microdynamic equation. The microdynamic equation thus becomes:

$$N_i(\mathbf{r}+\mathbf{c}_i, t+1) = N_i(\mathbf{r},t) + \sum_j \Phi_{ij}(N_j(\mathbf{r},t) - N_j^{eq}(\mathbf{r},t)) + f_i(\mathbf{r},t) \,. \tag{28}$$

Mass conservation implies

$$\sum_i f_i = 0 \,. \tag{29}$$

In the case of constant density simulations, the net local body force $\mathbf{f}$

$$\sum_i f_i \mathbf{c}_i = \mathbf{f} \tag{30}$$

is related to the pressure gradient ∇P and to volume forces $\rho\mathbf{a}$ through the relation

$$\mathbf{f} = -\nabla P + \rho\mathbf{a} \,. \tag{31}$$

In lattice simulations, the body force $\mathbf{f}$ may depend on space and time, be independent of velocity (e.g. gravity) or linear in the velocity (e.g. the Coriolis force) or derive from a generic potential distribution acting on the fluid. In two-dimensional FHP simulations, by further taking into account the different row and column spacing, the relation between the lattice vector f_i and the vector $\mathbf{f} = (\mathrm{f}_1, \mathrm{f}_2)$ is given by

$$f_i(\mathbf{r},t) = \frac{1}{4}\left(\mathrm{f}_1 - \frac{\mathrm{f}_2}{\sqrt{3}}\right) f_i^{(1)} + \frac{\mathrm{f}_2}{2\sqrt{3}} f_i^{(2)} \,, \tag{32}$$

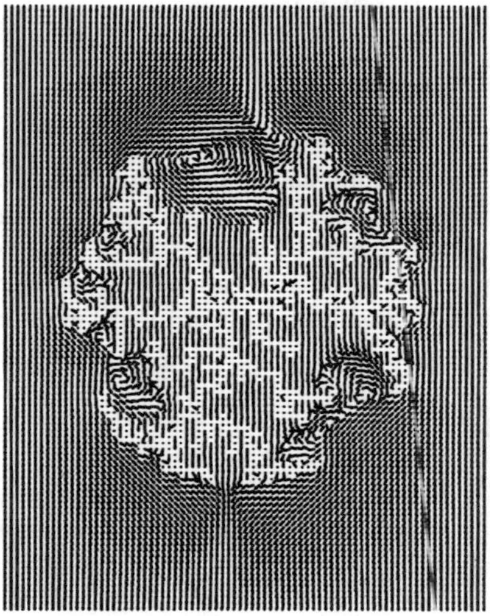

Figure 2: Velocity field of incompressible flow motion around a fractal object (such as a Diffusion Limited Aggregate (DLA)), for a linear pressure drop forcing the fluid in the y-direction. (DLA giration radius $R_g = 26$ lattice units, $Re = 5$).

where

$$f^{(1)} = \{1, 1, -1, -1, -1, 1\}, \quad f^{(2)} = \{1, 1, 1, -1, -1, -1\} \tag{33}$$

are lattice vector determining a net momentum transfer in the directions associated with the unit lattice vectors $\mathbf{c}_1$, $\mathbf{c}_2$ (see Figure 1).

5. Applications

Porous flow problems are usually solved by volume-averaged approaches. However in many applications the key point lies in understanding the functional relationship between volume-averaged parameters and the microscopic process occuring at the pore scale.

Because of the very complex boundaries occurring in pore structures, a macroscopic framework cannot provide exact solutions of the flow through porous media. LGA and the LBE are computational methods which easily handle arbitrary complex boundaries by representing the boundary conditions in terms of particle reflections and bouncing back [20,21,22,23]. For example, by modelling a complex pore medium as a random sequence of inaccessible and empty lattice sites on the surface of each blocked site, as well as on any lateral wall present, no-slip boundary conditions can be imposed: particles impinging in a given direction are reflected back into the fluid in the opposite direction. In this section we analyze several applications of body-

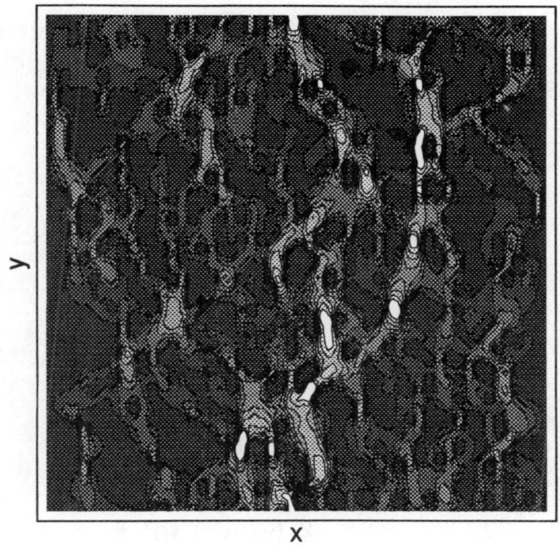

Figure 3: Velocity field contour plot of incompressible flow motion through a percolation lattice ($p = 0.75$) placed between two parallel plates; fluid is forced in the y direction by means of a linear pressure drop. The black clusters indicate regions with low values of velocity.

force representation associated with fluid dynamic problems in disordered and fractal structures, with natural convection phenomena and with flow of charged particles.

5.1. Incompressible flow through disordered media and fractal objects

As discussed above, the simplest way of generating a pressure gradient is to make use of a uniform body force throughout the whole lattice. As an example, figure 2 shows the velocity field of incompressible flow motion around a fractal object (such as a Diffusion Limited Aggregate (DLA)[24]), in the case of linear pressure drops.

Even at a very low Reynolds number ($Re = 5$ in the simulation), the complex structure of the object causes vortices and instability in fluid motion.

Detailed knowledge of the velocity field around fractal structures allows us to estimate the hydrodynamic radius and the friction factor and their dependence on the Reynolds number and on geometrical parameters. By considering different fractal objects (the simplest way of controlling the fractal dimensions is to generate deterministic fractal objects by means of Iterated Function Systems[25]), it is also possible to examine when the steady flow loses stability above the critical Reynolds number as a function of the fractal dimension.

The ease of handling complicated boundary conditions makes the LBE an excellent candidate to address long-standing problems in the physics of disordered and fractal

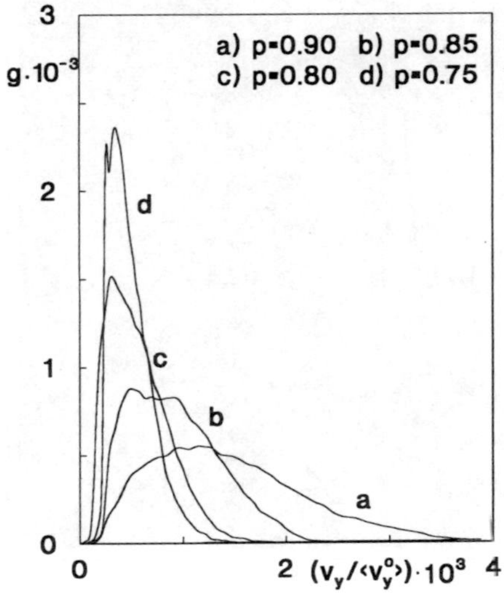

Figure 4: Probability distribution function $g(v_y/ < v_y^0 >)$ for 2D flow motion of an incompressible fluid through percolation lattices (with different percolation probability p) placed between two parallel plates. Fluid is forced along the y direction by means a linear pressure drop (the same for all the simulations).

media: the calculation of transport parameters and fluid mechanical properties in terms of the microscopic parameters characterizing the geometry of the structure.

In particular, the effect of disorder on convective-driven phenomena can be properly addressed by means of LBE simulations, possibly coupling this technique with other methods deriving from the stochastic theory of transport. Detailed analysis of the dependence of the permeability in twodimensional percolation lattices is developed elsewhere[26]. Here we focus on some statistical aspects of flow through porous media. Figure 3 shows the velocity-field contour plot of incompressible flow motion through a percolation lattice (generated by means of Leath's algorithm) placed between two parallel plates; fluid is forced in the y direction by means of a linear pressure drop. The black clusters indicate regions with low values of velocity.

Figure 3 clearly shows the slowing down of v_y in very large regions and a characteristic channeling phenomenon (clearest zone) in regions characterized by high accessibilty values[27]. The spatial irregularity of the velocity field does not disapperar even for high values of the percolation probability. It should also be noted that to this irregular spatial behaviour corresponds a smooth distribution of the mean velocity in the imposed direction $g(v_y/ < v_y^0 >)$, as can be seen from Figure 4. Here $< v_y^0 >$ is the mean velocity in the y direction in an empty twodimensional duct, for the same pressure drop.

It is possible to perform measurements of the local flow rate in packed beds by means of optical methods (laser Doppler measurements). The analysis of the local

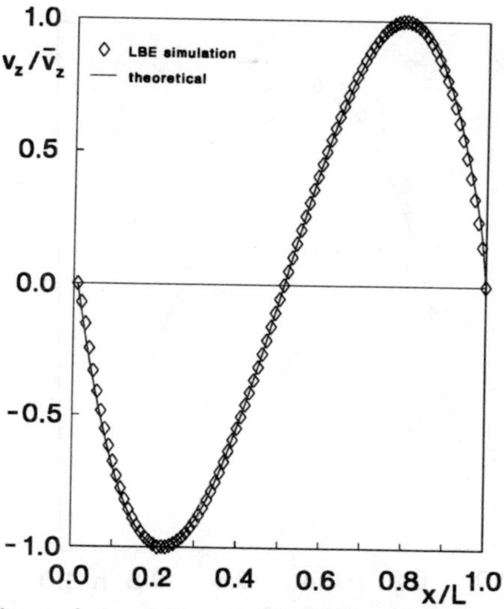

Figure 5: Comparison of simulation results and closed form solution for the velocity profile of an incompressible flow between two vertical parallel pla tes, in the presence of a linear temperature profile ($T = T(x)$). In the figure z is the direction parallel to the plates, x the normal direction and L the distance between th e plates. v_z is rescaled with respect to its maximum value $\overline{v}_z$.

velocity fluctuations has been recently discussed by Hosseini et al.[28]. It is important to point out that there is a close analogy between the velocity fluctuations and the characterization of disordered porous structures by means of a pointwise distribution of the local porosity, which indicates that it should be possible to describe the geometrical properties of a porous medium influencing convective transport by means of the statistical characterization of the local porosity field[27].

The existence of velocity fluctuations in macroporous structures indicates the importance of a detailed representation of the porous structure. Models of unit porous cells based on the representation of the porous structure in terms of a few equally distributed spherical or rectangular obstacles are in many cases an oversimplified picture of the structure of the medium, especially in connection with the prediction of convective transport parameters[29].

5.2. LBE for natural convection problems

Although in one-speed lattice models mass conservation implies energy conservation and density has to be strictly constant, natural convection phenomena can be addressed by means of a suitable body force, which takes into account the dependence of the density on the temperature gradients (Boussinesq approximation). In order to

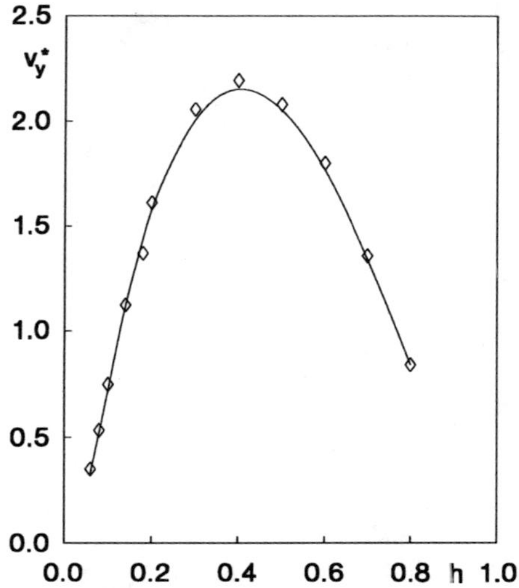

Figure 6: v_y^* vs. h ($\epsilon = 10^{-3}$, L=100 Lattice Units).

simulate natural convection phenomena, we can define a body force such that

$$\sum_i f_i(\mathbf{r}, t)\mathbf{c}_i = -\rho_o \beta_o \mathbf{g}(T(\mathbf{r}) - T_o) \ , \tag{34}$$

where β_o is the thermal expansion coefficient evaluated at a properly chosen mean temperature T_o, and T the pointwise temperature field, the solution of the heat balance equation (in stationary conditions)

$$k\nabla^2 T = v_\beta \partial_\beta T \ , \tag{35}$$

k being the thermal diffusivity .

As an example, we can consider the motion of an incompressible fluid between two parallel vertical plates, each mantained at a constant temperature T_1 and T_2 ($T_1 < T_2$). Figure 5 shows the LBE simulation results compared with the closed-form solution of the velocity profile. In its simplicity (uncoupling between heat and momentum balance equations), this example represents a test of the broad applicability of LBE to physics.

In the more general case in which coupling between heat and momentum transfer is considered, it is still possible to develop an integrated simulator (applicable also to complex geometries) in which, iteratively, the momentum balance is solved by means of the LBE and the heat balance by a finite difference scheme.

5.3. Body forces from potential field

By applying Eqs. (28) and (32), it is possible to simulate arbitrary distributions of potentials and pressure drops. As an example, it is a straightforward matter to study the fluid dynamic properties of a charged fluid in the presence of a "dipolar" $(1/r)$-potential on a periodic structure in isobaric conditions. The potential considered has the form

$$\phi(\mathbf{r}) = \phi(x,y) = \varepsilon(2\eta(y)-1)\int \frac{\delta(\mathbf{r}_o - R)}{|\mathbf{r}-\mathbf{r}_o|}dV(\mathbf{r}_o) \Big/ \int\int \frac{\delta(\mathbf{r}_o - R)}{|\mathbf{r}-\mathbf{r}_o|}dV(\mathbf{r}_o)dV(\mathbf{r}) \ , \tag{36}$$

where $dV(\mathbf{r})$ is the volume element, $\delta(\mathbf{r}_o - R)$ the Dirac delta function centered on the surface of a twodimensional ball of radius R (located in the middle of the lattice $\{(x,y)| - L \leq x, y \leq L\}$), $\eta(y)$ the step function ($\eta(y) = 1$ for $y > 0$ and $\eta(y) = 0$ for $y < 0$). Consequently, the "dipolar" potential is attractive for $y < 0$ and repulsive for $y > 0$, which produces a net momentum generation in the fluid in the y direction.

Figure 6 shows the behaviour of the dimensionless mean velocity $v_y^* = < \rho v_y R/\mu >$ as a function of the geometric parameter $h = R/L$ (the simulations were carried on a 100×100 exagonal lattice). The nonmonotonic behaviour observed for v_y^* is a consequence of the superposition of two opposite geometric and potential effects. For small h, the hindrance effect due to the presence of the obstacle is negligible. According to Eq. (36), $v_y^* \sim R$. For high values of h, the geometric effects prevail and v_y^* tends to zero. Therefore, there exists an intermediate value of h for which v_y^* attains a local maximum.

Many other problems involving charged fluids can be analyzed in similar fashion with the LBE (e.g. in connection with ionized plasma). It should be observed, however, that the introduction of a potential refers to the interaction of the whole fluid considered as a continuum. For this reason, in typical problems associated with solute motion (e.g. dispersion) in a stream of an inert carrier, it is not possible to follow such a direct approach and LBE methods should be extended.

References

1. U. Frisch, D. d'Humieres, B. Hasslacher, P. Lallemand, Y. Pomeau and J.P. Rivet, *Complex Systems* **1** (1987) 649.
2. M.A. Van Der Hoef and D. Frenkel, *Physica* **D 47** (1991) 191.
3. S. Chen, K. Diemer, G.D. Doolen, K. Eggert, C. Fu, S. Gutman and B.J. Travis, *Physica* **D 47** (1991) 72.
4. G.R. McNamara and G. Zanetti, *Phys. Rew. Lett.* **61** (1988) 2332.
5. S. Succi, R. Benzi and F. Higuera,*Physica* **D 47** (1991) 219.
6. F.J. Higuera and J. Himenez, *Europhys. Lett.***9** (1989) 663.
7. S. Succi, E. Foti and F. Higuera, *Europhys. Lett.***9** (1989) 433.
8. L. Kadanoff, G.R. McNamara and G. Zanetti, *Complex Systems* **1** (1987) 791.
9. R. Benzi and S. Succi, *J. Stat. Phys.* **60**, (1989) 69.

10. S. Wolfram *J. Stat. Phys.* **45** (1986) 471.
11. K. Diemer, K. Hunt, S. Chen, T. Shimomura and G. Doolen, *Lattice Gas Methods for Partial Differential Equations*, G. Doolen et al., eds. (Addison-Wesley, New York, 1990).
12. D. d'Humieres and P. Lallemand, *Complex Systems* **1** (1987) 599.
13. Y.H. Qian and S. Orszag, *Europhys. Lett.* **21** (1993) 255.
14. Y.H. Qian, D. d'Humieres and P. Lallemand, *Europhys. Lett.* **17** (1992) 479.
15. M. Vergassola, R. Benzi and S. Succi, *Europhys. Lett.* **13** (1990) 411.
16. G.R. McNamara, *Europhys. Lett* **12** (1990) 329.
17. J.T. Wells, D.R. Janecky and B.J. Travis, *Physica* **D 47** (1991) 115.
18. F.J. Higuera, S. Succi and R. Benzi, *Europhys. Lett.* **9** (1989) 345.
19. F.J. Higuera and S. Succi, *Europhys. Lett.* **8** (1989) 517
20. I. Ginzbourg and P.M. Adler, *J. Phys II France* **4** (1994) 191.
21. P. Papatzacos, *Complex Systems* **3** (1989) 383.
22. H. Rothman, *Geophys.* **33** (1988), 309.
23. S. Succi, E. Foti and F. Higuera, *Europhys. Lett.* **10** (1989) 433.
24. T. Vicsek, *Fractal Growth Phenomena* (World Sci., Singapore, 1989).
25. M. Barnsley, *Fractals Everywhere* (Academic Press, New York, 1989).
26. M. Giona and A. Adrover, this book.
27. M. Giona and A. Adrover, in *Characterization of Porous Solids III*, J. Rouquerol, F. Rodriguez-Renoso, K.S.W. Sing and K.K. Unger, eds. (Elsevier, Amsterdam, 1994), pp. 197-206.
28. M. E. Hosseini-Ashrafi and U. Tüzun, *Chem. Eng. Sci.* **48** (1993) 53.
29. O. A. Plumb and S. Whitaker, in *Dynamics of Fluids in Hierarchical Porous Media*, J. H. Cushman ed., (Academic Press, New York, 1990), pp. 97-176.

PARAMETER ESTIMATION IN TRANSPORT PHENOMENA THROUGH DISORDERED SYSTEMS

ANDREA REVERBERI
ISTIC, Universitá di Genova
Via Opera Pia 15, 16145 Genova, ITALY

LEONARDO ACEVEDO DUARTE
Departemento de Ingenieria Quimica, UIS
Bucaramanga, COLOMBIA

and

VINCENZO DOVI'
ISTIC, Universitá di Genova
Via Opera Pia 15, 16145 Genova, ITALY

ABSTRACT

An application of the theory of estimate in case of a model connecting experimental variables to parameters in the space of Laplace transforms by means of an antitrasformation technique is proposed. The model is the 3-d diffusion equation applied to transport phenomena in disordered systems. Simulated data are generated by a random number generation routine with 10% relative stochastic error. Percent error estimates of the anomalous diffusion exponent show a good reproducibility of test data and a satisfying robustness of the model.

1. Introduction

Nowadays a wide literature shows that there is a need for a new method in order to tackle some inadequacies that characterize the traditional approach when it is applied to fractal media. Many authors (e.g .Sahimi et al.[1]) stated that it seemed even impossible to successfully describe transport phenomena in fractals by means of the standard mathematical techniques applied for regular Euclidean structures. In fact, the use of a standard diffusion equation though implemented by adding both a volume scaling and an effective distance-depending diffusion coefficient $D_e(r)$ did not proved successful in order to correctly fit the shaped Gaussian distribution of the average probability density $P(r,t)$ referred to a walker diffusing in a random media, whose asymptotic behaviour is

$$P(r,t) = P_o(t)\left(\frac{r}{R(t)}\right)^{\alpha} \exp\left[-b\left(\frac{r}{R(t)}\right)^{u}\right] \tag{1}$$

Where u is bound to the anomalous diffusion exponent D_w by the relation

$$u = \frac{D_w}{D_w - 1} \,. \tag{2}$$

Giona and Roman[2] observed that the use of a fractional diffusion equation (FDE) could circumvent the drawbacks connected to both singularities of the geometric structure and to the anomalous behaviour in the mean square displacement with time, say

$$< r^2(t) >= K t^{\frac{2}{D_w}} \qquad D_w > 2 \, . \tag{3}$$

They proposed a generalization of the constitutive equation on fractals in a three-dimensional case by use of a consistent fractional diffusion equation of type:

$$\partial_t^{\frac{1}{D_w}} P(r,t) + A[\partial_r P(r,t) + \frac{k}{r} P(r,t)] = 0 \tag{4}$$

where r is the geometric coordinate, and ∂_t^q is the fractional derivative of order q, i.e.,[3,4]

$$\frac{\partial^q P(r,t)}{\partial t^q} = \frac{1}{\Gamma(n(q)-q)} \frac{\partial^{n(q)}}{\partial t^{n(q)}} (\int_0^t \frac{P(r,t)d\tau}{(t-\tau)^{q-n(q)+1}}) \tag{5}$$

A solution of the equation (3) can be sought in the space of Laplace transforms L, where, for a given $f(r,t)$,$(t>0)$,

$$L[f(r,t)] = \hat{f}(r,s) = \int_0^\infty e^{-st} f(r,t) dt \tag{6}$$

If $f(r,t)$ is assumed to be piecewise continuous in $[t]$ and of exponential order, there exists a real number C, so that $\hat{f}(r,s)$ is an analitic function for $Re[s] > C$. The number C is the abscissa of convergence of $\hat{f}(r,s)$. Remembering (3), the final expression will be:

$$\hat{P}(r,s) = Q(s) \frac{1}{(r s^{\frac{1}{D_w}})^k} \exp(-\frac{s^{\frac{1}{D_w}} r}{A}) \tag{7}$$

with the conditions that:

- $Q(s) \cong s^{-(1-\frac{D_f}{D_w})}$ to ensure normalization , i.e. $\int_0^\infty \hat{P}(r,s) dV = \frac{1}{s}$
- $\int_0^\infty r^2 \hat{P}(r,s) dr \cong s^{-(1-\frac{D_f}{D_w})}$ to ensure the consistency with the anomalous behaviour. D_f, D_w and k are parameters to be estimated during the regression procedure.

There are different techniques for numerically inverting the Laplace transform. However, only a few methods are suitable for automated inversion. Piessens and Huysmans[5] suggested to use an implemented version of Durbin's formula[6], where the approximate value $\hat{f}(r,s)$ for the inverse Laplace transform is given by

$$\hat{f}(r,t) = \frac{e^{at}}{T} \left\{ \frac{Re[\hat{f}(r,a)]}{2} + \sum_{k=1}^{\infty} Re\left[\hat{f}\left(r, \left\{a + \frac{ik\pi}{T}\right\}\right)\right] \cos\left(\frac{k\pi t}{T}\right) + \right.$$
$$\left. - \sum_{k=1}^{\infty} Im\left[\hat{f}\left(r, \left\{a + \frac{ik\pi}{T}\right\}\right)\right] \sin\left(\frac{k\pi t}{T}\right) \right\} \tag{8}$$

The parameters a and T may be chosen freely (with some restrictions), but their values affect strongly the discretization error. In this paper, a suggested function relating the values of the free parameters in (8) with the values of the independent variable t and the abscissa of convergence is used. The strategy allows us to perform the optimal trade-off between the need of a little error estimate and a good reliability of the global method. In fact, large values of T reduce the discretization error of expression (8) but, on the other hand, the evaluation of the infinite series slows down the convergence. In other words, it is possible to avoid the drawbacks that affect the usual spline interpolation techniques which are based on the determination of optimal values of the free parameter a.

2. Results and Discussion

As noticed by Huysmans and Crump[7], the series in (8) can be efficiently evaluated using the ϵ-algorithm[8]. The calculations are stopped so that the estimated truncation error (which is much greater than the discretization one),is kept within the accuracy requirement. The criterion must be also modelled in view of the further application of antitransformation results to the generation of simulated data. The latter are produced by adding to antitransformation results a 10% relative stochastic error. A FA03A - Harwell Library random number generation routine supplies the relevant noise distribution according to $N(0, \sigma)$

A well known technique in parameter estimation is based on the minimization of a suitable objective function, i.e.

$$\Phi = \sum_{\mu=1}^{n} \mid \bar{P}_\mu - P(\theta, r_\mu, t) \mid^2 = \min \tag{9}$$

where θ are the parameters, $\bar{P}_\mu$ the simulated values and $P(\theta, r_\mu, t)$ the theoretical values. Unfortunately we can not make use of an analitical expression of $P(\theta, r, t)$ as a result of an antitransformation of $\hat{P}(r, s)$ in (6). A Gauss-Newton estimate[9] of θ in (8). will finally supply:

$$\theta^{i+1} = \theta^i - N^{-1} q \tag{10}$$

where

$$N_{ij} = 2\sum_{\mu=1}^{n} \left(\frac{\partial P_\mu}{\partial \theta_i} \frac{\partial P_\mu}{\partial \theta_j} \right) \tag{11}$$

$$q_i = \frac{\partial \Phi}{\partial \theta_i} = -\sum_{\mu=1}^{n} 2[\bar{P} - P(\theta, r, t)] L^{-1} \left[\frac{\partial \hat{P}}{\partial \theta_i}(\theta, r, s) \right] \tag{12}$$

The regression is performed by means of a nonlinear parameter estimation DAPHNE algorithm. Figure 1 shows the behaviour of the percent error estimate versus the values of the anomalous diffusion exponent D_w at constant values of fractal dimension.

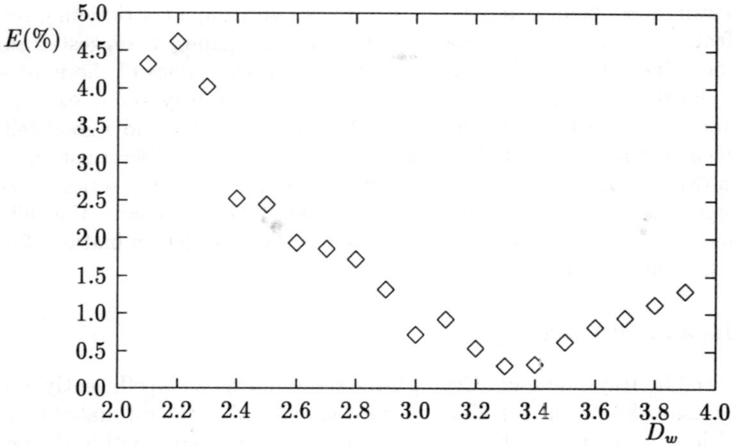

Figure 1: Percent error estimate $E(\%)$ of $P(r,t)$ vs the anomalous diffusion exponent D_w, for a fixed value of the fractal dimension $D_f = 2.5$.

The reader can observe a good reproducibility of test data. This is a satisfactory result that proves the robustness of the model on a significant range of parameter variations.

References

1. M. Sahimi, G. R. Gavalas and T.T. Tsotsis, *Chem. Eng. Sci.* **45** (1990) 1443.
2. M. Giona and H. E. Roman, *Chem. Eng. J.* **49** (1992) 1.
3. B. Ross, *Fractional Calculus and its Applications*, (Springer-Verlag, Berlin, 1974).
4. K. B. Oldham and J. Spanier, *The Fractional Calculus*, (Academic Press, New York, 1974).
5. R. Piessens and R. Huysmans, *ACS* **10** (1984) 348.
6. F. Durbin, *The Computer Journ.* **17** (1974) 371.
7. K.S. Crump *J. ACM* **23** (1976) 89.
8. P. Winn, *MTAC* **10** (1958) 91.
9. Y. Bard, *Non-linear Parameter Estimation*, (Academic Press, London, 1974).

DRUG RELEASE EFFECTS CONTROLLING PSEUDO NON FICKIAN DIFFUSION

ALESSANDRA ADROVER
Dipartimento di Ingegneria Chimica, Universitá di Roma "La Sapienza"
Via Eudossiana 18, 00184 Roma, ITALY

MASSIMILIANO GIONA
Dipartimento di Ingegneria Chimica, Universitá di Cagliari
Piazza d'Armi, 09123 Cagliari, ITALY

MARIO GRASSI, ROMANO LAPASIN and SABRINA PRICL
Dipartimento di Ingegneria Chimica, Universitá di Trieste
Piazza Europa 1, 34127 Trieste, ITALY

ABSTRACT

Controlled release from swollen gels is analyzed through numerical simulation of the release phenomenon from percolation lattices far from criticality. A two-phase model is also proposed to account for the apparent non-Fickian kinetics found experimentally.

1. Introduction

Controlled-release technology has found wide-ranging applications for the release of a variety of bioactive materials including pesticides, drugs, pheromones and plant regulators[1]. Conventionally, by focusing on the rate-controlling step in the release phenomenon, controlled-release delivery systems can be divided into: Diffusion controlled systems (reservoir or matrix devices), Chemically controlled systems (bioerodible or pendant chain), Swelling controlled systems and Magnetically controlled systems.

One of the open problems in the analysis of controlled release from swollen hydrogels is the interpretation of the experimentally found pseudo non-Fickian kinetics[2,3] expressed by a semiempirical power-law behaviour in the fractional solute release M_t/M_∞

$$M_t/M_\infty \sim t^n \qquad M_t/M_\infty < 0.5 \tag{1}$$

with an exponent n usually greater than 1/2, which means a deviation from the classical release models of Fickian diffusion[4]. In this article we show, by means of numerical simulation on percolation lattices, that the disordered and highly complex nature of the gel structure cannot be responsible for the anomalies found in sorption kinetics, underlining the differences between anomalous transport in disordered and fractal structures and the notion of "anomalous" transport in swollen gels.

The phenomenological origin of this anomaly should be sought in the presence of correlations in diffusive motion deriving from trapping or field effects which can be macroscopically described by means a two-phase percolation model able to predict an exponent $n > 1/2$, depending on an internal release kinetic coefficient. The predictions of the two-phase model proposed are checked against experimental results for the release of theophylline from water-swollen scleroglucan matrices and from HEMA-GMA mixtures in an acidic medium.

2. Percolative analysis of "anomalous" release

Drug release from swollen gels is a diffusion-step controlled phenomenon related to steric and potential interaction between solute molecules and macromolecular chains. An approach to this kind of problem is grounded on the simulation of transport in disordered structures representing the porous medium: transport parameters are evaluated from reasonable models of disorder of the pore network[5,6]. In order to understand the origin of the "anomalous " exponent n, we analyze the release kinetics from disordered models of polymeric matrices, taking percolation clusters as the paradigm of a disordered structure. Let us consider solute release from a swollen gel in slab form, with burst operating conditions. In a one-dimensional schematization of the phenomenon, the transport equation reads as:

$$\frac{\partial c}{\partial t} = D\frac{\partial^2 c}{\partial x^2} , \tag{2}$$

to be solved with the intial and boundary conditions

$$c(x,0) = c_0 , \quad 0 \le x \le L \tag{3}$$

$$\left.\frac{\partial c}{\partial x}\right|_{x=0} = 0 , \quad c(L,t) = 0 . \tag{4}$$

The latter boundary condition derives from the assumption of perfect sink conditions in $x = L$. D is the effective diffusion coefficient, which takes into account solute/matrix steric interactions. It is possible to solve Eq. (2) numerically with the initial and boundary conditions (3) on a two-dimensional disordered lattice structure by means of the simple explicit algorithm[7,8]

$$\begin{aligned} c_{ij}^{(n+1)} &= c_{ij}^{(n)} + \beta' \sum_{(h,k)\in I(i,j)} \chi(h,k) J_{hk\to ij}^{(n)} \\ &= c_{ij}^{(n)} + \beta \sum_{(h,k)\in I(i,j)} \chi(h,k)[c_{hk}^{(n)} - c_{ij}^{(n)}] , \end{aligned} \tag{5}$$

where $\beta = \mathcal{D}\Delta t/\Delta x^2$, and $\chi(i,j)$ is the characteristic function of the porous medium $\mathcal{M}$

$$\chi(i,j) = \begin{cases} 1, & (i,j) \in \mathcal{M} \\ 0, & \text{otherwise} . \end{cases} \tag{6}$$

Figure 1 shows the log-log plot of sorption curves for percolation lattices with burst operating conditions. Simulations on percolation lattices give a value of n (slope of

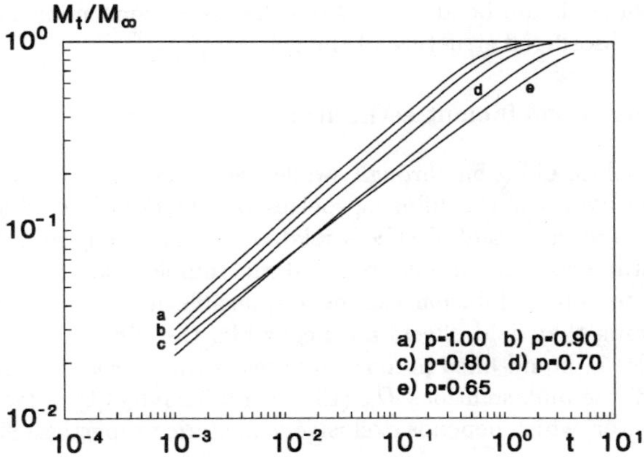

Figure 1: Log-log plot of M_t/M_∞ vs t [a.u.] (D=1 [a.u]) for burst-effect simulations on percolation lattices for different values of the percolation probability p.

the curves in Figure 1) equal to 0.5 for values of the percolation probability p far from the critical threshold (p_c=0.593). Only for $p = 0.65$ does the slope change slightly and attain the apparent value of $n = 0.38 \pm 0.03$. This result can be explained by considering that for p tending towards criticality the highly heterogeneous nature of the porous matrix emerges, as appears from anomalous characteristics in diffusive motion ($n \neq 1/2$). However, the resulting exponent n is always less than 1/2. Consequently, an anomalous exponent n greater than 1/2 (which implies an enhancement of the release kinetics) cannot possibly be related to non-Fickian diffusion, as in the case of fractal structures. In the theory of dynamic phenomena on fractals[9], anomalous transport means deviation from Fick's law in the specific sense that the mean square displacement of a Brownian particle, $< r^2(t) >$, deviates from a linear behaviour with time and follows a scaling law

$$< r^2(t) > \sim t^\beta \ , \tag{7}$$

with an exponent β less than unity. Eq. (7) is valid asymptotically, i.e. for $t \to \infty$ and β is a universal exponent characteristic of the fractal structure of the medium. Eq. (7) implies that the heterogeneity at all the lengthscales of a fractal structure induces a characteristic slowing down ($\beta < 1$) in the diffusional propagation for all the time scales. Anomalous transport in swollen polymers also shows an exponent n which deviates from the classical value $n = 1/2$ (which is strictly the counterpart of $\beta = 1$ for the scaling of the mean square displacement), but the physical origin of this deviation is completely different from the case of diffusion in disordered or fractal media. The scaling law of Eq. (1) is not valid asymptotically, as in the case of Eq. (7), and is not a manifestation of universal physical properties, as in the case of the exponent β. Starting from these considerations, the origin of the apparent non-Fickian behaviour should be attributed to other kinetic mechanisms rather than to the microscopic disordered structure of the gel.

3. Effective diffusion coefficient estimation

In the case of solute diffusion through swollen gels, the mechanism of transport depends on the relative size of the diffusing species, on the mesh formed by the macromolecular chains[10] and on potential effects related to solute/polymeric interactions. Morphological features associated with reduced macromolecular chain mobility and potential barriers to solute diffusion can be responsible for the decrease in solute diffusivity. Assuming that field effects are negligible, the diffusion coefficient D of the solute molecules in liquid-filled porous matrices deviates from the corresponding coefficient in the dilute bulk solutions D_E (Einstein diffusivity) by a factor G, called the accessibility factor, which depends exclusively on steric properties (Size Exclusion Effect, SEE),[11]

$$D = D_E G\,. \tag{8}$$

On the basis of equation (2) with burst operating conditions, Eqs.(3-4), the accessibility factor G and consequently the effective diffusion coefficient can be evaluated from the typical exponential relaxation associated with diffusive motion

$$\ln(1 - M_t/M_\infty) \sim -\frac{\pi^2}{4L^2} Dt\,. \tag{9}$$

Eq. (9) can be considered an important stage in the analysis of experimental data of controlled release from swollen gels. In fact, the assumption of a macroscopically constant effective diffusion coefficient also in the presence of erosion effects or swelling increasing during the experiment can be validated by an experimentally found exponential decay of $1-M_t/M_\infty$ with time. Many authors suggest using a time-dependent drug-diffusion coefficient to take into account the possible increasing of swelling during the release experiment

$$D(t) = D_\infty(1 - e^{-\gamma t})\,, \tag{10}$$

where D_∞ is the diffusion coefficient at the swelling equilibrium and γ a relaxation constant characteristic of the specific polymer-drug-solvent combination [4]. A dependence of D on t implies

$$1 - M_t/M_\infty \sim \exp[-{\lambda_1}^2 \int_0^t D(t')dt']\,, \tag{11}$$

λ_1 being the eigenvalue associated with the initial and boundary conditions (for burst operating conditions $\lambda_1 = \pi/(2L)$. If the experimental sorption curves seem to deviate from Eq. (9), then the swelling increase significantly affects the phenomenon: it cannot be neglected and a functional dependence of D on t should be considered and checked with Eq. (10). The analysis of diffusional relaxation in the presence of a concentration dependent diffusion coefficient is developed elsewhere[7].

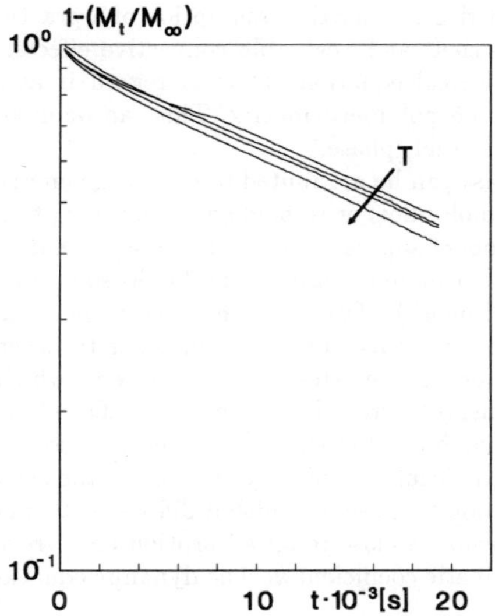

Figure 2: Log-normal plot of $1-(M_t/M_\infty)$ vs t for different values of the temperature T in the range 20° C $\leq T \leq 35^o$ C for the system thophylline/scleroglucan. The arrow indicates the increasing temperature.

Under the assumption of a pratically constant diffusion coefficient, Eq. (9) can be used to evaluate the diffusion coefficient for a specific solute diffusing through the gel without any information regarding the final porosity or the average pore radius of the gel. Figure 2 shows the behaviour of $1-M_t/M_\infty$ vs t (for different values of the temperature in the range 20° C $\leq T \leq 35^o$ C) for the system theophylline/water-swollen scleroglucan matrix (in slab form of thickness $L = 6$ mm). The initial loading of the matrix is c_0=1.0 [mg/gr H_2O] and the experimental boundary condition are well described by Eqs. (3-4). The swelling degree $s = 49$ [mg H_2O/mg polymer] is practically constant during the experiments. The value of D evaluated from equation (9) for the operating conditions described above is $6.0 \pm 0.7 \cdot 10^{-6}$ [cm^2/s]. It can be seen (from Figure 2) that the dependence of D on T may be neglected in practical terms by further considering that the variability range is of the same order of magnitude as the error in evaluating D itself.

4. Two-phase kinetic model

The results of the previous section indicate that an exponent n greater than 1/2 is not related to anomalous diffusion effects in disordered media. Deviation from $n = 1/2$ can be due to different phenomena : concentration dependent-diffusion coefficient, erosion of the gel matrix, convective contribution to mass transport, etc. In the case of non-erodible matrices, under the assumption of a practical concentration-independent diffusion coefficient and negligible convective effects, a simple way of explaining the phenomenological equation (1) is to regard it as a consequence of interaction between solute and polymeric matrix. This can be modeled by assuming the presence of an "trapped" (gel) phase.

The presence of this phase can be attributed to generic geometric trapping effects (especially for large solute molecules) or to field effects deriving from potential interactions. In a macroscopic model approach, both these phenomena lead us to consider the presence of two phases: a mobile phase in the liquid solution (i.e. a sol phase) and a stationary phase (gel phase). Two-phase models in the analysis of controlled release have been considered by other authors[3,12] in order to describe the effects of interfacial transfer kinetics on release rates for those systems which show significant free-energy barriers for transport across liquid-liquid interfaces[3] or to model the release condition above the solubility threshold[12]. In the proposed model we assume that a fraction ϵ_s of solute molecules is initially trapped in the stationary phase and irreversibly released in the mobile phase, in which it diffuses with an effective diffusion coefficient D. The phenomenon is close to an adsorption and irreversible desorption first-order kinetics with a kinetic coefficient k. The dynamic equations describing the phenomenon are

$$\frac{\partial c}{\partial t} = D\frac{\partial^2 c}{\partial x^2} - \frac{\partial q}{\partial t} \tag{12}$$

$$\frac{\partial q}{\partial t} = -kq \, , \tag{13}$$

where $c(x,t)$ and $q(x,t)$ are drug concentrations in the stationary and mobile phases respectively. Equations (12-13) are to be solved with the initial and boundary conditions (burst effect)

$$c(x,0) = (1-\epsilon_s)c_0 \quad 0 < x < L \, , \qquad q(x,0) = \epsilon_s c_0 \, , \tag{14}$$

$$\left.\frac{\partial c}{\partial x}\right|_{x=0} = 0 \, , \quad c(L,t) = 0 \, . \tag{15}$$

Even in this case, the asymptotic relaxation behaviour of M_t/M_∞ is given by Eq. (9) if $kL^2/D > \pi^2/4$, which means that the kinetic rate k is much larger than the diffusive rate D/L^2. In the simulations on percolation lattices, the stationary phase is represented by the first nearest neighbouring site of each site representing the polymeric matrix (see Figure 3) and is therefore implicitly a function of position since it is defined at the matrix boundary sites. Figure 4A shows the simulation results of the two-phase kinetics on the percolation cluster for different values of the kinetic constant k [a.u.], keeping constant the values of the initial partition coefficient ε_s and of the percolation probability p ($\varepsilon_s = 1$, p=0.80). The model furnishes values of the exponent n (evaluated at intermediate time scales) which cover all the "non-Fickian" range. This is more evident from Figure 4B, which shows the behaviour of n vs k for different values of the percolation probability. All the curves satisfy the condition

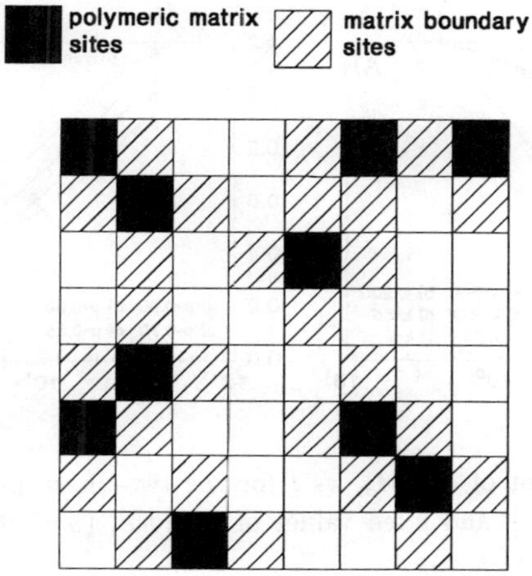

Figure 3: Schematic representation of two-phase simulation in percolation lattices. In the simulations the concentration q is defined only on the nearest neighbours of a matrix site (matrix boundary site) while c is defined at all the sites not belonging to the polymeric matrix.

that, in the limit of large values of the ratio kL^2/D , the internal release kinetics is so fast with respect to the diffusive motion in the sol phase that the presence of the gel phase can be neglected in practice. Consequently, the two-phase model tends to the classical diffusion model and n converges towards $1/2$.

Therefore, we may conclude that a simple kinetics model which takes into account the internal-release kinetics between a stationary and a mobile phase is able to predict a release exponent n greater than $1/2$. As the simulations show, the value of n depends on the kinetic parameters introduced in the model.

5. Experimental validation of the two-phase model

Figure 5 shows the comparison of the experimental release curve of theophylline from scleroglucan (for the experimental conditions discussed above) with the prediction of the two-phase model with $D = 6.0 \cdot 10^{-6}$ [cm^2/s] (see section 3) and $k = 1.5 \cdot 10^{-3}$ [s^{-1}] (evaluated as a best-fit parameter for $M_t/M_\infty < 0.5$. The agreement is fairly good. It should also be observed that, having no information about the initial division of the solute molecules between the two phases, we have assumed $\varepsilon_s = 1$. The agreement between the theoretical model and the experimental results would probably be closer if the value of ε_s could be estimated a priori. This would involve a better understanding of the nature of the drug/matrix interactions and will be one of the future steps in our investigation. The two-phase model can also be

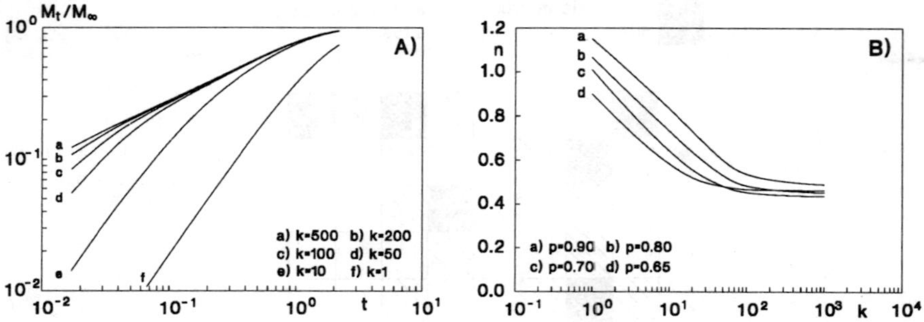

Figure 4: A) Log-log plot of M_t/M_∞ vs t for the two-phase percolation model for different values of k [a.u.] and fixed values of p and ε_s (p=0.80, ε_s=1); B) n vs k [a.u.] for different values of p, ε_s=1.

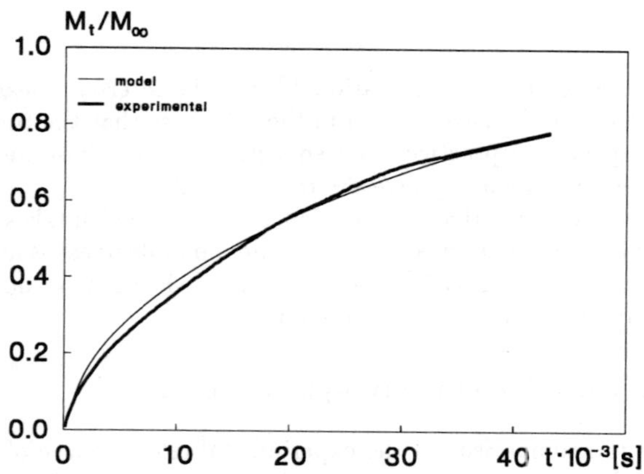

Figure 5: Comparison of the experimental release curve of theophylline from scleroglucan ($T = 30^o$ C, $c_o = 1$ [mg/g H_2O]) with the two-phase model predictions ($D = 6.0 \cdot 10^{-6}$ [cm^2/s], $k = 1.5 \cdot 10^{-3}$ [s^{-1}], $\varepsilon_s = 1$).

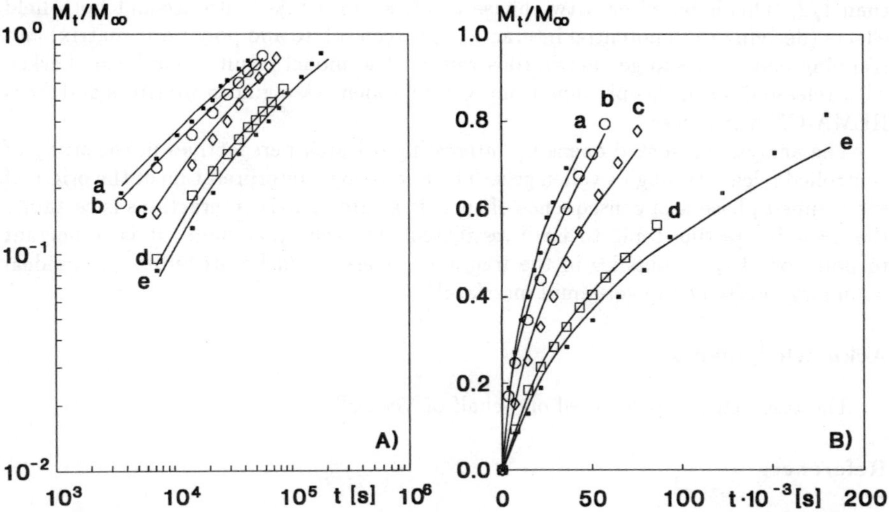

Figure 6: A) Comparison of the experimental data (point) of M_t/M_∞ taken from Vyavahare et al.[14] for theophylline in 0.05 N H_2SO_4 at 37 °C with the two-phase model predictions (curves) in log-log scale: a) p(HEMA-15%GMA) copolymer; b) p(HEMA-20%GMA) copolymer; c) p(HEMA-30%GMA) copolymer; d) p(HEMA-40%GMA) copolymer; e) p(HEMA-50%GMA) copolymer; B) Same data as Figure 6a in normal scale.

applied in controlled-release experiments in the presence of moderate swelling effects. Vyavahare et al.[13] present experimental data on the release of theophylline from 2-hydroxyethyl methacrylate-glycidil methacrylate (HEMA-GMA) mixtures in acidic medium. Figures 6A-B show satisfactory agreement between the fractional release data of Vyavahare et al. and the two-phase model.

The only case in which theory and experiment do not fit is that of Figure 6B-e, for which Vyavahare et al. found the maximal relative swelling degree increment $\Delta s = 0.1701$ [g H_2O /g polymer] during the release compared to the initial swelling degree $s_0 = 0.2340$ [g H_2O / g polymer].

6. Concluding remarks

The analysis of controlled-release kinetics in disordered models (percolation clusters) of polymeric matrices has shown that the "anomalous diffusion" experimentally found in the release kinetics from swollen gels cannot possibly be related to hindered diffusive motion in disordered media, i.e. is not a consequence of the disordered structure of the matrix. A kinetic model is proposed to account for an exponent n greater than 1/2. This is based on a two-phase kinetics which takes into account both field effects (deriving from potential interaction between solute and polymeric matrix) and trapping effects due to geometric constraints. The model results have been checked with release data of theophylline from water-swollen scleroglucan matrices and from HEMA-GMA matrices.

The analysis presented opens up interesting research perspectives in the study of controlled release through swollen gels. The microscopic interpretation of the origin of the trapped phase as a consequence of specific solute/matrix interactions is certainly the most interesting topic to be investigated. On the other hand, it is important to point out that anomalies in the fractional relase could be attibuted to nonideal boundary effects at the gel-liquid interface[14].

Acknowledgments

The research was performed on behalf of VectorPharma.

References

1. R. W. Baker and H. K. Lonsdal, in *Controlled Release of Biologically Active Agents*, eds. A. C. Tanquary and R. E. Lacey (Plenum Press, New York, 1974).
2. P. L. Ritger and N. A. Peppas, *J. Controlled Release* **5** (1987) 177.
3. J. Hadgraft, *Int. J. of Pharmaceutics* **2** (1979) 177.
4. J. Crank, *Mathematics of Diffusion* (Clarendon Press, Oxford, 1975).
5. R. A. Siegel, J. Kost and R. Langer, *J. Controlled Release* **8** (1989) 223.
6. M. Giona, A. Adrover and D. Spera, *Chem. Biochem. Eng. Q.* **7** (1994) 199.
7. M. Giona and A. Adrover, this book.
8. M. Giona, A. Adrover and A.R. Giona, in *Fractals in the Natural and Applied Sciences*, ed. M. M. Novak, (North Holland, Amsterdam, 1994), pp. 153-163.
9. A. Bunde and S. Havlin, *Fractals and Disordered Systems*, (Springer-Verlag, Berlin, 1992).
10. C. T. Reinhart and N. A Peppas, *J. Membrane Sci.* **68** (1992) 93.
11. W.M. Deen, *AIChE J.* **33** 1409-1425.
12. R. S. Harland, C. Dubernet, J. P. Benoit and N. A. Peppas, *J. Controlled Release* **7** (1988) 207.
13. N. R. Vyavahare, M. G. Kulkarni and R. A. Mashelkar, *J. Membrane Sci* **54** (1990) 221.
14. M. Grassi, R. Lapasin, S. Pricl (private communication).

FRACTAL ADSORPTION ISOTHERMS: A CRITICAL COMPARISON OF ENERGETIC AND GEOMETRIC DESCRIPTIONS OF HETEROGENEITY IN ADSORPTION

MANUELA GIUSTINIANI, LUIGI MARRELLI
Dipartimento di Ingegneria Chimica, Universitá di Roma "La Sapienza"
Via Eudossiana 18, 00184 Roma, ITALY

MASSIMILIANO GIONA and ANTONIO VIOLA
Dipartimento di Ingegneria Chimica, Universitá di Cagliari
Piazza d'Armi, 09123 Cagliari, ITALY

ABSTRACT

A critical analysis of thermodynamic models for describing adsorption equilibria on heterogeneous surfaces is presented, focusing attention on the thermodynamic consistency and on the geometric/energetic description of heterogeneity. The Keller model of adsorption isotherms is applied to literature data of adsorption equilibria on molecular sieves and the apparent fractal dimension of zeolitic materials is discussed.

1. Introduction

The application of fractal concepts to describe the heterogeneous structure of surface and microporous materials began with the work of Avnir, Farin and Pfeifer[1,2], who collected and analyzed adsorption equilibrium data for many porous solids and surfaces and found that the monolayer coverage n_m can be related to the geometric size of the molecules (the adsorption cross-section σ_i or the molecular radius r_i) through a fractal scaling law

$$n_m \sim \sigma_i^{-D/2} \sim r_i^{-D} , \tag{1}$$

where D is the fractal dimension of the surface. By applying Eq. (1) or similar scaling relationships deriving from a geometric approach to the monolayer covering, it has been found[3] that many surfaces exhibit a fractal structure with a characteristic fractal dimension between 2 and 3 at molecular scales.

These results have been followed by many other contributions, connecting the fractal structure of a surface with the thermodynamical properties[4], with polymer adsorption in a fractal distribution of sites[5], with the kinetics of adsorption processes[6] and with the relation existing between geometry and adsorption/desorption hysteresis[7]. Despite this highly diversified literature, the fractal approach to adsorption has been always considered a useful analytical chemistry technique for obtaining some characterization of rough surfaces.

The introduction of fractal models to describe rough surfaces had no direct impact on the characterization of adsorption for engineering purposes. The usual models widely applied in chemical engineering to characterize adsorption on heterogeneous surfaces overlook geometric effects and describe heterogeneity or by means of pure semiempirical correlations (just to fit the experimental data) or by introducing a continuous distribution of adsorption energies[8].

Outstanding work on the relationship between the thermodynamics of adsorption and the intrinsic geometric characterization of the adsorbent was developed by Keller[9], who proposed a class of thermodynamically consistent adsorption isotherms (AI), hereafter referred to *Keller isotherms*.

The scope of this article is twofold: to analyze the geometric and energetic description of adsorption equilibria, demonstrating that Keller analysis provides a fundamental tool for a rational evaluation of adsorption isotherms in heterogeneous surfaces and microporous materials; to address the problem of the fractal characterization of zeolitic microporous adsorbent.

The article is organized as follows. First we review the basic definition of the continuous description of heterogeneous adsorption. Then we analyze the properties of Keller AI and apply this class of AI to the adsorption equilibria on zeolites in order to ascertain the origin of the apparent fractal dimension observed experimentally.

2. Thermodynamic analysis

2.1. The continuous approach to heterogeneous adsorption

In modelling equilibria on heterogeneous surfaces it is customary to assume the existence of a continuous distribution $g(E)$ of adsorption energies E[10,11]. Therefore, if $n_e(P,T;E)$ indicates the adsorbed concentration ([mol/g Adsorbent]) for a given value of E, the overall AI, $n(P,T)$ is given by

$$n(P,T) = \int n_e(P,T;E)g(E)dE \ , \quad \int g(E)dE = 1 \ . \tag{2}$$

It has been shown by Sircar and Myers[8] that the choice of the elementary function n_e (which can be e.g. of Langmuir shape or of step shape) does not significantly influence the overall adsorption isotherm. The key problem in the continuous modelling is thus to obtain $g(E)$ from AI and from calorimetric measurements (isosteric heat of adsorption)[10,11].

However, the continuous approach to adsorption equilibria displays two fundamental shortcomings:

- the geometric properties of the surface are not considered;
- it is generally assumed that adsorbed particle-particle interactions are negligible and therefore that the adsorption energies are independent of the surface coverage (or of the partial pressure in the fluid phase).

Both these limitations can make it impossible to establish a correct theoretical position on fluid-solid adsorption equilibria and neglect of these points is one reason why AI-analysis is mostly reduced in application to a black-box parameter fitting.

Nevertheless, it is possible to demonstrate on purely thermodynamic grounds that the introduction of the two above-mentioned effects into AI models serves precisely to make the analysis thermodynamically consistent.

2.2. Keller adsorption isotherms: thermodynamic consistency and geometric scaling

The thermodynamics of equilibria between a fluid phase (f) and the adsorbed phase (a) in a multicomponent mixture can be developed by including in the expression of the Gibbs free energy the energetic contribution derived from the spreading of the adsorbate molecules over the surface[12]. This can be achieved by introducing a pair of intensive/extensive variables, the spreading pressure π and the extension area A, so that the Gibbs free energy G^a and its differential attain the form*

$$G^a = \sum_{i=1}^{N} \mu_i^a n_i \,, \tag{3}$$

$$dG^a = -S^a dT + A d\pi + \sum_{i=1}^{N} \mu_i^a dn_i \,, \tag{4}$$

where $\{n_i\}$ is the set of molar concentrations in the adsorbed phase per unit mass of the adsorbent. Equilibrium conditions imply that the chemical potentials of each component in the two phases are equal

$$\mu_i^a = \mu_i^f \qquad (i = 1,..,N) \,. \tag{5}$$

A complete set of thermodynamic variables describing the state of the adsorptive fluid is given by $\Sigma^f = (T, \{P_i\})$, where $\{P_i\}$ is the set of partial pressures in the fluid phase, while the adsorbed phase can be described by means of the set of variables $\Sigma^a = (T, \{n_i\})$.

By developing the differential of G^a from Eq. (3) and comparing the result with Eq. (4) (bearing in mind also that $\pi = \pi(T, \{n_i\})$) one obtains

$$A\left(\frac{\partial \pi}{\partial T}\right)_{\mathbf{n}} = S^a + \sum_{k=1}^{N} \left(\frac{\partial \mu_k^a}{\partial T}\right)_{\mathbf{n}} n_k \tag{6}$$

$$A\left(\frac{\partial \pi}{\partial n_i}\right)_{T,\mathbf{n}} = \sum_{k=1}^{N} \left(\frac{\partial \mu_k^a}{\partial n_i}\right)_{T,\mathbf{n}} n_k \qquad (i = 1,..,N) \,. \tag{7}$$

The last set of equations can be further developed by making use of Eq. (5) and by introducing the expression of the chemical potentials in terms of the fugacities f_i,

$$\mu_i^a = \mu_i^f = \mu_i^{of}(T) + RT \log\left(\frac{f_i}{P^o}\right), \quad (i = 1,..,N) \tag{8}$$

---

*the definition of the Gibbs energy (Eqs. (3) and (4)) is fully equivalent to the definition adopted by Ruthven[12], whereby $G^a = -A\pi + \sum_{i=1}^{N} \mu_i^a n_i$ and consequently $dG^a = -S^a dT - \pi dA + \sum_{i=1}^{N} \mu_i^a dn_i$

in order to obtain the relations

$$A\left(\frac{\partial \pi}{\partial n_i}\right)_{T,\mathbf{n}} = RT\sum_{k=1}^{N}\frac{n_k}{f_k}\left(\frac{\partial f_k}{\partial n_i}\right)_{T,\mathbf{n}} \qquad (i=1,..,N)\,. \tag{9}$$

Eq. (9) is obtained by considering that the thermodynamic state of the system can be referred to the adsorbed-phase variables Σ^a. Equivalently, by regarding Σ^f as state variables, the following equations hold true

$$A\left(\frac{\partial \pi}{\partial T}\right)_{\mathbf{P}} = S^a + \sum_{k=1}^{N}\left(\frac{\partial \mu_k^a}{\partial T}\right)_{\mathbf{P}} n_k \tag{10}$$

$$A\left(\frac{\partial \pi}{\partial P_i}\right)_{T,\mathbf{P}} = \sum_{k=1}^{N}\left(\frac{\partial \mu_k^f}{\partial P_i}\right)_{T,\mathbf{P}} n_k = RT\sum_{k=1}^{N}\frac{n_k}{f_k}\left(\frac{\partial f_k}{\partial P_i}\right)_{T,\mathbf{P}} \qquad (i=1,..,N)\,. \tag{11}$$

From Eq. (10), by considering the definition of the specific entropies $s_i^f = -(\partial\mu_i^f/\partial T)_{\mathbf{P}}$ and the definition of G^a, one obtains

$$\Delta S = S^f - S^a = -A\left(\frac{\partial \pi}{\partial T}\right)_{\mathbf{P}} \quad ; \quad \Delta H = H^f - H^a = -AT\left(\frac{\partial \pi}{\partial T}\right)_{\mathbf{P}} \,. \tag{12}$$

By starting from Eq. (4) and developing the differential in terms of the state variables Σ^f, the following closure conditions (Maxwell equations) are recovered

$$\left(\frac{\partial A}{\partial P_i}\right)_{T,\mathbf{P}}\left(\frac{\partial \pi}{\partial P_j}\right)_{T,\mathbf{P}} + \sum_{k=1}^{N}\frac{RT}{f_k}\left(\frac{\partial f_k}{\partial P_i}\right)_{T,\mathbf{P}}\left(\frac{\partial n_k}{\partial P_j}\right)_{T,\mathbf{P}} =$$
$$\left(\frac{\partial A}{\partial P_j}\right)_{T,\mathbf{P}}\left(\frac{\partial \pi}{\partial P_i}\right)_{T,\mathbf{P}} + \sum_{k=1}^{N}\frac{RT}{f_k}\left(\frac{\partial f_k}{\partial P_j}\right)_{T,\mathbf{P}}\left(\frac{\partial n_k}{\partial P_i}\right)_{T,\mathbf{P}} \,, \quad (i\neq j=1,..,N)\,, \tag{13}$$

which represent a set of $N(N-1)/2$ equations to be identically satisfied by a *thermodynamically consistent* model of AI, $n_i = n_i(T,\{P_i\})$. The Maxwell relations, Eq. (13), reduce to simple expressions under the assumption that the extension area does not depend on the partial pressures and that the fluid phase is an ideal gas phase, $f_i = P_i$

$$\frac{1}{P_i}\left(\frac{\partial n_i}{\partial P_j}\right)_{T,\mathbf{P}} = \frac{1}{P_j}\left(\frac{\partial n_j}{\partial P_i}\right)_{T,\mathbf{P}} \qquad (i\neq j=1,..,N)\,. \tag{14}$$

Keller[9,13] has shown that a class of adsorption isotherms satisfying the Maxwell closure equations are given by

$$n_i(T,\{P_i\}) = n_\infty(T)\Phi(f^*)\left[\alpha_i c_i f_i^{\alpha_i} + \frac{f_i}{RT}\sum_{k=1}^{N}\alpha_k c_k f_k^{\alpha_k}\left(\frac{\partial E_k}{\partial f_i}\right)_T\right] \,, \tag{15}$$

where Φ is a continuous differentiable function of its argument (e.g. of Langmuir shape $\Phi(x) = (1+x)^{-1}$), $E_k = E_k(T, \{f_i\})$ are the adsorption energies, and the quantities c_k, f^* are defined as

$$\begin{aligned} c_k &= [P_{k\infty} \exp(-E_k/RT)]^{-\alpha_k} , \\ f^* &= \sum_{k=1}^{N} c_k f_k^{\alpha_k} . \end{aligned} \tag{16}$$

Moreover, the enthalpy difference ΔH, Eq. (12), is given, for an ideal fluid phase, by

$$\Delta H = \sum_{i=1}^{N} \left\{ E_i n_i - \frac{n_\infty(T)\Phi(f^*) f_i E_i}{RT} \sum_{k=1}^{N} \alpha_k c_k f_k^{\alpha_k} \left(\frac{\partial E_k}{\partial f_i} \right)_T \right\} . \tag{17}$$

To complete the picture,

$$\pi = \Psi(f^*) , \tag{18}$$

where Ψ is a generic differentiable and increasing function of its argument and

$$A = n_\infty(T) RT \frac{\Phi(f^*)}{\Psi'(f^*)} , \tag{19}$$

where Ψ' indicates the derivative of Ψ with respect to its argument. In the next section we discuss how fractality comes into play and what theoretical and practical improvement is offered by Keller AI.

2.3. Fractal dimension as closure condition

Let us consider Eq. (15) in the case of monolayer covering, and let us assume Φ of Langmuir form, $\Phi(x) = (1+x)^{-1}$. Hereafter we will always consider the ideal gas phase, $f_i = P_i$. In general, $E_i(T, \{P_i\})$ are bounded decreasing functions of P_i, so that

$$\lim_{P_i \to \infty} n_i(T, \{P_i\}) = n_\infty(T)\alpha_i \quad (i = 1, .., N) . \tag{20}$$

By comparing Eq. (20) with Eq. (1), it follows that

$$n_\infty(T)\alpha_i \sim r_i^{-D} \qquad (i = 1, .., N) . \tag{21}$$

Therefore, given one reference species of radius r_r and exponent α_r, all the other exponents α_i can be obtained from Eq. (21) as

$$\alpha_i = \alpha_r \left(\frac{r_r}{r_i} \right)^D \qquad (i = 1, .., N) . \tag{22}$$

In other words, the exponents α_i are not independent of each other but, insofar as the scaling law Eq. (1) holds true, are obliged to satisfy Eq. (22). In this framework, the application of the fractal scaling law, Eq. (1), represents a closure condition to be imposed on the Langmuir-Freundlich exponents α_i.

According to these observations, the effects of surface roughness are expressed by three parameters: the reference exponent α_r, related to the monolayer coverage of the reference species; the extensive value $n_\infty(T)$, which does not depend on the adsorbate and is a function uniquely of the temperature; and the intensive variable D, the fractal dimension of the adsorbent solid, describing the scaling of the surface roughness with the yardstick size.

It should be observed that Keller AIs enable us to distinguish the geometric effects of surface heterogeneity (expressed by α_r, n_∞, D) from energetic heterogeneity (represented in the model by the functional dependence of the adsorption energies E_i on the partial pressures). The decoupling of these two contributions cannot be obtained by approaching the thermodynamic analysis of adsorption within the continuous framework (see section 2.1). Indeed, as the results of Sircar and Myers[8] show, the overall AIs are practically independent of the choice of the elementary function n_e and all the effects associated with geometric/energetic heterogeneity are concentrated on the distribution function $g(E)$. Moreover, and this is the crucial theoretical point, continuous AI models are thermodynamically inconsistent, while Keller AIs are strictly derived from thermodynamic arguments.

2.4. Some practical observations

Let us first consider the case of constant adsorption energies, E_i independent of $\{P_i\}$, and of a Langmuir-shaped $\Phi(x)$. In this case, Eq. (15) reduces to the Langmuir-Freundlich equation

$$n_i(T,\{P_i\}) = n_\infty(T)\alpha_i c_i P_i^{\alpha_i}\left[1+\sum_{k=1}^{N} c_k P_k^{\alpha_k}\right]^{-1} \qquad (i=1,..,N)\,, \tag{23}$$

where the exponents α_i are given by Eq. (22). Langmuir-Freundlich isotherms have been widely used as three-parameter fitting curves of adsorption equilibria[14], $n_i = n_{\infty i}k_i P_i^{\beta_i}/(1+k_i P_i^{\beta_i})$ in the case of single-component adsorption. For arbitrary $n_{\infty i}$, k_i, β_i, these expressions do not satisfy Eq. (14).

However, there is a profound difference between Eq. (23) and the Langmuir-Freundlich equation regarded as a simple three-parameter expression. Let us examine this aspect in the case of the prediction of single-component equilibria of N different species. The number of parameters N_p needed to apply Langmuir-Freundlich expressions in the form discussed e.g. by Carrà and Morbidelli[14] is $N_p = 3N$ ($n_{\infty i}$, k_i, β_i for each species). In the case of Eq. (23), this number reduces to $N_p = 3 + N$ for $N > 1$ (in the case of $N = 1$, $N_p = 3$, but the concept of fractal dimension makes no sense). These parameters are n_∞, α_r, D, which characterize the surface properties and the adsorption capacity of the adsorbent, and the N quantities c_i (or rather $[P_{i\infty}\exp(-E_i/RT)]^{-\alpha_i}$), which depend on the adsorbed species.

This means that for a sufficiently large number N of adsorbates, the Keller AI model reduces in practice to a single-parameter model and the number of parameters needed to interpret the experimental data is definitely much smaller than in the brute-force application of three-parameter Langmuir-Freundlich AI. This result is exclusively a consequence of the thermodynamic consistency of Eq. (23), and of the fractal hypothesis Eq. (22).

There are two other practical problems in the application of Eq. (15) that need to be discussed with some care: estimation of the radii r_i and the functional form of the adsorption energies.

The choice of the values for the molecular radii r_i gives rise to a subtle question since the radii r_i may depend on the energetic interactions in the adsorbed phase. This is equivalent to saying that the adsorption cross section can be different from the molecular cross section deriving from force-constant estimation of the Lennard-Jones potential determined from viscosity or second virial coefficients[15]. Indeed, Keller[9] has observed that the values of r_i should be considerably modified in order to obtain a better fitting of the experimental data. An ad-hoc modification of the value of the molecular radii drastically weakens the approach proposed by Keller: if we remove the hypothesis that the molecular radii are known a priori and independent of adsorption, Eq. (22) no longer represents a closure condition and the estimation of a fractal dimension in adsorption loses its meaning.

In principle, the adsorption cross sections could be obtained from a statistical thermodynamic approach. A simple statistical thermodynamic AI model has been proposed[16,17] for zeolitic adsorbents: the materials are represented by a series of M cavities which can contain at most m molecules. Within this framework it is possible to introduce the effects of intermolecular potential energies (e.g. of Lennard-Jones type) and thus estimate the section σ. However, even this approach will represent an a posteriori fitting of the adsorption data if it is further considered that the structure parameters M and m are also to be estimated from adsorption experiments.

For these reasons, we prefer to consider the radii r_i as purely geometrical parameters associated with the steric properties of gas molecules. Consequently, the data for r_i were obtained from force constants deriving from kinetic predictions of viscosity[15]. It should be noted in passing that the values for r_i obtained in this way do not differ significantly from the corresponding values obtained from statistical mechanical analysis of adsorption (see the data of Costa et al.[18] compared with the table of force constants reported by Hirschfelder et al.[15]).

The last point is related to the functional dependence of the adsorption energies $E_i(T, \{P_i\})$ on the partial pressures. This dependence can be regarded as a manifestation of the deviation from ideality in the adsorbed phase.[†] The introduction of pressure-dependent E_i does not modify the asymptotic behaviour for $P_i \to \infty$ (monolayer coverage). In general, due to screening effects, E_i is a decreasing function of P_i. A non-constant value for the E_i should be considered in the case of multicomponent AIs, due to the presence of energetic interactions between different molecules.

[†]In this context, we define an adsorbed phase as nonideal if the interaction between adsorbed molecules is purely steric and the adsorption energies do not therefore depend on the composition. This is the counterpart of the definition of ideality in a gas mixture, where the only energetic interactions considered are hard-sphere collisions.

In this work we assume the functional expression proposed by Keller[9]

$$E_i(T,\{P_i\}) = E_{o,i}(T)\left[1+\sum_{j=1}^{N} a_{ij}(T)P_j\right] \Big/ \left[1+\sum_{j=1}^{N} b_{ij}(T)P_j\right] \qquad (i=1,..,N)\,, \quad (24)$$

where a_{ij}, b_{ij}, $E_{o,i}$ are in general functions of the temperature T. Alternative expressions can also be considered assuming Lennard-Jones models of interaction[19].

3. Application to zeolites and other molecular sieves

3.1. Analysis of experimental results

Literature adsorption data [23,24,25,26,27,28,29] for pure gases and their binary mixtures on zeolites and activated carbon at different temperatures have been correlated by making use of Keller AI. The parameters entering into the AI are obtained by means of optimization procedure[22]. In the Keller AI model we have assumed a Langmuir form for Φ and adsorption energies given by Eq. (24). Moreover, in order to analyze the effect of non-constant adsorption energies we have compared the results deriving from the simplified model with E_i=constant, Eq. (23), with the more general model including $E_i = E_i(T,\{P_i\})$ given by Eq. (24).

The reference component is always taken as the species with the smallest radius in each experimental series.

The simplified model, in which the energetic parameters are independent of the number of molecules in the adsorbed phase, is expected to be valid only when molecular interactions are negligible (i.e. in a very low pressure range and at high temperature or in the case of a small amount of adsorbed molecules with respect to the capacity of the adsorbent). Close agreement between this model and experimental data is found in few cases and not under the most interesting gas-adsorption conditions (for instance at low temperature).

Figure 1 shows the correlation of literature data for CH_4, C_2H_4, C_2H_6, C_3H_6 obtained at 293.15 K and pressures of up to 100 kPa on activated carbon[27]. In this case, the simplified model furnishes good results. This is not true for the description of zeolitic systems.

The predictive capacity of the Langmuir-Freudlich model with regard to this data stems from the fact that the operating conditions lie inside the region in which the interactions can be neglected (the amount adsorbed compared to the capacity of the surface, represented by $n_\infty(T)\alpha_r$, is indicative in this sense).

Comparison of the two AI expressions is interesting in evaluating the nonideal behaviour deriving from the dependence of the adsorption energies on the amount adsorbed. Some examples are shown in figures 2,3. Figure 2 shows adsorption data of H_2S, CO_2, C_3H_8 obtained at 303.15 K, in a pressure range up to 206.8 kPa, on a H-Mordenite zeolite molecular sieve[26]. Figure 3 shows the same analysis for the adsorption equilibrium data of CO_2, C_2H_4, C_2H_6, $i-C_4H_{10}$ at 298.15 K, with pressures up to 137.8 kPa, on a 13X molecular sieve[25].

The results obtained with the more general AIs are shown in figures 4,5. Figure 4 shows the adsorption data of the pure gases O_2, N_2, CO at 144.26 K, with pressures

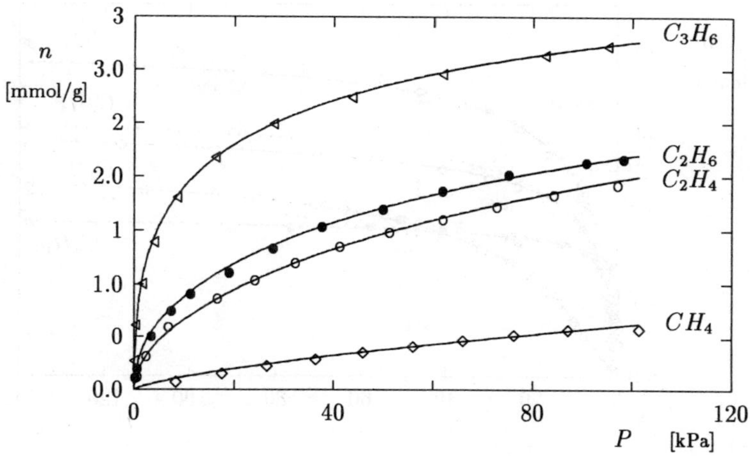

Figure 1: $n(T,P)$ vs P in activated carbon at T=293.15 K (data of Costa et al.[27]).

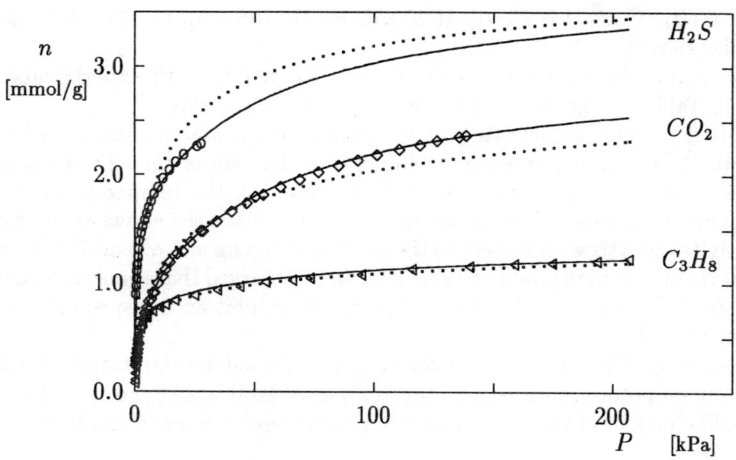

Figure 2: $n(T,P)$ vs P in H-mordenite zeolite at T=303.15 K (data of Talu and Zwiebel[26]). Solid line: AI including the effect of the dependence of adsorption energies on the pressure; dotted line: AI with E_i=constant.

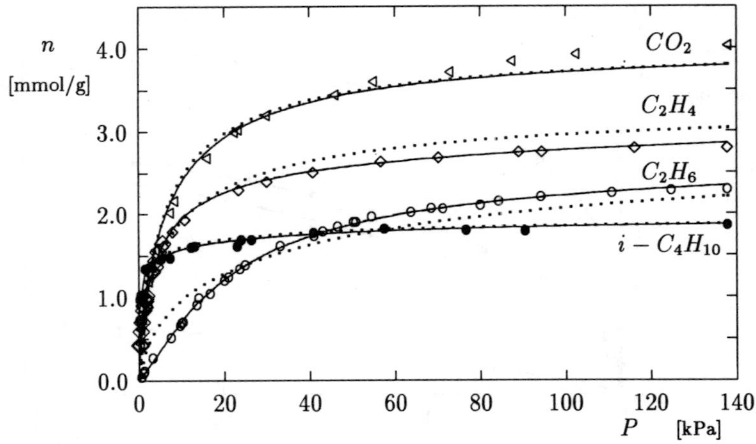

Figure 3: $n(T,P)$ vs P in a 13X molecular sieve at T=298.15 K (data of Hyun and Danner[25]). Solid line: AI including the effect of the dependence of adsorption energies on the pressure; dotted line: AI with E_i=constant.

up to 178.1 kPa , on a 10X molecular sieve[24]. Figure 5 shows the pure component data of C_2H_4, C_2H_6, $i-C_4H_{10}$ at 373.15 K, pressures up to 137.8 kPa, on a 13X molecular sieve[25].

The quantitative comparison of the two AI models with experimental data is shown in Table 1, together with the average percentage error.

Table 2 presents an overview of the parameters characterizing the microporous material. As can be observed from Table 2, in the case of the 13X molecular sieve there is a small change in the fractal dimension with the temperature. A detailed analysis of the influence of temperature on D will be discussed elsewhere by developing a sensitivity analysis with respect to the three parameters n_∞, α_r and D characterizing the adsorbent. From the data of Table 2 it can be observed that some zeolitic materials have a fractal dimension close to 2 while others exhibit an unexpectedly high value of D (close to 3).

To sum up, the Keller AI model furnishes a good interpretation of adsorption data on microporous materials. In the analysis of zeolitic structures, inclusion of the nonideal behaviour of the adsorption energies represents a considerable improvement.

3.2. Prediction of multicomponent equilibria

The Keller function for $E_i = E_i(T,\{P_i\})$ (Eq. (24)), written for a two-component mixture, introduces four binary interaction parameters, a_{ij}, b_{ij}, with $i \neq j$ taking into account the interactions between molecules of different species. The other energetic parameters entering into the model (a_{ii}, b_{ii}, E_{oi}) and the geometric ones derive from

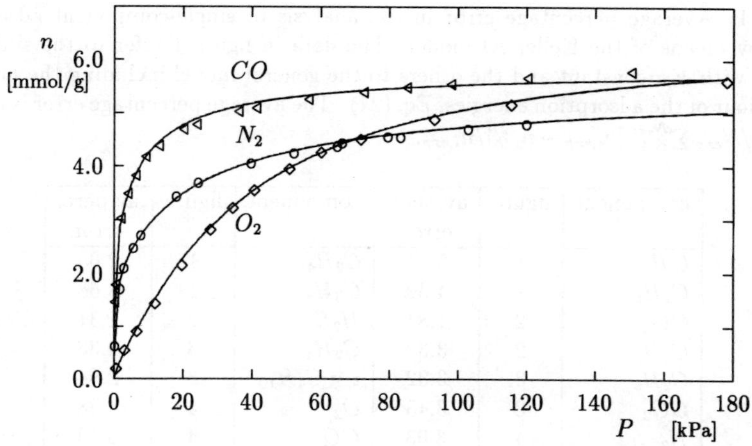

Figure 4: $n(T, P)$ vs P in a 10X molecular sieve at T=144.26 K (data of Danner and Wenzel[24]).

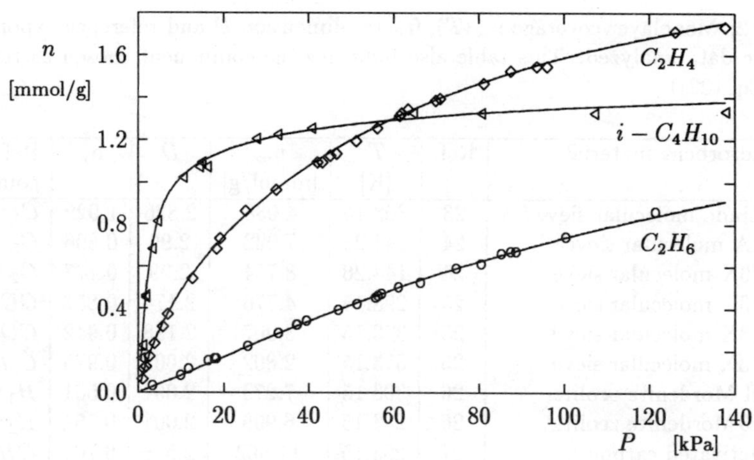

Figure 5: $n(T, P)$ vs P in a 13X molecular sieve at T=373.15 K (data of Hyun and Danner[25]).

Table 1: Average percentage error in the analysis of single-component adsorption data by means of the Keller AI model. The data in figure 1 refer to the simplified model with E_i=constant and the others to the general model including the nonideal behaviour of the adsorption energies, Eq. (24). The average percentage error is defined as $100/N_{exp}\sum_{i=1}^{N_{exp}}|n_{i,exp}-n_{i,th}|/n_{i,exp}$.

component	figure	av.perc. error	component	figure	av.perc. error
CH_4	1	5.71	C_2H_4	1	2.05
C_2H_6	1	4.32	C_3H_6	1	3.68
CO_2	2	1.84	H_2S	2	1.31
C_3H_8	2	3.32	C_2H_4	3	4.33
C_2H_6	3	3.32	$i-C_4H_{10}$	3	2.94
CO_2	3	3.43	O_2	4	4.38
N_2	4	3.03	CO	4	3.13
C_2H_4	5	2.18	C_2H_6	5	2.46
$i-C_4H_{10}$	5	2.85			

Table 2: Monolayer coverage $n_\infty(T)$, fractal dimension D and reference exponent α_r for the data analyzed. This table also indicates the component chosen as reference (see Eq. (22)).

adsorbent material	Ref.	T [K]	n_∞ [mmol/g]	D	α_r	Ref. comp.
Linde molecular sieve 5A	23	308.15	4.038	2.826	1.029	CO_2
5A molecular sieve	24	144.26	7.022	2.99	0.806	O_2
10X molecular sieve	24	144.26	8.754	2.99	0.777	O_2
13X molecular sieve	25	298.15	4.776	2.174	0.851	CO_2
13X molecular sieve	25	323.15	3.907	2.198	0.942	CO_2
13X molecular sieve	25	373.15	2.802	2.000	0.975	C_2H_4
H-Mordenite zeolite	26	303.15	7.273	2.001	0.561	H_2S
H-Mordenite zeolite	26	283.15	6.999	2.001	0.551	H_2S
activated carbon	27	293.15	11.962	2.556	0.764	CH_4
activated carbon	28	301.40	17.234	2.719	0.695	CH_4
5A molecular sieve	29	275.55	1.847	2.00	1.083	Ar

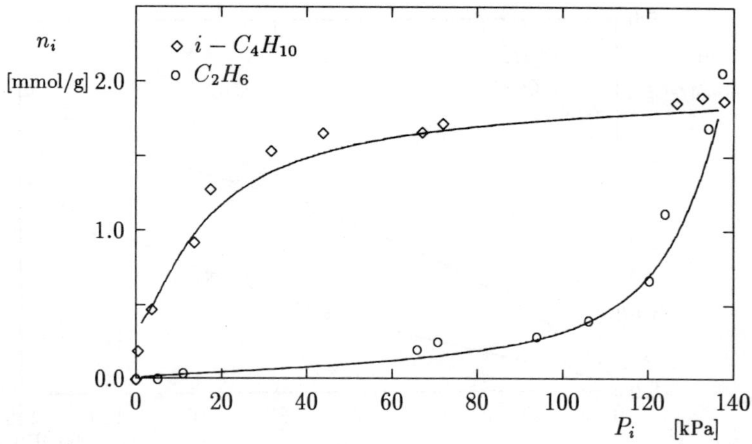

Figure 6: $n_i(T, \{P_i\})$ vs the partial pressures P_i for binary mixtures: $i - C_4H_{10}/C_2H_6$ on a 13X molecular sieve at T=298.15 K and P=137.8 kPa (data of Hyun and Danner[25]).

the analysis of pure component data.

Predictions for the binary gas-mixture adsorption data reported by Hyun and Danner[25] are shown in Figures 6-8. The adsorption data were obtained at a temperature of 298.15 K and at a constant total pressure of 137.8 kPa on a 13X molecular sieve. Figure 6 shows the $i - C_4H_{10}/C_2H_6$ system, figure 7 the C_2H_4/CO_2 mixture and figure 8 the system $i - C_4H_{10}/C_2H_4$. As discussed by Hyun and Danner, it should be observed that the $i - C_4H_{10}/C_2H_4$ system (figure 8) displays highly nonideal behaviour, whereas the C_2H_4/CO_2 mixture exhibits slight nonideality. With the inclusion of binary interaction parameters, the Keller AI model proves experimentally useful for the prediction of multicomponent adsorption equilibria. We can also observe that the AI model including only pure-component contributions is not able to furnish accurate predictions of binary data.

3.3. On the fractal nature of zeolites

Zeolitic materials are fairly regular porous structures characterized by well defined porous cages connected by straight or zig-zag channels[20]. For this reason, it could be intuitively expected that such microporous structures will not exhibit a fractal scaling behaviour.

The results on synthetic faujasite (taking N_2 and $n - alkenes$ as adsorbates) reported by Avnir et al.[3] give a fractal dimension $D = 2.02 \pm 0.05$, as is to be expected given the regular structure of the materials.

Sulikowski[21] recently undertook an experimental analysis of the structure and the

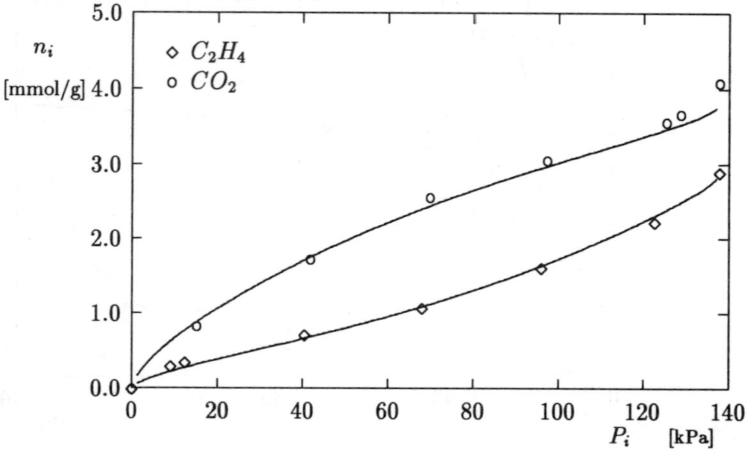

Figure 7: $n_i(T, \{P_i\})$ vs the partial pressures P_i for binary mixtures: C_2H_4/CO_2 on a 13X molecular sieve at T=298.15 K and P=137.8 kPa (data of Hyun and Danner[25]).

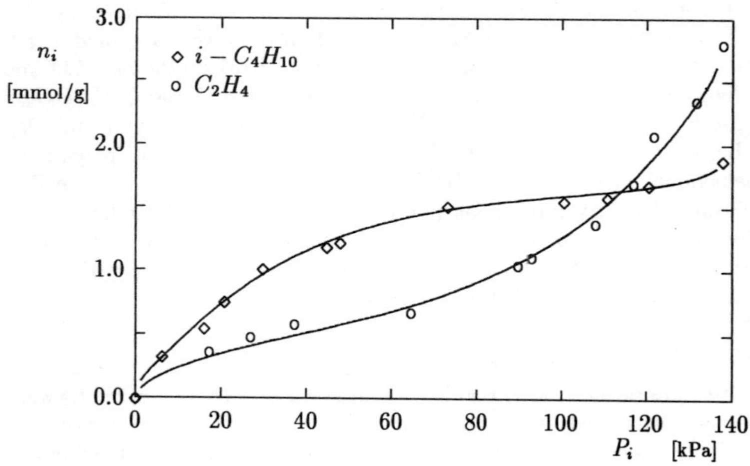

Figure 8: $n_i(T, \{P_i\})$ vs the partial pressures P_i for binary mixtures: $i-C_4H_{10}/C_2H_4$ on a 13X molecular sieve at T=298.15 K and P=137.8 kPa (data of Hyun and Danner[25]).

fractal dimension deriving from adsorption measurements (using methanol, ethanol, 2-propanol and 2-methyl-1-propanol as adsorbates) on synthetic faujasite (Na-Y) under different preparation conditions: e.g. by applying alumination and dealumination procedures. Depending on the preparation process, Sulikowski found fractal dimensions ranging from $D = 2.02 \pm 0.05$ for the parent faujasite up to $D = 2.43 \pm 0.04$ for modified faujasite. This wide range of fractal dimensions is probably due the distribution of extra-framework aluminum cations and by the corresponding vacancies generated in the solid at atomic scales.

Table 2 reviews the values for the fractal dimensions found by AI analysis on molecular sieves. Some of the structures (see Table 2) are characterized by high values of the fractal dimension D (much greater than two) while other structures exhibit a fractal dimension close to 2. The reason for the apparently high fractal dimension is probably connected with two factors: the three-dimensional nature of the microporous zeolitic material and the substitutional disorder in the microporous cages. Substitutional changes in cage structure may be the cause of a certain level of disorder causing the fractal dimension D to be greater than 2 at least for the considered range of molecule lengthtscales.

A natural objection to the interpretation given of fractality in microporous materials is that the apparent fractal dimension found by the AI model could be a fitting parameter accounting for energetic interactions not properly described by means of the functional expression adopted for the adsorption energies, Eq. (24). If this criticism were correct, one would expect a different estimate of the fractal dimension for different values of the temperature, since the energy contribution enters into the equations as the ratio E_i/RT between adsorption energy and thermal fluctuation energy: the higher the temperature, the lower the relative strength of E_i compared with RT. This aspect will be analyzed elsewhere.

It is impossible to proceed any further in the investigation of this problem with purely thermodynamic arguments. However, research on this topic should be oriented in two main directions: a detailed structural characterization of the material with particular emphasis on the effects of substitutional disorder induced by imperfectly regular localization of the cations; and Monte Carlo (or Molecular) simulation of adsorption/desorption processes on lattice models of zeolitic materials. This can be done with the method proposed by Seri-Levy and Avnir[7].

4. Concluding remarks

In this article we have discussed a thermodynamically consistent approach to adsorption equilibria first proposed by Keller, focusing attention on the closure role played by the fractal characterization of the adsorbent. The resulting AIs are thermodynamically consistent and the approach followed enables us to discriminate between the roles of geometric and energetic heterogeneity.

Analysis of literature data indicates that the resulting AIs provide a good description of adsorption equilibria on molecular sieves for single component and multicomponent mixtures. In the latter case the dependence of the adsorption energies on the partial pressures should necessarily be taken into account.

Thermodynamic analysis indicates that many zeolitic materials seem to possess a fractal dimension between 2 and 3 (and sometime much larger than 2). The origin of the fractality observed in zeolitic materials should be interpreted as the effect of structural disorder on the distribution of cations at the molecular level. A possible tool for analyzing this problem is Monte Carlo simulation in simple lattice models of the microporous structure. However, leaving aside the structural study of the origin of a value of D greater than two, it should be stressed that the inclusion of an intensive geometric parameter (D) ensures that the analysis of adsorption equilibria is well-framed from the thermodynamic standpoint.

Acknowledgments

The authors grateful to Douglas Ludlow, Mizuho Schwalm and William Schwalm for stimulating discussions.

References

1. P. Pfeifer and D. Avnir, *J. Chem. Phys.* **79** (1983) 3558.
2. P. Pfeifer, D. Avnir and D. Farin, *J. Stat. Phys.* **36** (1984) 699.
3. D. Avnir, D. Farin and P. Pfeifer, *Nature* **308** (1984) 261.
4. A. V. Neimark, *JETP Lett.* **51** (1990) 607.
5. M. Urbakh and L. Daikhin, *Surf. Sci.* **287/288** (1993) 847.
6. A. Seri-Levy and D. Avnir, J. Phys. Chem. **97** (1993) 10380.
7. A. Seri-Levy and D. Avnir, *Langmuir* **9** (1993) 3067.
8. S. Sircar and A. L. Myers, *AIChE J.* **32** (1986) 650.
9. J. U. Keller, *Physica* **A 166** (1990) 180.
10. G. F. Cerofolini and N. Re, *Rivista del Nuovo Cimento* **16** (1993) 1.
11. A. W. Adamson, *Physical Chemistry of Surfaces* (J. Wiley, New York, 1982).
12. D. M. Ruthven, *Principles of Adsorption and Adsorption Processes* (J. Wiley, New York, 1984).
13. J. U. Keller, *Ber. Bunsenges. Phys. Chem.* **92** (1988) 1510.
14. S. Carrá and M. Morbidelli, *Int. Rev. Phys. Chem.* **6** (1987) 351.
15. J. O. Hirschfelder, C. F. Curtiss and R. B. Bird, *Molecular Theory of Gases and Liquids* (John Wiley, New York, 1954).
16. D. M. Ruthven, K. F. Loughlin and K. A. Holborow, *Chem. Engng. Sci.* **28** (1973) 701.
17. D. M. Ruthven, *AIChE J.* **22** (1976) 753.
18. E. Costa, G. Calleja and L. Cabra, in *First International Conference on Fundamentals of Adsorption* (Engineering Found., New York, 1983).
19. J. U. Keller, R. Staudt and M. Tomalla, *Proc. AIChE Annual Meeting*, St Louis November 7-12 (1993) 187b.

20. D. W. Breck, *Zeolite Molecular Sieves* (J. Wiley, New York, 1974).
21. B. Sulikowski, *J. Phys. Chem.* **97** (1993) 1420.
22. G. Buzzi Ferraris, *Quaderni dell'ingegnere chimico italiano* **4** (1968) 171.
23. A. J. Glessner and A. L. Myers, *Chem. Eng. Progress Symp. Series* **65** (1969) 73.
24. R. P. Danner and L. A. Wenzel, *AIChE J.* **15** (1969) 515.
25. S. A. Hyun and R. P. Danner, *J. Chem. Eng. Data* **27** (1982) 196.
26. O. Talu and I. Zwiebel, *AIChE J.* **32** (1986) 1263.
27. E. Costa, J. L. Sotelo, G. Calleja and C. Marron, *AIChE J.* **27** (1981) 5.
28. R. Reich, W. T. Ziegler and K. A. Rogers, *Ind. Eng. Chem. Process Des. Dev.* **19** (1980) 336.
29. H. Vereist and G. V. Baron, *J. Chem. Eng. Data* **30** (1985) 66.

THEORETICAL MODEL FOR PHOSPHOLIPID AGGREGATION INTO CLOSED VESICLES

MARIA CRISTINA ANNESINI, ANDREA PIRRI
Dipartimento di Ingegneria Chimica,
Università di Roma "La Sapienza"
via Eudossiana 18, 00184 Roma, ITALY

and

GIUSEPPE MOSSA
Istituto di Medicina Sperimentale,
Consiglio Nazionale delle Ricerche
via Marx 15-43, 00137 Roma, ITALY

ABSTRACT

Spherical closed structures of phospholipids (liposomes) are very important in many different areas of science and technology, but the mechanisms of vesicle formation and the factors that determine the size distribution are not yet well understood. Two theoretical models for liposome formation are considered. The first one assumes that lipids aggregate spontaneously into small unilamellar vesicles that represent an equilibrium state. Size and shape of the aggregates are determined by energetic and geometrical parameters. The second model depicts the closed vesicles as inherently unstable structures, formed by distortion of a planar membrane. Interaction between growth and closure of lipid bilayer controls the properties of liposomes.

1. Introduction

Amphiphilic molecules - i.e., molecules composed of a polar, hydrophilic group (head) and a non polar hydrophobic (tail) - self associate, in aqueous environment, into a variety of structure in order to segregate their hydrophobic region from water. Different supramolecular structures as micelles, inverted micelles, planar lamellae, vesicles can appear depending on the molecular shape of amphiphilic compounds, their concentration, temperature, pH, ionic strength, etc.. Furthermore aggregates often exhibit a broad distribution of sizes. The macroscopic behaviour of these systems results in a surprising rich phase diagram, whose complexity is increased by the presence of metastable phases.

The particular class of supramolecular structures we are interested in are the closed vesicles, made up of one or several concentric bilayers. We refer mainly to phospholipid vesicles or liposomes, which are interesting both from a theoretical and a practical point of view. The main topics about liposome structure and properties and about their applications in several fields have been recently reviewed by Lasic[1]. The liposome resemblance to biological membranes makes them an extraordinary biomimetic system to study cell topology during cellular processes, cell-cell interaction (adhesion, fusion), molecular recognition, membrane enzymology. Recently, also fascinating works have appeared[2,3] dealing with liposomes as autopoietic systems to study the primitive life forms. Several large-scale applications of liposomes have been reported in the last ten years, mainly in biomedicine, as drug-delivery system, as carrier of genetic material, in diagnostic, etc.

A large amount of data is now available concerning different liposome preparation methods, the influence of preparation conditions on the liposome properties (lamellarity, size) or their stability in different environmenst. On the other hand, little is understood of their complex characteristics and behaviour. While much work has been done about forces between amphiphilic molecules and between aggregates, about surfactant aggregation in micelles or phospholipid aggregation in flat bilayer, little work has been done to explain the mechanism of liposome formation, the size distribution, the stability. Fundamental questions about the spontaneity of liposome formation, their thermodynamical stability, the transition from open aggregates to closed structures are still open. Experimental evidence that slightly different conditions result in liposomes with a wide variety of properties or that liposomes do not change their size upon varying the concentration, supports the conclusion that liposomes are not in a thermodynamic equilibrium state but in metastable "kinetic trap"[4]. Thermodynamic models can only get insight on the stability of different structures, but fail to predict which is the real state of the system. Thus, a kinetic model must be considered.

The aim of this work is to study the theoretical models for liposome formation in order to point out their shortcomings and possible developments. We mainly refer to liposomes produced by detergent depletion techniques, but similar arguments hold also for different preparation methods. A short description of the phenomena related to vesiculation precedes the theoretical sections, in order to clarify the behaviour that should be described. Two different approaches are considered. The first one, which may be defined as a molecular model, is based on self-aggregation thermodynamics, improved by taking into account of geometrical constraints. The second one considers the bilayer as an elastic membrane and depicts the liposome formation process as the distortion and closure of a flat structure. This paper reports the preliminary results of our studies in this field. We show some numerical simulations carried out to probe the ability of these simple models to describe liposome formation and to predict the liposome properties.

2. Liposome preparation and solubilization

Liposomes can be produced by different techniques as sonication, injection of organic lipid solution into an aqueous system, reverse-phase evaporation, detergent depletion, extrusion, etc. This work deals with liposomes formed after detergent removal by dialysis, dilution, chromatography or adsorption. These methods have the advantage of yielding relatively homogeneous small unilamellar vesicles with good reproducibility. Furthermore they can be used also to incorporate sensitive biomolecules with a proper choice of detergent. The main disadvantage is the presence of unremoved detergent in the liposome bilayer.

Different lipid cocktails (mainly egg or synthetic lecithins) with different additives (cholesterol, antioxidants) and different detergents (alkylglycosides, cholates) are used. The properties (lamellarity, size, size distribution) depend not only on thermodynamic parameters as lipid composition, pH, temperatures, but also on the history of preparation. There is a mess of data on this subject, sometimes controversial and not always reproducible and reliable. Here, we summarize some interesting results.

a) Different liposome sizes are obtained with different lipid cocktails[5]. Cholesterol inclusion increases the vesicle size (at least at low pH)[6];

b) Sodium cholate produces smaller liposomes than alkylglycosides. The length of glycoside sidechain has also a significant effect on the vesicle size[7];

c) The vesicle size increases with increasing the lipid concentration, but too high lipid concentration results in multilamellar vesicles[8].

d) The initial detergent/lipid ratio seems to be significant, but controversial results have been obtained[5].

e) Faster detergent removal rate produces smaller liposomes[9,10].

Liposome formation by detergent removal was the first method for which a mechanism of vesicle formation was proposed[11]. A schema of this mechanism is reported in figure 1: mixed detergent/lipid micelles grow through fusion into disk-like flat micelles, with small detergent molecules segregated at the disk edge until there is not

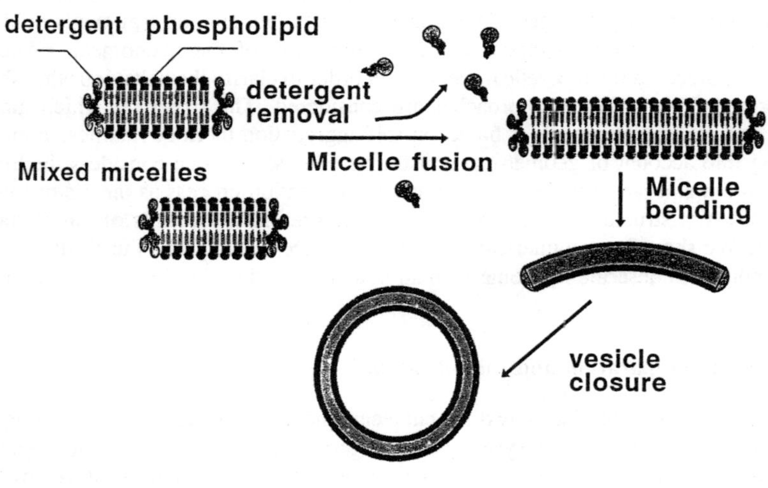

Figure 1 Mechanism of liposome formation by detergent depletion.

enough detergent to shield the edge and micelles collapse into closed vesicles. Since vesiculation does not occur instantly, it was possible to obtain some experimental evidence of these disk-like micelles[12,13].

It is also interesting considering the reverse (but not symmetric) process of liposome formation, i.e. solubilization of preformed vesicles by detergent addition. This process is widely studied by biologists and biophysicist dealing with membrane solubilization and reconstitution. Many Authors agree that the process occurs in three subsequent steps, as determined from turbidity, NMR or electron microscopy data[14,15]. In the first step detergent is incorporated into the bilayer, until liposome saturation; in the second step liposomes and mixed micelles coexist, each structure with a constant detergent to lipid molar ratio; in the third step all phospholipids are present as mixed micelles with increasing detergent to lipid molar ratio. A schema of the solublization process is reported in Figure 2.

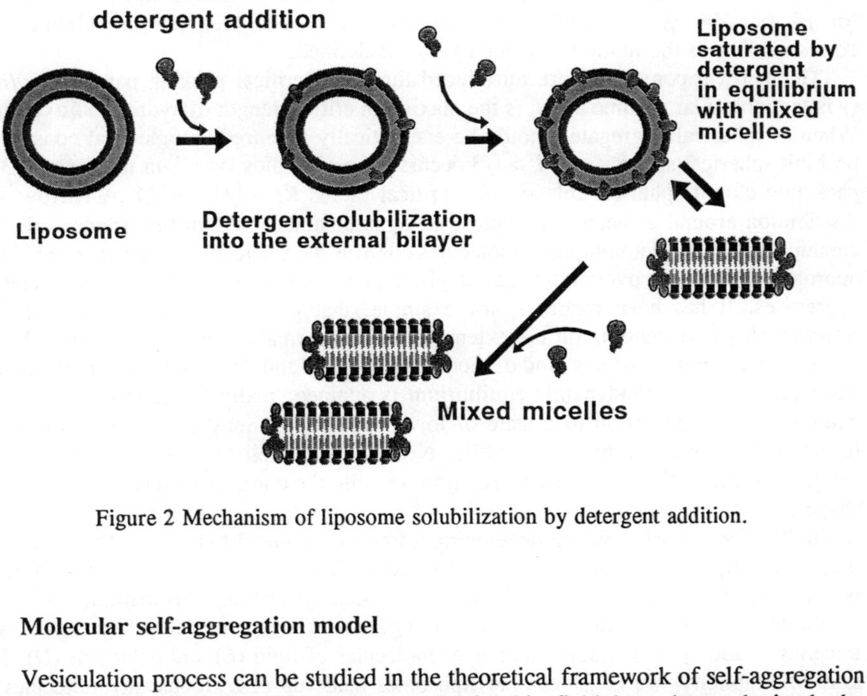

Figure 2 Mechanism of liposome solubilization by detergent addition.

3. Molecular self-aggregation model

Vesiculation process can be studied in the theoretical framework of self-aggregation thermodynamics. The most interesting results in this field have been obtained by Israelachvili[16,17], who discussed also the geometric aspects of the phenomenon.

Many studies deal with micelles of a single amphiphilic compound, but little attention has been devoted to mixed micelles or to vesicles.

Equilibrium thermodynamics between aggregates made up of N amphiphilic molecules results in the following size distribution:

$$X_N = N\left\{\frac{X_M}{M}\exp[M(\mu_M^o - \mu_N^o/N)/kT]\right\}^{M/N} \tag{1}$$

where X_N is the density of molecules incorporated into aggregates made up of N molecules and μ_N° is the reference chemical potential of these aggregates (M is any reference aggregation number). The effective size distribution depends on the sequence $\{\mu_N^\circ\}$: large aggregates appear only if μ_N°/N attains a minimum value at some finite value of N. From a physical point of view, this behaviour can be due to the competition effect between attractive hydrophobic interaction of amphiphilic tails and repulsive interaction of polar heads. A functional form in terms of molecular surface area, a, is often used for μ_N° ($\mu_N^\circ/N= \gamma a + K_r/a$). Thus, an optimal surface area per amphiphile, a°, corresponding to the minimum value of μ_N°, is defined.

Geometrical constraints are introduced through a critical packing parameter $v/a^\circ l_c$ (v is the molecular volume and l_c is the maximum critical length of hydrophobic chains). Whereas spherical aggregates should be energetically favoured, geometrical constraints prohibit spherical shape if $v/a^\circ l_c > 1/3$ occurs. Phospholipids ($v/a^\circ l_c$ in the range 2/3-1) pack into closed spherical bilayer with critical radius $R_c \sim l_c(1-v/a^\circ l_c)$. A narrow size distribution around a mean radius close to R_c is obtained. Within this framework, the chemical structure of amphiphilic molecules controls the shape of aggregates: even if this approach may seem oversimplified, it gives good results in predicting the shape of aggregates. It has been reported, for example, that gangliosides self aggregate into different shapes depending on the extension and conformation of the head group[18].

The shortcomings of this kind of models are evident and derive from the fundamental assumption that thermodynamic equilibrium is attained: really liposomes are far from equilibrium and transition to a state of lower free energy may be kinetically blocked (often by liposomal morphology itself). Nevertheless, these models are useful to get insight into energetics of different structures and into the domains of stability of different shapes.

In this framework, we are developing a first-order model to describe the process of preformed liposome solubilization by detergent addition. We believe that this process is not so far from equilibrium as the reverse process of liposome formation.

In this model two different types of aggregates are considered: liposomes and micelles, made up of variable number of molecules of lipid (L) and detergent (D). For sake of simplicity, we consider only liposomes made up of L_o (constant) molecules of lipids; L_o depends on the preparation procedure. Equilibrium between different aggregates and free detergent molecules in solution results in the following expression for numerical density of liposomes, $X(D)$, and micelles, $Y(L,D)$:

$$X(D)=X(0)\exp\left\{\frac{1}{kT}(\mu_L^o(0)-\mu_L^o(D)+D\mu_D^o)\right\}c_D^D \tag{2}$$

$$Y(L,D)=X(0)^{L/L_o}\exp\left\{\frac{1}{kT}\left(\frac{L}{L_o}\mu_L^o(0)-\mu_M^o(L,D)+D\mu_D^o\right)\right\}c_D^D \tag{3}$$

where $\mu_L^\circ(D)$ and $\mu_M^\circ(L,D)$ are the reference chemical potential of liposomes (made up of L_o lipid molecules and D molecules of detergent) and micelles (made up of L lipid molecules and D molecules of detergent), μ_D° is the reference potential of detergent and c_D is the detergent mole fraction in the aqueous phase.

Eq.(2) and (3), together with the mass balance equations, allow us to obtain the liposome and micelle distribution for different total detergent to lipid ratio. Of course, the effective distribution depends on the sequences $\{\mu_L^\circ(D)\}$ and $\{\mu_M^\circ(L,D)\}$ that contain most of the physics of molecular interactions between lipid and detergent within the aggregates and between the aggregate and the solvent. In order to derive physically motivated expressions for $\{\mu_L^\circ(D)\}$ and $\{\mu_M^\circ(L,D)\}$, we consider that swollen liposomes contain detergent molecules only in the external layer. Detergent is prevented from penetrating into the internal layer by the external layer itself, according to some experimental observations[19], and this results in a substantial difference between liposome solubilization and liposome formation processes. Furthermore, disk-like micelles with lipid molecules in the inner core and detergent molecules segregated at the edge are assumed. Physical considerations in terms of counterbalance of hydrophobic attractive and repulsive forces lead to the following forms of $\{\mu_L^\circ(D)\}$ and $\{\mu_M^\circ(L,D)\}$:

$$\mu_L^o(D) = a'+b'D+cD^2 \tag{4}$$

$$\mu_M^o(L,D) = aL+b\sqrt{L}+c\,\frac{D^2}{\sqrt{L}}-eD \tag{5}$$

It is likely that Eq.(5) does not hold at high detergent to lipid ratio, when transition to spherical micelles occurs.From now on, we use Eqs.(4) and (5), but they must be considered only as working hypothesis.

A further simplification arises if only a constant number of lipid molecules, L_o, is considered also in micelles. Some preliminary tests, carried out with this simplifying assumption, show that the model predictions qualitatively agree with the experimental behaviour previously described. The content of detergent in the liposomes increases with detergent concentration until a critical value of D/L is reached. At this point, a further increase of detergent concentration yields a transition from vesicles to micelles whereas the detergent to lipid molar ratios in the two different aggregates remains almost constant. Liposome and micelle distributions at different detergent concentrations in the second stage of solubilization (i.e. when liposomes and micelles coexist) are reported in Figure 3.

If the restriction of L_o lipid molecules in micelles is relaxed, a bivariate micelle distribution appears. Figure 4 shows some typical distributions obtained in this case.

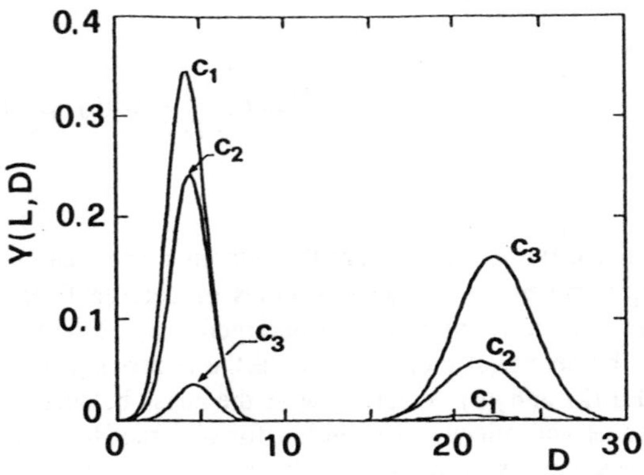

Figure 3 Distribution curves of detergent molecules in liposomes (left) and micelles (right): $c_1 < c_2 < c_3$. L_o lipid molecules in both liposomes and micelles.

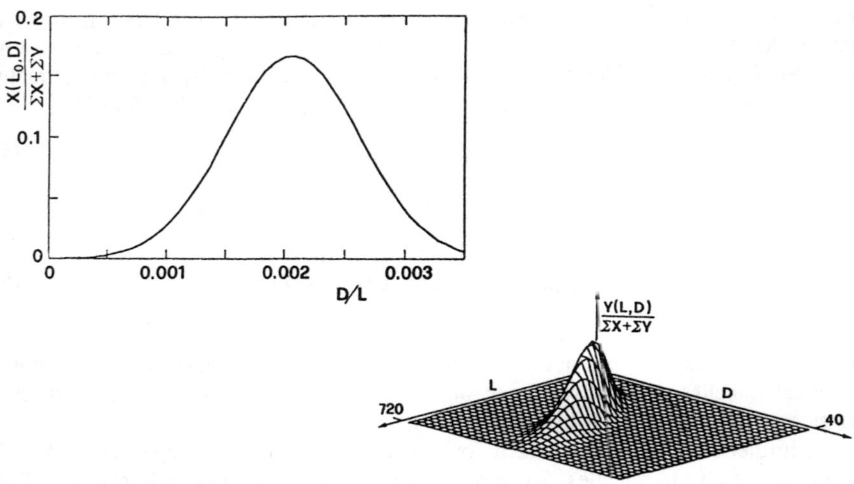

Figure 4. Lipid and detergent distribution in liposomes (left) and micelles (right). L_o=2000 lipid molecules in liposomes. a'/kT=-100, b'/kT=0.1, c'/kT=0.4, a/kT=0.00075, a/kT= 0.00075, b/kT=0.067, c/kT=4.47, e/kT=2, $c_D/exp(-\mu^{\circ}_D/kT)$=30.

4. Membrane bending model

From a different point of view, the liposome bilayer may be regarded as a continuous membrane with elastic properties, without attention to details of the molecular structure. In the first approximation, a non stretchable membrane is considered and only bending deformation, with constant surface area, is allowed. The elastic energy of a closed structure is given by[20]:

$$H=\int dA\{\tfrac{1}{2}\kappa(C_1+C_2-C_0)^2+\kappa' C_1C_2\} \tag{6}$$

where C_1 and C_2 are the two principal curvatures, C_0 is the spontaneous curvature, κ and κ' are the elastic rigidity and Gaussian curvature modulus (κ' contribution is often neglected). Typical values of κ for phospholipid bilayers are in the range 10^{-19}-10^{-20} J.

For spherical shape with zero spontaneous curvature (as required for symmetric bilayer) $H=4\pi(2\kappa+\kappa')$ independently of the radius. In general, a vesicle ensemble with a fixed amount of lipids (and therefore with fixed area) is thermodynamically unstable against fusion, since the entropic term cannot overcome the high elastic energy: liposomes are therefore metastable structures. Spontaneous vesiculation can occur only for very low κ values[18] or for a mixture of surfactants with different composition in the inner and outer bilayer, which results in a non zero spontaneous curvature[21]. A substantial improvement of the model accounts for the effective rigidity that decreases logarithmically with the increasing of the vesicle size[22,23].

This approach has been widely used to predict vesicle shapes[24-26] and vesicle size distribution[23], with interesting fall out also in biological studies (erythrocyte morphology, cell budding, etc.[27-29]). An interesting model for liposome formation by detergent depletion has been suggested[12,30] in this context. The main idea of this model is that liposome formation proceeds through fusion of fragments of planar bilayer with detergent segregated at the edge. The superimposition of edge tension, γ, and elastic energy results in the instability against curvature of disk-like micelles above a critical radius ($r_c= 8\kappa/\gamma$)[31]. In details, the Gibbs free energy of a fragment is given by:

$$\Delta G=\Delta G_s[\Omega^2+V_F(1-\Omega^2)]^2 \tag{7}$$

where ΔG_s is the energy of a spherical fragment, Ω is the shape parameter ($\Omega=0$ for planar disk and $\Omega=\pm1$ for a closed sphere) and V_F is the vesiculation index equal to $R\gamma/2\kappa$. It is easy to show that: *a)* for $0<V_F<1$ the minimum of free energy occurs at $\Omega=0$ and planar disks are stable; *b)* for $1<V_F<2$ a local minimum exists at $\Omega=0$, but the lowest energy occurs at $\Omega=\pm1$. Transition from a disk-like micelle to a closed vesicle proceeds over the activation barrier. Fokker-Plank equation gives the following rate for pre-critical transition from disk to sphere:

$$k_c = k^\circ_c \exp\left\{-(\Delta G_s/kT)(1-\tfrac{1}{2}V_F)^2\right\} \tag{8}$$

and that for the reverse sphere burst:

$$k'_c = k'^\circ_c \exp\left\{-(\Delta G_s/kT)(\tfrac{1}{2}V_F)^2\right\} \tag{9}$$

The model has been applied to determine the stability lines for vesicle closure and vesicle burst micellization[32].

The appealing aspect of this approach is that the edge tension, which is modulated by detergent concentration, controls the vesiculation kinetics. In turn, the interaction between kinetics of micelle growth by fusion and kinetics of disk closure controls the size distribution of the obtained vesicles.

Therefore, a theoretical study is in progress to develop a vesiculation model that is based on these physical considerations and that allows us to analyse the dynamic behaviour of the system. The vesiculation at constant edge tension (therefore at constant detergent concentration) has been studied as a first example. In this model, two populations, i.e. micelles and vesicles, are considered. Micelles are growing as a result of binary collisions with other micelles; simultaneously micelles can undergo vesiculation with a probability strongly dependant on their size.

Dynamics of concentrations of micelles (x_j) and vesicles (y_j) made up of j primary fragments is given by:

$$\frac{dx_j}{d\tau} = \left[-x_j\sum_{i=1}^{\infty} K_{i,j}x_i + \frac{1}{2}\sum_{i=1}^{j-1} K_{i,j-i}x_i x_{j-i}\right] - v_j x_j \tag{10}$$

$$\frac{dy_j}{dt} = v_j x_j \tag{11}$$

The first two terms on the right-hand side of eq.(10) derive from classical Smoluchowski model for irreversible aggregation. The last term accounts for vesiculation rate; this term is equivalent to a sink that eliminates micelles from the dynamic system. Micelle fragmentation, vesicle burst or vesicle fusion have not been considered in this first approximation model.

Kinetics of such a system is substantially different from that of classical irreversible aggregation[33,34] and a variety of dynamic behaviour can be observed, depending on the aggregation kernel $K_{i,j}$, on the function of the vesiculation rate, v_j. as well as on the initial conditions. A first approach to the system assumes an initial monodisperse suspension of primary fragments ($x_j=\delta_{j1}$ for $t=0$), a constant aggregation kernel $K_{i,j}=1$

and a vesiculation rate given by:

$$
\begin{aligned}
&v_j=0 && j\le j_{\min}\\
&v_j=\beta\exp\left[-\alpha(1-\sqrt{j/j_{\max}})^2\right] && j_{\min}<j<j_{\max}\\
&v_j=\infty && j\ge j_{\max}
\end{aligned}
\tag{12}
$$

where j_{min} and j_{max} are the number of primary particles that result in $V_f=1$ and 2 respectively.

Figure 5 shows the numerical result of the micelle and vesicle distributions ($x_j/\Sigma x_j$ and $y_j/\Sigma y_j$) at different times. β value used in this numerical calculation corresponds to a fast vesiculation process: micelles grow until they reach a critical size, then vesiculation is almost instantaneous. The micelle distribution is therefore cut off at some large j value and this cut off decreases with time as the ratio between aggregation and vesiculation rate decreases. In this condition, liposome distribution is almost constant with time (but this is not true for different β values). In order to determine the scaling law of the system, we report the moments of both distributions vs. time, determined numerically for different β values. Figure 6 shows that for large t micelle distribution moments decrease as $t^{-\alpha}$ with α near to 1. Vesicle distribution moments approach a limiting value, M_∞, with the same temporal scaling ($M-M_\infty \sim t^{-1}$), as shown in Figure 7.

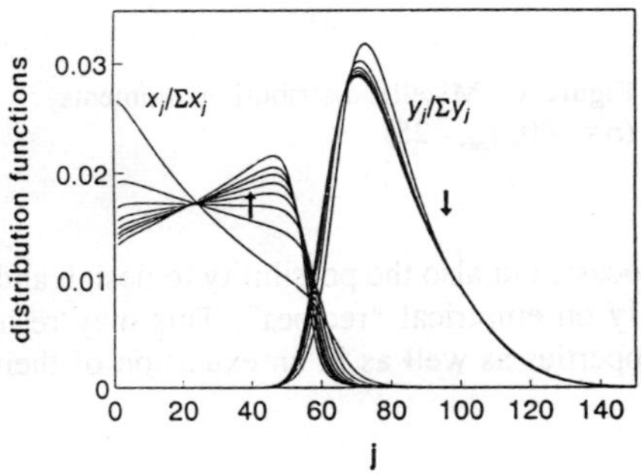

Figure 5. Micelle and liposome distribution at different times for $\alpha=100$, $\beta=1$, $j_{min}=25$.

5. Concluding remarks

The models reported, based on the classical theories of self-aggregation thermodynamics or on the membrane elasticity, may be useful to understand some aspects of the liposome formation and related properties. Of course, the true significance of these models must be judged by their success in describing the real behaviour of such systems. First attempts to compare, at least qualitatively, the model predictions with experimental evidence gave promising results, although all the models are very simplified with respect to the complexity of the system. We believe that is worth going on both in modeling the liposome formation process and in obtaining reliable experimental data to validate the models. The final goal is not only a deeper understanding of an interesting

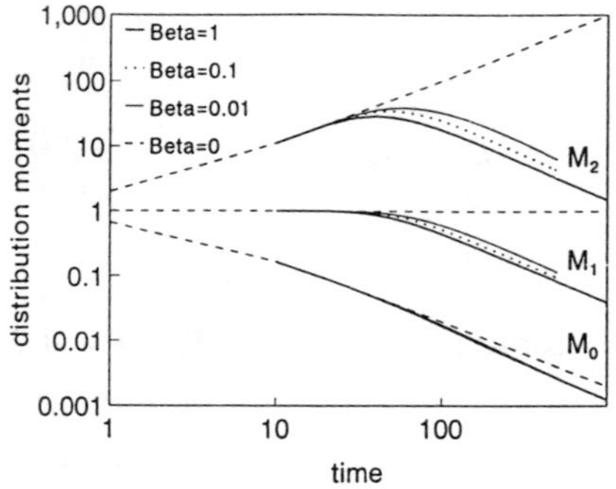

Figure 6. Micelle distribution moments ($\alpha=100$, $j_{min}=25$).

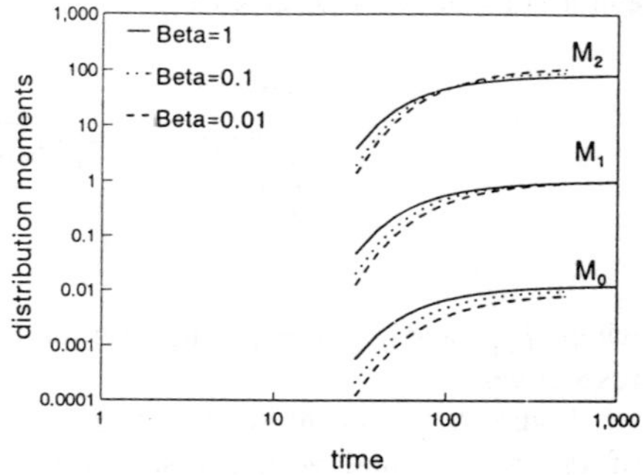

Figure 7. Liposome distribution moments ($\alpha=100$, $j_{min}=25$).

process, but also the possibility to design and control liposome formation without relying only on empirical "recipes". This may result in a significant improve of the liposome properties as well as in an extension of their application field.

Acknowledgements

This work was partially supported by a financial contribution of Ministero dell'Università e della Ricerca Scientifica e Tecnologica.

References

1. D.D. Lasic, *Liposomes: from Physics to Applications* (Elsevier Science Publishers B.V., Amsterdam, 1993).
2. P.K. Schmidly, P. Schuntenberger, and P.L. Luisi, *J. Am. Chem. Soc.* **113** (1991) 8127.
3. P.L. Luisi, in *Thinking about biology,* eds. W. Stein and F.J. Varela (Addison-Wesley, 1993).
4. D.D. Lasic, *J. Colloid Interface Sci.* **140** (1990) 302.
5. O. Zumbuehl and H.G. Weder, *Biochim. Biophys. Acta* **640** (1981) 252.
6. V. Rhoden and S.M. Goldin, *Biochemistry* **18** (1979) 4173.
7. R.A. Schwendener, M. Asanger, and H.G. Weder, *Biochem. Biophys. Res. Comm.* **100** (1981) 1055.

8. J.S. Tauskela, M. Akler, and M. Thompson, *Analytical Biochemistry* **201** (1992) 282.
9. A. Sabouni, W.A. Ritschel, and M. Ashraf, *Pharmaz. Ind.* **54** (1992) 74.
10. J.M. Wrigglesworth, M.S. Wooster, J. Elsden, and H.J. Danneel, *Biochem. J.* **246** (1987) 737.
11. D.D. Lasic, *Biochem. J.* **256** (1988) 1.
12. P. Fromherz and D. Ruppel, *FEBS Letters* **179** (1985) 155.
13. C.Y. Shiau and C.C. Chang, *Chem. Eng. Commun.* **125** (1993) 227.
14. T. Inoue, T. Yamahata, and R. Shimozawa, *J. Colloid Interface Sci.* **149** (1991) 345.
15. S. Almog, B.J. Litman, W. Wimley et al., *Biochemistry* **29** (1990) 4582.
16. J.N. Israelachvili, D.J. Mitchell, and B.W. Ninham, *J. Chem. Soc. Faraday Trans. 2* **72** (1976) 15225.
17. J.N. Israelachvili, S. Marcelja, and R.G. Horn, *Q. Rev. Biophys.* **13** (1980) 121.
18. M. Corti and L. Cantu', *J. Liposome Research* **3** (1993) 611.
19. M. Ueno, *Biochim. Biophys. Acta* **904** (1987) 140.
20. W. Helfrich, *Z. Naturforsch.* **28c** (1973) 693.
21. S.A. Safran, P. Pincus, and D. Andelman, *Science* **248** (1990) 354.
22. W. Helfrich, *J. de Physique* **46** (1985) 1263.
23. W. Helfrich, *J. de Physique* **47** (1986) 321.
24. H.J. Deuling and W. Helfrich, *J. de Physique* **37** (1976) 1335.
25. R. Lipowsky, *Nature* **349** (1991) 475.
26. T. Sekimura and H. Hotani, *J. Theor. Biol.* **149** (1991) 325.
27. A. Alonso, M.A. Urbaneja, F.M. Goni, F.G. Carmona, F.G. Canovas, and J.C. Gomez-Fernandez, *Biochim. Biophys. Acta* **902** (1987) 237.
28. J. Kas and E. Sackmann, *Biophys. J.* **60** (1991) 821.
29. T.M. Fischer, *Biophys. J.* **65** (1993) 687.
30. P. Fromherz, C. Rocker, and D. Ruppel, *Faraday Discuss. Chem. Soc.* **81** (1986) 39.
31. W. Helfrich, *Physics Letters* **50A** (1974) 115.
32. P. Fromherz, *Chemical Physics Letters* **94** (1982) 259.
33. M.H. Ernst, in *Fractals in Physics,* eds. L. Pietronero and E. Tosatti (North-Holland, Amsterdam, 1986).
34. Z. Racz, in *Fractals in Physics,* eds. L. Pietronero and E. Tosatti (North-Holland, Amsterdam, 1986).

MONTE CARLO SIMULATION OF AGGREGATION PROCESSES: KINETIC PROPERTIES OF DEPOSITIONAL AGGREGATES

ORESTE PATIERNO, ALESSANDRA MENICOCCI, ALESSANDRO R. GIONA
Dipartimento di Ingegneria Chimica, Universitá di Roma "La Sapienza"
Via Eudossiana 18, 00184 Roma, ITALY

and

MASSIMILIANO GIONA
Dipartimento di Ingegneria Chimica, Universitá di Cagliari
Piazza d'Armi, 09123 Cagliari, ITALY

ABSTRACT

Aggregation kinetics in the formation of deposits is investigated by analyzing the temporal behaviour of the mean coverage and of the distribution of layer-aggregation probabilities. Monte Carlo simulation results are compared with stochastic models of growth (KPZ equation). The meaning of layer aggregation probabilities is investigated in details. The heterogeneous structure of the deposit kinetics is described by introducing the concept of aggregation time and aggregation time distribution.

1. Introduction

Research on aggregation phenomena encompasses many fields of chemical engineering science and technology, electrochemistry, vapour deposition technology, and the kinetics of surface growth. Classical analytical techniques for describing these phenomena have been developed[1]. However, the macroscopic approach based on deterministic kinetic models neither explain nor predict the structural properties of the aggregates. For this reason Monte Carlo (MC) simulation is the simplest tool for determining structural and kinetic properties and relating them to macroscopic measurable parameters characterizing the aggregation process. The analysis of the structure of depositional aggregates has been developed elsewhere[2,3,4].

In the recent years the study of surface growth kinetics gained an increasing interest, especially because stochastic models of surface formation were proposed[5] and compared with MC simulations and experimental results[6]. A comprehensive review on this topic can be found in the book of Family and Vicsek[7].

In this article we analyze aggregation kinetics of depositional aggregates in order to determine the relationship existing between the macroscopic kinetics of deposit formation and their structural properties. The dinamic properties of surface growth of aggregates can be investigated by looking at some macroscopic quantities: temporal evolution of the layer coverage, aggregation probability distribution, etc. In this sense the approach followed in the present analysis differs from dynamic scaling analysis of surface formation[7] because it is more oriented to macroscopically mesurable quantities and not to the spatio-temporal propagation of surface instabilities.

Moreover, in order to understand the heterogeneous nature of aggregation processes the concept of particle age (aggregation time) is introduced and related to the fluid dynamic condition of aggregation.

2. Deposit growth and dynamics

2.1. Simulation condition

The simplest way of generating depositional aggregates is to make use of MC simulation. By means of such simulation techniques it is possible to take into account fluid dynamic conditions and kinetic phenomena by tuning the walk statistics of aggregating particles and the probability of aggregation once a particle reaches an aggregation site (sticking probability). The details of the MC simulation employed were discussed by Tassopoulos et al.[3] and by Giona and Patierno[4]. In the present work we considered two-dimensional aggregates growing on a 1000×1000 square lattice with periodic boundary conditions. Moreover, we limit ourselves to growth processes with sticking probability equal to unity, so that the kinetic of deposit formation is controlled by the Peclet number $Pe = vL/D$ and by the occupation probability f_m. Following Tassopulos et al.[3], the ***downward probability*** p, i.e. the probability of a particle to move in the direction of growing aggregate, is related in twodimensional simulations to the Peclet number through the relation

$$Pe = \frac{3\,(4p-1)}{2\,(1-p^2)}\,. \tag{1}$$

The ***occupation probability*** f_m represents the average number of sites in the non-aggregate phase occupied by moving particles.

2.2. Stochastic models of growth

The dynamics of surface growth can be modelled by means of stochastic differential equations[7]. Let $h(t, \underline{x})$ be the height of the deposit at time t and at position $\underline{x}$ ($\underline{x}$ is the position vector in the plane transverse to the surface growth direction; it is a $(d-1)$-dimensional vector if the growth evolves in a d-dimensional space). It can be shown[5] that the surface growth can be modelled by means of the stochastic differential equation

$$\frac{\partial h(t,\underline{x})}{\partial t} = v + \alpha|\nabla h|^2 + \nu\nabla^2 h + \eta(t,\underline{x}) \tag{2}$$

where v is the growth velocity in the orthogonal direction to the surface, η is a stochastic Wiener process[8] uncorrelated both in space and in time

$$<\eta(t,\underline{x})>=0 \qquad <\eta(t,\underline{x})\eta(t',\underline{x}')>=\sigma^2\delta(t-t')\delta(\underline{x}-\underline{x}')\,. \tag{3}$$

The term $\alpha|\nabla h|^2$ takes into account the effect of the lateral growth, while the Laplacian contribution $\nu\nabla^2 h$ stands for the smoothening action deriving from surface tension. By considering a reference frame moving uniformly with the growth velocity v, and by putting $\tilde{h} = h - vt$, Eq. (1) becomes

$$\frac{\partial \tilde{h}(t, \underline{x})}{\partial t} = \alpha |\nabla \tilde{h}|^2 + \nu \nabla^2 \tilde{h} + \eta(t, \underline{x}) \; . \tag{4}$$

Last equation can be made dimensionless by defining

$$\underline{\xi} = \frac{\underline{x}}{L} \, , \quad \tau = \frac{\nu t}{L^2} \, , \quad z = \frac{\tilde{h}}{L} \, , \quad \tilde{\eta} = \frac{L\eta}{\nu} \, , \tag{5}$$

where L is the characteristic lenght in the orthogonal direction to the growth process, so that Eq. (4) becomes

$$\frac{\partial z}{\partial \tau} = a |\nabla_\xi z|^2 + \nabla_\xi^2 z + \tilde{\eta} \; , \tag{6}$$

where $a = \alpha/\nu$. This equation is known as Kardar-Parisi-Zhang (KPZ) equation. For KPZ equation a detailed dynamic scaling theory has been presented, focusing attention on the propagation of fluctuations in the height dynamics as a function of time t and of deposit size L[7].

3. Monte Carlo results and mean field approach

Besides dynamic scaling theory it is interesting to investigate deposit kinetics in terms of other global parameters related to macroscopic quantities: the mean layer coverage and the layer-aggregation probability density function (pdf).

3.1. Mean layer coverage

A quantity characterizing the deposit formation is the mean layer coverage, which is the counterpart in aggregation process of sorption curve in diffusion/adsorption phenomena[9].

Let us consider a two-dimensional model and let y be a generic height (in MC lattice simulation it corresponds to the lattice coordinate of a row orthogonal to deposit growth direction). The layer coverage $c(\theta, y)$ represents the fraction of sites belonging to the deposit aggregated at height y and at time $\theta = t - t_y$, where t_y is the time instant the first particle aggregates on layer y. The mean layer coverage $< c(\theta) >$ is the average of $c(\theta, y)$ over the height y

$$< c(\theta) > = \frac{1}{Y} \int_{y_o}^{y_o + Y} c(\theta, y) dy \; , \tag{7}$$

where the height y_o is sufficiently hight to ensure fully developed growth conditions. Particular attention should be posed on the definition of the time t. In MC simulations, the time t is consider equal to the number of aggregated particles. This definition of the time scales has a direct physical explanation, since by recalling Eq. (2) the height h and therefore the number of aggregated particles is proportional to time.

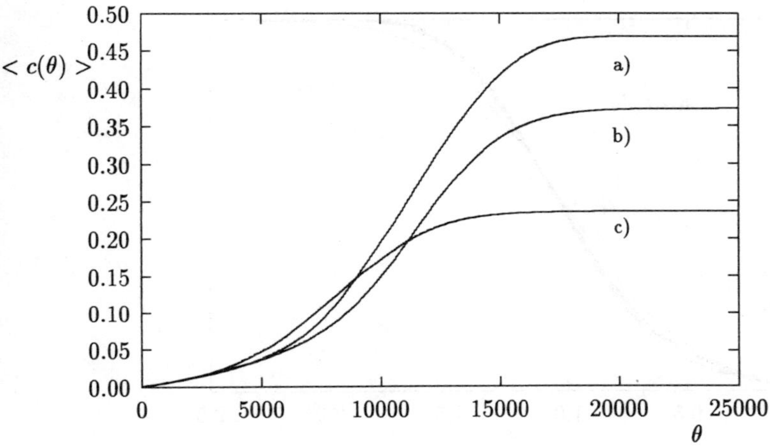

Figure 1: Mean layer coverage $< c(\theta) >$ vs θ in single particle deposition for different Peclet numbers. a)$Pe = 0$; b)$Pe = 2.0$; c)$Pe = \infty$.

Figure 1 shows the behaviour of $< c(\theta) >$ obtained from MC simulations in single particle deposition (diluted growth condition) for different values of the Peclet number. As it should be expected from the analogy with sorption curve, $< c(\theta) >$ shows a monotically increasing behaviour which asymptotically saturate to the corresponding value of the density[4] $c_\infty(Pe)$. It is convenient to normalize these quantities, by defining the normalized coverage $\Gamma =< c > /c_\infty$ (so that for all the aggregation condition $0 \leq \Gamma \leq 1$) and the dimensionless time s rescaled with respect to the first order moment of Γ (which can be considered as a probability distribution function)

$$s = \theta/m_1 \ , \quad m_1 = \int_0^\infty t \, d\Gamma(t) \ . \tag{8}$$

The normalized coverage obtained from MC simulations can be directly compared with the corresponding quantity evaluated starting from the mean field KPZ-equation. Figure 2 shows the single particle simulation results for in two limit case of Peclet numbers $Pe \in (0, \infty)$ ($Pe = 0$ corresponds to diffusion limited aggregates, $Pe = \infty$ to pure ballistic deposition) compared with the KPZ equation with $a = 1.5$ and $a = 2.75$. As can be observed all the MC simulation results for the normalized coverage kinetics are well interpreted by means of the KPZ equation with a value of the parameter a ranging in a small interval. A similar result has been found in the case of multiparticle simulation ($f_m > 0$).

In particular simulation results indicates that the normalized coverage $\Gamma(s)$ is weakly dependent on the occupation probability f_m, as shown in figure 3.

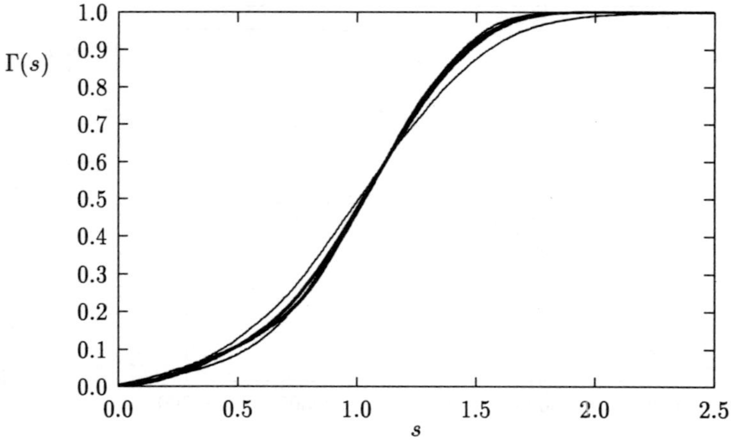

Figure 2: Normalized coverage $\Gamma(s)$ vs the dimensionless time s: The solid line show $\Gamma(s)$ deriving from KPZ model with $a = 1.5$ and $a = 2.75$ ($\sigma L/\nu = 0.707$) compared with MC result from single particle simulation (heavy lines) with $Pe = 0$ and $Pe = \infty$.

3.2. Layer aggregation pdf

Another interesting physical parameter is the layer aggregation pdf $\pi(y)$, defined as the probability that a particle will aggregate at a distance y from the current maximal height reached by the deposit.

The layer aggregation pdf $\pi(y)$ describes the spatial screening effects due to the competition between the various layers in the overall growth of the structure. Figure 4 shows the typical behaviour of $\pi(y)$ for a depositional aggregate. As it clearly appear from this figure the growth kinetics is limited to a narrow layer (the aggregation layer) close to the top of the structure. The width w of the aggregation layer can be estimated from the first order moment of $\pi(y)$

$$w = \int_0^\infty y\pi(y)dy \,. \tag{9}$$

The width w is a monotonically decreasing function of Peclet number.

3.3. Generalized layer growth models

In a macroscopic perspective, the layer aggregation pdf $\pi(y)$ characterizes the spatial effect in the growth, assuming that the kinetics depends exclusively on the coordinate y in the growth direction. Therefore, given $\pi(y)$ it is possible to simulate a practical unlimited variety of aggregation processes. For this reason we define as *generalized layer growth models*, the class of growth algorithms which can be characterized by specifying exclusively the layer aggregation pdf $\pi(y)$.

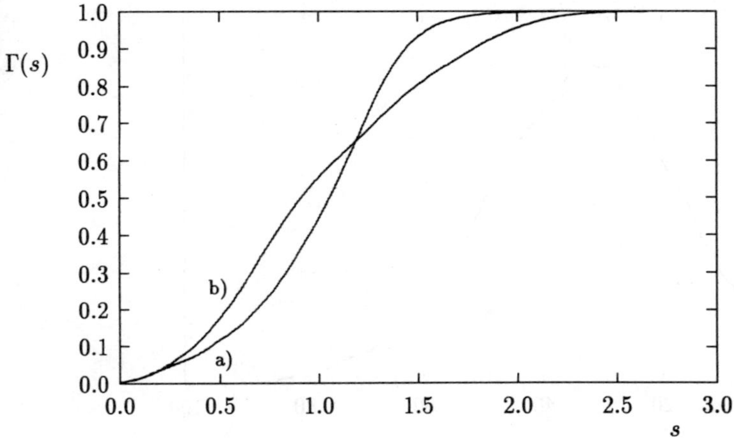

Figure 3: $\Gamma(s)$ vs s for multiparticle simulation for $Pe = 2.0$ and $f_m = 0.1$ (a) and $f_m = 1.0$ (b).

The lattice simulation of these model is straightforward and can be considered as a modification of the Leath algorithm of spreading percolation[10]. Starting from the actual surface profile, the first nearest neighbours of a surface sites represent the active sites for growth. At each time, the aggregation of a new particle on a site of lattice coordinate (i, j) (j is the coordinate in the direction of growth) is a random process whose statistics is given by $\pi(y)$, where $y = h_{max} - j$, h_{max} being the actual maximal height reached by the deposit. The aggregation process in each site is a temporally and spatially uncorrelated process. This kind of models can be improved by avoidng that the growth will proceed on a site inside a close loop of the aggregate. Of course this effect will induce nonlocal correlation (self-avoiding growth inside closed loops).

For instance, figure5 and 6 depict two deposits generated in this way by imposing a layer aggregation pdf of exponential shape

$$\pi(y) = \frac{1}{L_a} \exp(-y/L_a) , \tag{10}$$

where L_a is the width of the aggregation layer. Depending on the value of L_a, the resulting deposits exhibit a transition from very compact structures (L_a large, implies a more uniform probability of growth) to highly ramified structures (L_a small, implies almost exclusively a growth on the tips). The detailed characterization of the deposits as a function of the structure of $\pi(y)$ lies outside the scope of the present work. However, it is worthwhile to point out some general aspects of these kind of models.

The definition of growth models based on the assignment of a spatially dependent pdf $\pi(y)$ is the one of the simplest way to describe a growth process which takes into account of the spatial screening effect due to the layer competition. These kind

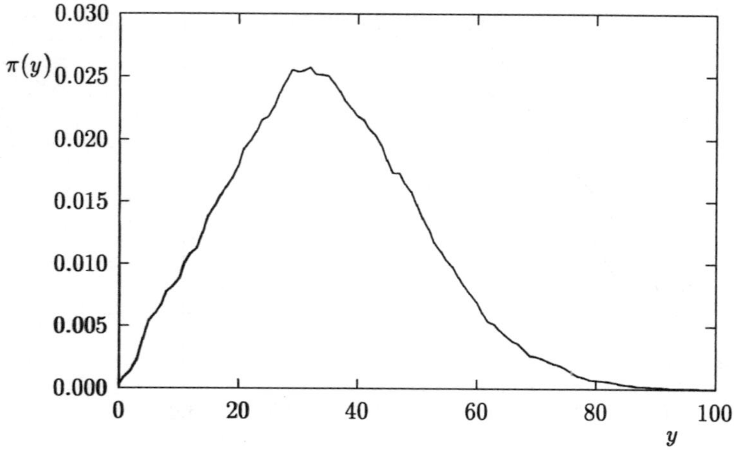

Figure 4: $\pi(y)$ vs y for single particle deposition with $Pe = 1.071$.

Figure 5: Depositional aggregate generated from the layer aggregation probability given by Eq. (10) with $L_a = 4$.

Figure 6: Depositional aggregate generated from the layer aggregation probabilty given by Eq. (10) with $L_a = 30$.

of models are physically well posed in comparison to other macroscopic models of deposit kinetics which a priori consider time dependent kinetic rates.

The models based on aggregation probabilities depending exclusively on the coordinate in the growth direction represent a simplified version of more general models of growth (like e.g. dielectric breakdown model[11,12] in which the aggregation probabilities depend on the position vector). Therefore, it is possible to obtain more refined models of deposit growth by considering that the aggregation probabilities will depends not only on the position but also on the local value of the tangential stress field induced by the fluid dynamic conditions of the process. In lattice simulations the stress field distribution can be obtained e.g. by solving the Navier-Stokes equation on the lattice by making use of Lattice Boltzmann Equation[13].

4. Aggregation time and kinetic heterogeneity of the aggregates

The analysis developed in the previous sections focused attention on the kinetic properties of deposit formation. Another interesting point that should be addressed is: how heterogeneous is the aggregation process as a function of the fluid dynamic conditions of particle movement? This problem can be approached by introducing age-distribution functions associated with the growth process similarly to what usually adopted in describing dispersion properties in chemical reactor dynamic[14].

Let us focus attention on a single aggregating particle. We define the *particle age* α (which can be called also the *aggregation time*) the time spent by a particle from the launch to its aggregation.

This concept is particularly interesting in multiparticle simulations in which the unit time is associated to the movement of each particle of one unit lenght (in the lattice simulation the unit length corresponds to a lattice site). In the simulations, aggregating particles are lauched from an height of 40 unit length from the actual top of the aggregate.

Once the aggregation time α has been introduced, the corresponding aggregation time pdf $p(\alpha)$, and the spatial distribution of the aggregation times $\alpha(\underline{x})$ can be defined. Figure 7 shows the behaviour of $p(\tilde{\alpha})$ vs $\tilde{\alpha} = \alpha/m_1(\alpha)$ ($m_1(\alpha)$ is the first order moment of $p(\alpha)$, i.e. the average aggregation time) for a fixed value of the Peclet number and for different values of the occupation probability f_m.

As can be seen all the pdfs collapse in a single curve. It means that only the mean aggregation time characterizes the distribution.

The average aggregation time $m_1(\alpha)$ of $p(\alpha)$ is shown in figure 8 as a function of the Peclet number for different values of the occupation probability. As it is to be expected $m_1(\alpha)$ decreases with the Peclet number (the dynamics of aggregating particles changes from purely diffusive to convective) and with the occupation probability f_m.

Figure 9 shows a particular snapshot of the spatial distribution of the aggregation time $\alpha(\underline{x})$ on a deposit with $Pe = \infty$ and $f_m = 0.1$. As can be noticed the spatial distribution of the aggregation time is an highly heterogeneous function of the position. Its scaling properties can be characterized by making use of the multifractal

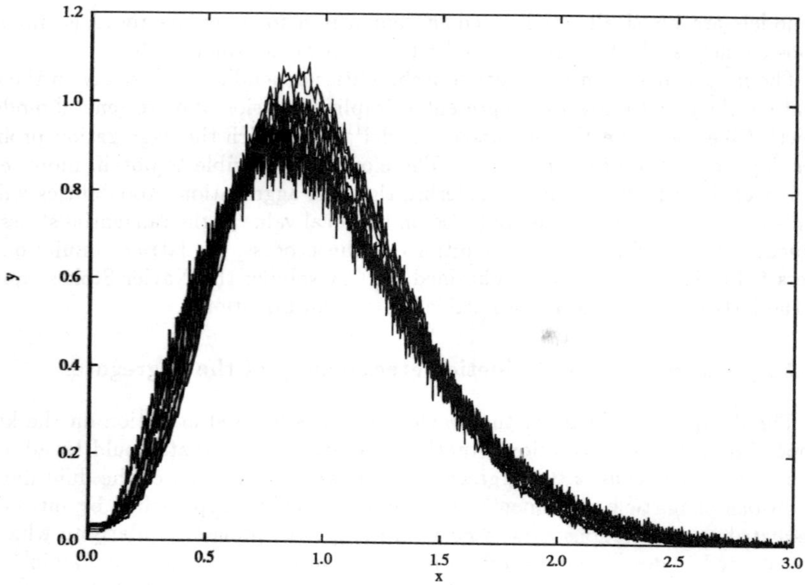

Figure 7: Aggregation time distribution $p(\tilde{\alpha})$ vs $\tilde{\alpha}$ ($Pe = 0$) for different values of the occupation probability f_m ranging from 0.0 to 1.0.

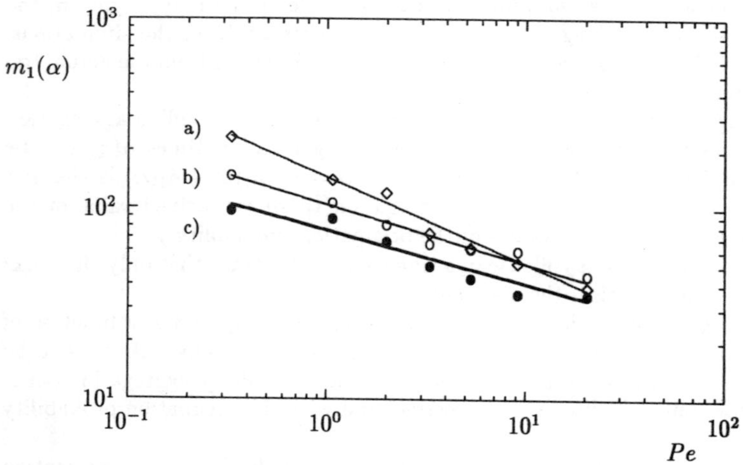

Figure 8: $m_1(\alpha)$ vs Pe for different values of f_m: a) $f_m = 0.1$; b) $f_m = 0.5$, c) $f_m = 1.0$.

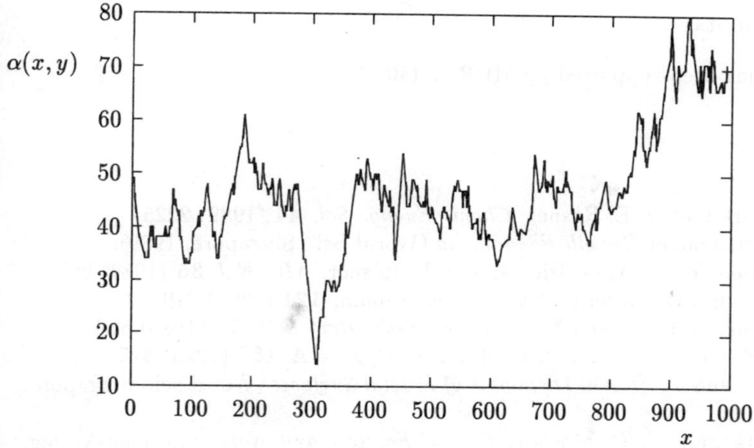

Figure 9: Snapshot of $\alpha(\underline{x}) = \alpha(x, y)$ vs x for afixed values of y (y is the coordinate in the growth direction) on a depositional aggregates ($Pe = \infty$, $f_m = 0.1$). For typographic reasons we have not indicated the points not belonging to the aggregate.

formalism[15].

In particular, a preliminary analysis of the spatial behaviour of $\alpha(\underline{x})$ indicates that, for low Peclet values, $\alpha(\underline{x})$ is statistically fairly homogeneous, close to a Wiener process. With the increase of the Peclet number, the distribution of $\alpha(\underline{x})$ undergoes structural changes originated by the presence of cluster of nearest neighbouring particles having the same age. This effect seems to be associated with some kind of channeling phenomenon similar to an aggregation avalanche of particle having the same age.

5. Concluding remarks

In this article we have discussed various methods to analyze the kinetics of deposit formation. The kinetics of layer coverage in depositional aggregation indicates that there is an acceptable agreement between the results of KPZ model and MC simulations.

The analysis of layer growth pdf suggests that the investigations of many aggregation phenomena of industrial interest could be treated within this approach, by defining a suitable expression for $\pi(y)$ eventually encompassing the effect of the tangential shear stress in particle disaggregation.

We have also indicated that a proper way to analyze kinetic heterogeneity could be represented by particle-age distribution and its spatial fluctuation behaviour. The detailed analysis of this aspect is left to future investigations.

Acknowledgments

This work has been supported by MURST (40%).

References

1. H. M. Park and D. E. Rosner, *Chem. Engng. Sci.* **44** (1989) 2225.
2. T. Vicsek, *Fractal Growth Phenomena* (World Sci., Singapore, 1989).
3. M. Tassopoulos, J. A. O'Brien and D. E. Rosner, *AIChE J.* **35** (1989) 967.
4. M. Giona and O. Patierno, *Chem. Eng. Comm.* **121** (1993) 219.
5. M. Kardar, G. Parisi and Y-C. Zhang, *Phys. Rev. Lett.* **56** (1986) 889.
6. T. Vicsek, M. Cserzö and V. K. Horvath, *Physica* **A 167** (1990) 315.
7. F. Family and T. Vicsek, *Dynamics of Fractal Surfaces* (World Sci., Singapore, 1991).
8. A. Lasota and M. C. Mackey, *Chaos, Fractals and Noise* (springer-Verlag, Berlin, 1994).
9. J. Crank, *The Mathematics of Diffusion* (Claredon Press, Oxford, 1975).
10. P. L. Leath, *Phys. Rev.* **B 14** (1976) 5046.
11. L. Niemeyer, L. Pietronero and H. Wiesman, *Phys. Rev. Lett.* **52** (1984) 1033.
12. F. Argoul, A. Arneodo, J. Elezgaray and H. L. Swinney, in *Growth and Form - Nonlinear Aspects* ed. by M. Ben Amar, P. Pelce and P. Tabeling (Plenum Press, New York, 1991).
13. A. Adrover and M. Giustiniani, this book.
14. C. Y. Wen and L. T. Fan, *Models for Flow Systems and Chemical Reactors* (Marcel Dekker, New York, 1975).
15. T. C. Halsey, M. H. Jensen, L. P. Kadanoff, I. Procaccia and B. I. Shraiman, *Phys. Rev.* **A 33** (1986) 1141.

FRACTAL CONCEPTS IN MODELING OF SUSPENSION RHEOLOGY

ROMANO LAPASIN, MARIO GRASSI and SABRINA PRICL
Department of Chemical, Environmental and Raw Materials Engineering
University of Trieste, Piazzale Europa 1, Trieste, I-34127 Italy

ABSTRACT

Modeling the rheology of aggregated suspensions essentially implies a rigorous definition of the structure of the disperse phase and the modifications it undergoes when the system is subjected to a deformation or a velocity field. In the present paper we develop a model able to describe and predict the shear viscosity in a wide range of structural conditions of the disperse phase, from fractal aggregates to primary particles.

1. Introduction

Modeling of suspension rheology is a subject of great interest both at the academic and industrial level. Suspensions of solid particles in a liquid phase are found in many industrial branches and applications, e.g. paints, printing inks, paper coatings, dyestuffs, pesticides, cement pastes, ceramic slips, foods etc. The knowledge of the rheological behavior of suspensions under controlled deformation or flow conditions often plays a crucial role in the analysis and control of preparation processes, as well as in the further handling and usage of these systems.

At present, no theoretical approach can be suitably used for a reliable and accurate prediction of shear behavior of suspensions in those ranges of disperse phase concentration usually occurring in practical applications (which include concentrated suspensions); therefore, the common practice relies on experimental tests in equivalent shear flow conditions and on empirical equations for correlating data. A quick glance at the relevant literature immediately reveals all the difficulties that can be encountered when attempting the formulation of a rheological model, even of a phenomenological nature, having general validity. Indeed, as the disperse phase concentration increases, the initial Newtonian behavior of suspensions changes into shear-thinning and/or plastic; furthermore, at high concentrations, a shear-thickening behavior is often observed over a more or less extended shear rate range. Finally, it should be noted that non-Newtonian behavior may also appear at low concentration, particularly when the solid particles are fine and anisometric.

A major implication of these considerations is that, even when recourse is made to phenomenological models in fitting the experimental data, objective difficulties often arise in trying to collect all data within a homogenous frame and, *a posteriori*, in finding suitable correlations among the model parameters and the characteristics of the disperse phase.

The definition of a theoretical model implies the macroscopic properties of a system to derive directly from the characteristics - both on micro- or mesoscale - exhibited by the system itself under the action of a well defined force or velocity field. Several theoretical models have been successfully formulated in the case of polymeric systems whereas, when suspensions are concerned, the actual modeling possibilities remain confined to those systems made up by stabilized (and hence, isolated) particles. Under these circumstances, some relationships have been derived by applying the basic concepts of the mean field theory, which hold in a dilute regime, and in the absence of hydrodynamic interactions among the particles. At greater disperse phase concentrations, interactions among the particles come into play and this factor, coupled with the high numerical density of the particles, rules the rheological behavior of the system; here, the recent theoretical developments based on the application of non-equilibrium molecular dynamics (*NEMD*) techniques have led to encouraging results in the simulation of real non-Newtonian behaviors[1,2].

Modeling the rheology of aggregated suspensions appears to be a much more difficult task, since the development of the model depends essentially on a rigorous definition of the structure of the disperse phase and the modifications it undergoes when the system is subjected to a deformation or a velocity field[3-8]. In the present paper, we shall examine the main problems and difficulties encountered in the formulation of a microrheological model, and present a case history of model derivation.

2. Aggregation Structures

The disperse phase of real aggregated suspensions, both dilute and concentrated, may present a wide variety of structures, depending on the specific characteristics of the particles and the different factors which govern the aggregation state. Nevertheless, it is possible to divide all disperse phase structures into two main classes: the *homogeneous quasi-crystal lattices*, which may occur under rest conditions and span the entire space from wall to wall, thus conferring to the system those mechanical characteristics typical of a gel state, and the *clusters or flocs aggregates*, having shapes and dimensions which change from system to system and with imposed flow conditions.

The aggregation state of the disperse phase is determined by a number of factors, essentially bound to the geometric and surface features of the particles (dimensions, shape, presence of charges or adsorbed components), to their numerical density and to the specific field condition (temperature, hydrodynamic conditions, presence of other components in the fluid phase). In the absence of an effective stabilizing action (of electrostatic and/or steric nature), particles do aggregate among themselves to an extent which is mainly a function of the particle dimensions. Thus, high degrees of aggregation are encountered in colloidal systems, where the formation of very large particle clusters occurs. Even in coarse particle suspensions, however, a temporary aggregation state can eventually be observed, depending on particle concentration and intensity of the applied velocity field[9].

When the aggregation state is so extended that a continuous three-dimensional network is set up, the rheological properties of the suspension are quite similar to those peculiar to weak gel systems. Under continuous shear flow conditions, when both the degree of particle aggregation and its variations with imposed flow conditions are no longer negligible, the viscosity of the suspension varies markedly with the imposed shear rate. Such variation may be reversible or present some characteristics of irreversibility, when the preceding deformation history causes a substantial and irreversible change in the structural features of the disperse phase.

Indeed, different shear flow conditions may originate quite different phenomena, which may coexists within the same system and result in either cooperative or antagonistic effects on the structural characteristics of the disperse phase and, hence, on the rheological properties of the suspension. In the vast majority of cases, the main effect induced by the applied shear conditions results in the breakdown of particle aggregates into smaller units; the greater the imposed shear rate or shear stress, the greater the extent of this breakdown process. When only aggregation phenomena due to the always present Brownian motions (*perikinetic aggregation*) oppose the shear destructive process, the macroscopic behavior of the suspension is shear-thinning and reversible. In some cases, the aggregate breakdown process can be accompanied by a structural rearrangement, leading to new aggregates having more compact features than the original ones. In other cases, the opposite situation is observed: the imposed flow field leads to an increased degree of aggregation among the particles (*orthokinetic aggregation*). This particular effect may appear during the flow and deformation history of the suspension by virtue of some changes induced at the particle/continuous phase interface. In both situations, the rheological behavior of the suspension begins to show some irreversible features, the entity of which is proportional to the relative weight of such concomitant phenomena.

Particle aggregations have been the subject of many experimental and theoretical studies, since they are of utmost importance in many branches of science and technology, especially in the colloid science area[10-23]. Much attention has been devoted to the simulation of different aggregation mechanisms and to the analysis of the resulting structures. One of the earliest models for the computer simulation of aggregate formation in three dimensions was introduced by Vold[10]. In this model, particles with random linear trajectories are added to a growing cluster of particles at the position where they first contact the cluster. Reorganization of the cluster is not allowed. One of the most important results of Vold's simulation was that the number of particles $N(l)$ within a length l of the center of gravity is given by $N(l) \sim l^{2.33}$, for $N(l) < 40\text{-}60\%$ of the total number of particles in the cluster. Even if these results were criticized and corrected by Sutherland[11], Vold's work opens a long series of simulation studies in two- and three-dimensional space, based upon different aggregation mechanisms.

All these analyses have been extended to very large clusters, having colloidal aggregates as reference systems; consequently, the concept of fractal dimensionality has been conveniently used to interpret the results of the computer simulations and characterize the aggregate structure. This idea is a particularly valuable way of describing many of the geometrical properties of structures which are self-similar or statistically self-similar. For sufficiently large cluster sizes, the dependence of the radius of gyration R_g on the number of particles in the cluster n can be expressed as $R_g \sim n^{1/D}$, where D is the fractal dimension. Similar scaling laws can be written for other geometric quantities: among them, an important quantity which characterizes self-similar objects is the density correlation function $C(r)$, where r is the distance from the center of mass of the cluster. If the object has a fractal dimension D and an ordinary Euclidean dimension d, then $C(r) \sim r^{(D-d)}$. The fractal dimension has important implications for colloidal systems: if $D < d$, the density of the cluster will become smaller and smaller as the cluster grows larger and larger. The mechanisms of aggregation and the results obtained from computer simulation, expressed in terms of fractal geometry, have been extensively reviewed and discussed by Meakin[21,22].

Table 1
Values of the fractal dimension calculated according to different mechanisms of aggregation

PROCESS[1]	*SUB-PROCESS*	*D*	*REF.*
DLCCA	constant reaction kernel	1.8	22
	hierarchical	1.78	22
BCCA	BCCA	1.95	22
RLCCA	Jullien and Kolb	1.98	14,15
	Brown and Ball (monodisperse)	1.94	16
	Brown and Ball (polydisperse)	2.11	16
	Leyvraz	2.09	19
Polydisperse DL	IR = 0	1.8	22
	IR = 1	2.09	
	IR = 2	2.17	
	IR = 3	2.18	
Polydisperse BALLISTIC	IR = 0	1.95	22
	IR = 1	2.13	
	IR = 2	2.18	
	IR = 3	2.19	
Polydisperse RL	IR = 0	2.09	22
	IR = 1	2.18	
	IR = 2	2.24	
	IR = 3	2.25	

[1] DLCCA = Diffusion Limited Cluster Cluster Aggregation; BCCA = Ballistic Cluster Cluster Aggregation; RLCCA = Reaction Limited Cluster Cluster Aggregation; DL = Diffusion Limited; RL = Reaction Limited

Table 1 reports a list of D values obtained from different mechanisms of aggregation: they range between 1.78 and 2.25, and converge to similar values when aggregation is combined with re-ordering processes. In many cases particle clusters are held together by relatively weak van der Waals forces, which may be disrupted by thermal fluctuations, external fields or by the relative momentum of two colloding clusters, leading to fission or to structural reorganization. Table 1 shows that more compact aggregates are obtained when three consecutive stages of structural reorganization (IR = 1, 2, 3), designated as "bending", "folding" and "twisting", are simulated[22].

It should be noted that in real systems the property of self-similarity may extend over only a limited range of length scales, so that scaling laws hold between some lower cutoff length, sensibly higher than the radius a of a primary particle, and an upper cutoff length approaching the overall size of the cluster. Moreover, different scaling laws can be derived from the same mechanism of aggregation, only by adjusting some parameters of the model. This is the case of clustering of clusters (CC), which is a basic aggregation model applicable to several growth process in colloid science.. A sign change of the mobility exponent which controls the diffusivity of the clusters relative to their size, results into a marked change of the cluster size distribution during aggregate growth and marks the boundary between a 'flocculation like' and a 'gelation like' regime. Another important parameter is the bonding probability, which permits the model to crossover from chemical or reactive CC to diffusive CC, with a corresponding change of fractal dimensionality. Finally, it must be emphasized that a crossover is expected from the flocculation regime at low concentration to the gelation regime at high concentration, with a corresponding gradual change of the fractal dimension from D to d.

All the above scaling laws cannot be extended to smaller degrees of aggregation, which are typical of noncolloidal systems or coarse suspensions in any flow conditions or may be encountered even in colloidal systems at high shear rates. Different criteria must be adopted to analyze the state of aggregation of the disperse phase observed in real systems and to characterize its dependence on the disperse phase concentration and the rate of shear, in order to formulate an adequate rheological model.

3. Rheological Modeling of Aggregated Suspensions

The fundamental objective of the rheological modeling is the definition of an equation able to describe the dependence of the viscosity on the shear rate and on the disperse phase concentration. The modeling procedure can be divided into several steps, the first concerning the choice of an adequate functional dependence of viscosity on disperse phase concentration.

Several models have been developed for concentrated suspensions of unaggregated spherical particles, taking into account the hydrodynamic interaction between spheres in different ways[24,25]. Their common feature is that the shear viscosity of the suspension is expressed as:

$$\eta = \eta_s \eta_r(\Phi) \tag{1}$$

where η_s is the viscosity of the continuous phase, and η_r is the relative viscosity, which characterizes the contribution of the disperse phase and depends on the disperse phase volume fraction Φ.

Particular mention is made of the Quemada model[26], which was derived starting from the optimization of the viscous energy dissipation. It can be written in the two equivalent forms:

$$\eta_r = \frac{1}{(1-\Phi/\Phi_m)^2} = \frac{1}{(1-\frac{[\eta]}{2}\Phi)^2} \tag{2}$$

where $[\eta]$ is the intrinsic viscosity, depending on particle shape and charge (for electrically neutral spherical particles, $[\eta] = 2.5$). The Quemada model appears to be the most interesting among the different models proposed, since it can provide a satisfactory correlation of viscosity data for many real suspensions of spherical and anisometric particles over a extended Φ range, and this is the reason why it can be properly written in the two equivalent forms reported above.

As mentioned above, the disperse phase of aggregated suspensions can be made up of different structural units (particles, flocs, clusters) having different shape, size and size distribution. A simple model suitable to take into account for all these factors can be hardly conceived and some simplifying assumptions are compulsory.

At any shear rate, the viscosity of an aggregated suspension is assumed to depend primarily on the effective volume of the disperse phase, which in its turn depends on the degree of aggregation of the solid particles. The main effect resulting from the aggregation processes is an increase in volume of the disperse phase with respect to its nominal value, i.e. the sum of the volumes of the constituent particles. A global measure of the volume increase of the disperse phase is introduced by the parameter C_{fp}, which is defined by:

$$C_{fp} = \frac{\Phi_{eff}}{\Phi} \tag{3}$$

where Φ_{eff} and Φ are the effective and the nominal volume fractions, respectively[5].

Assuming disperse phase aggregation and breakdown processes as dominant is equivalent to neglecting the effects bound to aggregate anisometry and their subsequent orientation in the flow field. Another assumption useful to model development is concerned with the shape and nature of the aggregates, which may be regarded as impenetrable spheres. Even if permeability of aggregates can be taken into account, however, the depth of liquid penetration into the aggregate is relatively small, so that the assumption of permeable aggregates exceeds the precision of the assumption regarding their form[6]. As a consequence, we can reasonably extend the validity of the Quemada model to the case of aggregated suspensions, provided we substitute Φ with Φ_{eff}:

$$\eta = \eta_s \eta_r(\Phi_{eff}) = \eta_s \eta_r(C_{fp}\Phi) \tag{4}$$

For a cluster made up by n_i monosized spherical particles of radius a, the dependence of C_{fp} on n_i can be written as follows:

$$C_{\mathrm{fp_i}} = \frac{V_\mathrm{i}}{n_\mathrm{i} V_\mathrm{p}} = \frac{R_\mathrm{i}^3}{n_\mathrm{i} a^3} \tag{5}$$

where V_i and V_p are the volume of the cluster and the volume of the particle, respectively, and R_i is the equivalent hydrodynamic radius of the cluster.

In the case of fractal aggregates, if we assume the hydrodynamic radius R_i to be proportional to the cluster radius of gyration R_g, we obtain:

$$R_\mathrm{i} \sim n_\mathrm{i}^{1/D}, \quad C_{\mathrm{fp_i}} \sim n_\mathrm{i}^{\beta} \tag{6}$$

with $\beta = 3/D-1$. When $D < 3$, the density of the floc decreases with increasing n_i.

When aggregate size ranges from that of primary particles to that of fractal objects, the variation of C_fp with n_i for each aggregate can be expressed through the following relationship:

$$C_{\mathrm{fp_i}} = \frac{B + A\left(n_\mathrm{i}^{\beta+1} - 1\right)}{B + n_\mathrm{i} - 1} \tag{7}$$

Figures 1(a) and 1(b) show the good quality of the fitting achieved with Eq. (7) in two extreme cases, relative to real dense aggregates of sintered glass particles[5] and to aggregates obtained by computer simulation according to the ballistic cluster-cluster mechanism[22], respectively. In the first case, $\beta = 0$; consequently, A represents the limiting value of C_fp for $n_\mathrm{i} \to \infty$. In the second case, for n_i large enough ($n_\mathrm{i} > 50$), C_fp tends to the same scaling law indicated by Eq. (6), with $\beta = 0.533$. We can sensibly conclude that the functional dependence expressed by Eq. (7) is valid independently of the particular aggregation mechanism involved.

In the case of monosized spherical aggregates, the value of C_fp for the global suspension coincides with that of each single cluster. Nevertheless, the real situation is characterized by an aggregate size distribution, variable with imposed flow conditions, which must satisfy the mass balance:

$$N = \sum_{n_\mathrm{i}=N_\mathrm{min}}^{N_\mathrm{max}} \nu_\mathrm{i} n_\mathrm{i} \tag{8}$$

Accordingly, we get:

$$C_\mathrm{fp} = \frac{\sum_{n_\mathrm{i}=N_\mathrm{min}}^{N_\mathrm{max}} \nu_\mathrm{i} V_\mathrm{i}}{N V_\mathrm{p}} = \frac{1}{N} \sum_{n_\mathrm{i}=N_\mathrm{min}}^{N_\mathrm{max}} \nu_\mathrm{i} \left(\frac{R_\mathrm{i}}{a}\right)^3 = \frac{1}{N} \sum_{n_\mathrm{i}=N_\mathrm{min}}^{N_\mathrm{max}} \nu_\mathrm{i} n_\mathrm{i} C_{\mathrm{fp_i}} \tag{9}$$

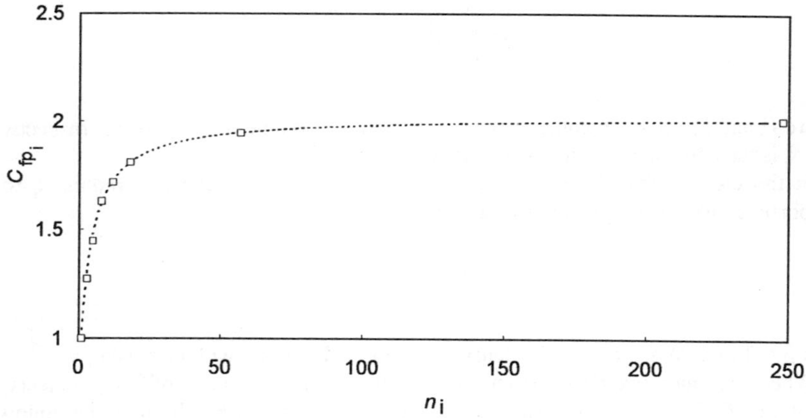

Figure 1(a). Comparison between experimental data for dense aggregates[5] (□) and model prediction (solid line) according to Eq. (7) (β = 0, A = 2.024, B = 4.544).

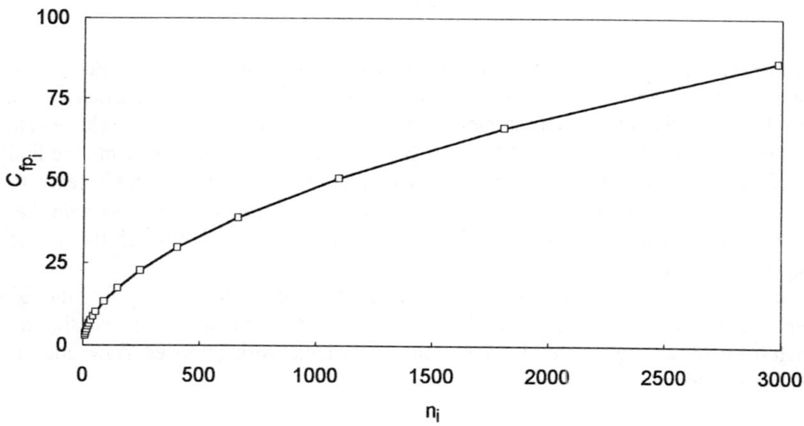

Figure 1(b). Comparison between computer simulated data for ballistic clustering aggregation[22] (□) and model prediction (solid line) according to Eq. (7) (β = 0.533, A = 1.21, B = 2.43).

where N is the total number of the particles in the system. For fractal aggregates, if the aggregate size distribution is wide and reasonably approximated (at least to a good extent) by a box-type distribution, ($\nu_i \cong \text{cost} \cong \overline{\nu}$, per $N_{min} \le n_i \le N_{max}$), we have:

$$C_{fp} \sim \bar{n}^{3/D} \sim N_{max}^{3/D} \tag{10}$$

For non colloidal suspensions, as well as for colloidal fractal systems at the early stages of growth, we can assume a hyperbolic size distribution $\nu_i(n_i)$ such as:

$$\nu_i = \frac{N}{N_{max} - N_{min} + 1} \frac{1}{n_i} \tag{11}$$

which satisfies the condition set by Eq. (8).

From Eqs. (9) and (11) we obtain, in the case of maximum generality ($N_{min} = 1$):

$$C_{fp} = \frac{1}{N_{max}} \sum_{n_i=1}^{N_{max}} C_{fp_i} \tag{12}$$

Combining Eqs. (7) and (12) yields:

$$C_{fp} = \frac{1}{N_{max}} \sum_{n_i=1}^{N_{max}} \frac{B + A\left(n_i^{\beta+1} - 1\right)}{B + n_i - 1} \tag{13}$$

To simplify the successive steps involved in model development, we can resort to an approximated expression of Eq. (13), containing either N_{max} or $\bar{n}$ as the independent variable and having the same functional dependence of Eq. (7):

$$C_{fp} = \frac{\bar{B} + \bar{A}\left(N_{max}^{\beta+1} - 1\right)}{\bar{B} + N_{max} - 1} \tag{14}$$

where the values of $\bar{B}, \bar{A}$ and β are 11.76, 1.98 and 0 for dense aggregates and 0.392, 1.495 and 0.44 for fractals aggregates, respectively.

In order to complete the rheological modeling is now sufficient to define the dependence of $\bar{n}$ (or of N_{max}) on the shear rate or the shear stress, on condition that we assume the conservation of the hyperbolic size distribution law upon variation of the imposed flow conditions. The maximum or the average aggregate dimension can be derived from a rate equation which expresses the balance between the particle aggregation and aggregate breakdown processes[5] or, alternatively, from experimental observations on aggregates after these have been subjected to different shearing conditions[27]. A plain relationship, which synthesizes all these considerations, is the following:

$$N_{max} = N_{max_l} + k\dot{\gamma}^{-p} \tag{15}$$

where $N_{\max_l}$ corresponds to the state of minimum degree of aggregation and is, of course, equal to 1 in case of complete structural breakdown. The second term in Eq. (15) derives from the ratio of the rate constants for the particle aggregation and aggregate breakdown processes (the rate constant of the aggregation process is assumed to be dependent on temperature and independent of the applied shear rate whereas the rate constant for the breakdown process increases with increasing shear rate, with a functional dependence of the power-law type). According to its definition, Eq. (15) includes, as a special case, the shear dependence observed for fractals aggregates ($N_{\max} \propto \dot{\gamma}^{-p}$) in a narrow range of $\dot{\gamma}$.

Combining Eqs. (4), (14) and (15) leads to the final expressions for the viscosity as a function of the disperse phase volume fraction and the shear rate, respectively:

$$\eta_r = \left(1 - 1.25 C_{fp}(N_{\max})\Phi\right)^{-2} = \left[1 - 1.25\Phi \frac{\bar{B} + \bar{A}\left(N_{\max}^{\beta+1} - 1\right)}{\bar{B} + N_{\max} - 1}\right]^{-2} \tag{16'}$$

$$\eta_r = \left\{1 - 1.25\Phi \frac{\bar{B} + \bar{A}\left[\left(N_{\max_l} + k\dot{\gamma}^{-p}\right)^{\beta+1} - 1\right]}{\bar{B} + N_{\max_l} + k\dot{\gamma}^{-p} - 1}\right\}^{-2} \tag{16''}$$

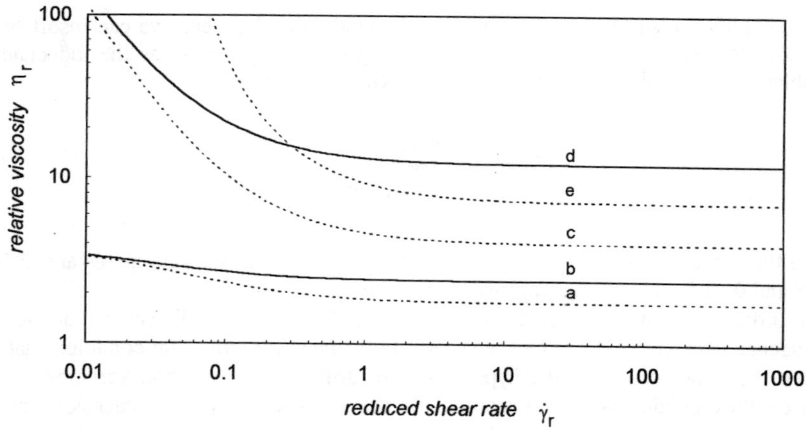

Figure 2. Relative viscosity curves for dense aggregates predicted from Eq. (16") (a: $\Phi = 0.2$, $N_{\max_l} = 1$; b: $\Phi = 0.2$, $N_{\max_l} = 10$; c: $\Phi = 0.4$, $N_{\max_l} = 1$; d: $\Phi = 0.4$, $N_{\max_l} = 10$; e: $\Phi = 0.5$, $N_{\max_l} = 1$).

Figure 2 shows the profiles of the relative viscosity as a function of the reduced shear rate $\dot{\gamma}_R = \dot{\gamma} / \dot{\gamma}_c$ (where $\dot{\gamma}_c = k^{1/p}$) for dense aggregate suspensions ($\beta = 0$). For low value of the volume fraction Φ we can observe only a slight deviation from Newtonian behavior and the value of $N_{\max_l}$ does not sensibly affect the global response of the systems; on the contrary, at high disperse phase concentration $N_{\max_l}$ plays a crucial role in governing the shear dependent behavior and, particularly, in determining the asymptotic viscosity value at high rates of shear.

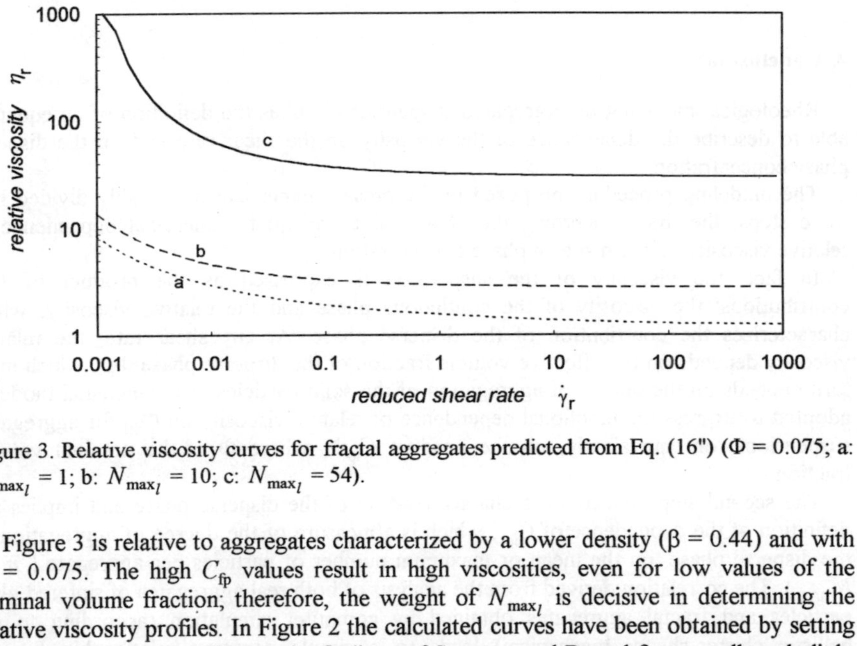

Figure 3. Relative viscosity curves for fractal aggregates predicted from Eq. (16") ($\Phi = 0.075$; a: $N_{\max_l} = 1$; b: $N_{\max_l} = 10$; c: $N_{\max_l} = 54$).

Figure 3 is relative to aggregates characterized by a lower density ($\beta = 0.44$) and with $\Phi = 0.075$. The high C_{fp} values result in high viscosities, even for low values of the nominal volume fraction; therefore, the weight of $N_{\max_l}$ is decisive in determining the relative viscosity profiles. In Figure 2 the calculated curves have been obtained by setting p equal to 0.878, according to the findings of Sonntag and Russel from small-angle light scattering experiments, whereas those reported in Figure 3 have been obtained by setting $p = 0.5$. Quite obviously, increasing p results in a stronger dependence of η_r on $\dot{\gamma}$, all other factors being equal.

All the profiles calculated from the model, both for dense and fractal aggregates, closely resembles those observed in real colloidal and noncolloidal suspensions. Moreover, when β reduces to zero, the resulting expression for the relative viscosity is equivalent to the Quemada model[5,26]:

$$\eta_r = \left[1 - 1.25\Phi \frac{C_{fp,0} + C_{fp,\infty}(\dot{\gamma}/\dot{\gamma}_c)^{-p}}{1 + (\dot{\gamma}/\dot{\gamma}_c)^{-p}}\right]^{-2} \tag{17}$$

The difference between the limiting Newtonian viscosities (for $\dot{\gamma} \to 0$ and for $\dot{\gamma} \to \infty$, respectively) depends on the corresponding parameters $C_{fp,0}$ and $C_{fp,\infty}$. The above model is able to describe shear-thinning behavior, for $C_{fp,0} > C_{fp,\infty}$, and plastic behavior, for $C_{fp,0} \rightarrow 1/1.25\Phi$. Finally, by modifying the assumptions concerning the rate constants in the kinetic equation leading to Eq. (14), we obtain $C_{fp,0} < C_{fp,\infty}$ and, accordingly, the model predicts a shear-thickening behavior. Several examples are given in literature, which confirm the suitability of the model in describing the rheological behavior of aggregated suspensions under shear flow conditions.

4. Conclusions

Rheological modeling of aggregated suspensions implies the definition of an equation able to describe the dependence of the viscosity on the shear rate and on the disperse phase concentration.

The modeling procedure proposed in the present paper can be suitably divided into three steps, the first concerning the choice of an adequate functional dependence of relative viscosity on the disperse phase concentration.

In fact, the viscosity of the suspension is expressed as the product of two contributions: the viscosity of the continuous phase and the relative viscosity, which characterizes the contribution of the disperse phase. At any shear rate, the relative viscosity depends on the effective volume fraction of the disperse phase Φ_{eff}, which in its turn depends on the degree of aggregation of the solid particles. The Quemada model is adopted to express the functional dependence of relative viscosity on Φ_{eff} for aggregated suspensions, or equivalently, on the product $C_{fp}\Phi$, where Φ is the nominal volume fraction.

The second step concerns the characterization of the disperse phase and implies the definition of the dependence of C_{fp} , which is a measure of the degree of aggregation of the disperse phase, on the mean or maximum number of particles per aggregate ($\bar{n}$ or N_{max}). The correlation derived from the analysis of both real aggregates of sintered glass particles and fractal aggregates obtained by computer simulation (according to the ballistic cluster-cluster mechanism) leads to a simple common relationship for the variation of C_{fp} with n_i (number of particles per aggregate) which is valid also when aggregate size ranges from that of primary particles to that of fractal objects. In the case of monosized spherical aggregates, the value of C_{fp} for the global suspension coincides with that of each single cluster. Nevertheless, the real situation is characterized by an aggregate size distribution, variable with imposed flow conditions. By assuming a hyperbolic size distribution $\nu_i(n_i)$, an expression is obtained for the dependence of C_{fp} on $\bar{n}$ or N_{max} , which contains a parameter β related to the density and, hence, to the fractal dimension of the aggregates. In real systems the structural conditions of the disperse phase strongly depend on the shear conditions and history experienced by the suspensions. At low shear no significant structural rearrangement is promoted by motion so that aggregates maintain their original structure and fractal dimension, which, in turn, depend on the aggregation mechanism. Conversely, the structural breakdown occurring at high shear is frequently accompanied by reordering and rearrangement processes and, then, particle aggregates tend to loose their fractal nature.

In the final step the dependence of $\bar{n}$ (or of N_{max}) on the shear rate or the shear stress is derived from a rate equation which expresses the balance between the particle aggregation and aggregate breakdown processes or, alternatively, from experimental observations on aggregates.

All the viscosity curves calculated from the model, both for dense and fractal aggregates, closely resembles those observed in real colloidal and noncolloidal suspensions. Thus, the proposed model appears to be suitable to describe and predict the shear viscosity in a wide range of structural conditions of the disperse phase, from fractal aggregates to primary particles.

References

1. H.O. Barnes, M.F. Edwards and L.V. Woodcock, *Chem. Eng. Sci.* **42** (1987) 591.
2. A.I. Jomha, A. Merrington, L.V. Woodcock, H.A.Barnes and A.Lips, *Powder Technol.* **63** (1991) 343.
3. B.A. Firth and R.J. Hunter, *J. Colloid Interface Sci.* **57** (1976) 266.
4. W.B. Russel, *Powder Technol.* **51** (1987) 15.
5. R. Lapasin, *Chem. Biochem. Eng. Q.* **1** (1987) 143.
6. A. Potanin and N.B.Uriev, *J. Colloid Interface Sci.* **142** (1990) 385.
7. A. Potanin, *J. Chem. Phys.* **96**(12) (1992) 9191.
8. A. Potanin, *J. Colloid Interface Sci.* **145** (1992) 140.
9. A.L. Graham, R.D.Steele and R.B.Bird, *Ind. Eng. Chem. Fundam.* **23** (1984) 420.
10. M.J. Vold, *J. Colloid Sci.* 18 (1963) 684.
11. D.N. Sutherland, *J. Colloid Interface Sci.* **25** (1967) 373.
12. Y.A. Witten and L.M. Sander, *Phys. Rev. Lett.* **47** (1981) 1400.
13. P. Meakin, *J. Colloid Interface Sci.* **96** (1983) 415.
14. M.Kolb and R. Jullien, *J. Phys. Lett.* **45** (1984) L977.
15. R. Jullien and M. Kolb, *J. Phys.* **A17** (1984) L639.
16. W.D. Brown and R.C. Ball, *J. Phys.* **A18** (1985) L517.
17. M. Kolb, *Physica* **140A** (1986) 416
18. R. Jullien and R. Botet, *Aggregation and Fractal Aggregates* (World Scientific, Singapore, 1987).
19. P. Meakin and F. Family, *Phys. Rev.* **A36** (1987) 5498.
20. H.M. Lindsay, M.Y. Lin, D.A. Weitz, P. Sheng, Z. Chen, R. Klein and P. Meakin, *Faraday Discuss. Chem. Soc.* **83** (1987) 153.
21. P. Meakin, *Ann. Rev. Phys. Chem.* **39** (1988) 237.
22. P. Meakin, *Adv. Colloid Interface Sci.* **28** (1988) 249.
23. M.Y. Lin, R. Klein, H.M. Lindsay, D.A. Weitz, R.C. Ball and P. Meakin, *J. Colloid Interface Sci.* **137** (1990) 263.
24. R. Lapasin, *ICP* **12** (1984) 14.
25. D. Quemada, *Journal de Mecanique* **special issue** (1985), 267.
26. D. Quemada, *Rheol. Acta* **16** (1977) 82; *ibidem* **17** (1978) 632.
27. R.C. Sonntag and W.B. Russel, *J. Colloid Interface Sci.* **115** (1987) 378.

Multiphase and Turbulent Flows

The numerical analysis of turbulent flows is closely connected with the development of supercomputers and parallel architectures. In the last few years, numerical codes for solving the Navier-Stokes equations began to be applied also in chemical process design and it can be expected that the design of reacting systems and separation units in future years will benefit from detailed fluid-dynamic simulations of mixing and turbulence.

Two articles in this section cover methodological aspects related to the direct simulation of turbulent flow and to wavelet analysis. The other two refer to the identification of multiphase-flow regimes by means of fluctuation analysis.

The topic of multiphase-flow identification and characterization is particularly interesting for any researcher involved with chaos, fractals and fluctuation theory. Upon examination of the recent literature on this field, two different *schools* can be sharply recognized: those who regard multiphase flow dynamics as a typically chaotic phenomenon and consequently analyze regime transitions by applying the classical methods of chaos theory (correlation dimension, embedding techniques); and those who treat multiphase-flow signals as a stochastic process and investigate characteristic scaling properties of the fluctuations (rescaled-range analysis, diffusional methods). Future work in the field will establish which approach is to be preferred.

DIRECT NUMERICAL SIMULATION OF TURBULENT DISPERSION UNDER STEADY DRIFT

ALFREDO SOLDATI, PAOLO ANDREUSSI

Dipartimento di Scienze e Tecnologie Chimiche, Università di Udine, via Cotonificio 108 Udine, 33100, ITALY

and

SANJOY BANERJEE

Department of Chemical and Nuclear Engineering, University of California at Santa Barbara Santa Barbara, CA, 93106, U.S.A.

ABSTRACT

The problem of turbulent particle transport in a closed channel flow is described. A pseudospectral method was used to simulate the temporally and spatially varying turbulent flow field. Then, the behaviour of dispersed charged particles was analyzed following a swarm of small particles in the flow field. Particles were under the effect of the flow field, the gravity force and the action of an applied steady and uniform electrostatic force. The qualitative features of the dispersion mechanisms are presented. The classical approach to evaluate the diffusion properties of the flow field in homogeneous turbulence, first presented by Taylor (1921) is discussed. An attempt is made to employ this method to estimate the diffusion coefficient of particles in the outer region of the channel, where the non homogeneity of turbulence is weaker.

1. Introduction

The dispersion of particles or aerosols in turbulent flow is certainly one of the most challenging problems in fluid mechanics. However, when dealing with turbulence related problems, the strong nonlinearity of the Navier-Stokes equations makes theoretical analysis extremely difficult and excludes the usual perturbative treatments. Physically, at large Reynolds number, a very large range of scales of motion is excited. These different scales interact together in a complex way that is not yet clarified as pointed out by Hussain[1,2]. Usually, turbulent transport phenomena have been solved by the use of quantities describing the mean features of the flow field. This approach, that may be termed *statistical* since only the statistical properties of the flow field are required in order to describe the phenomenon, supplies the overall transport characteristics. However, instantaneous properties may not be given.

In recent years, since the increased capabilities of large computers made it feasible, a different approach based on the direct solution of turbulent flow fields is becoming one of the main tools for fundamental and applied research in turbulence related phenomena. The Direct Numerical Simulation (DNS) allows the description of the temporally and spatially varying turbulent flow field without the use of constants and adjustable parameters. Usually, either finite differences or pseudospectral methods are used for DNS. When finite-differentiating a function, the accuracy of the results depends on the number of neighboring points on which the differentiation is based (order of the method). The accuracy of a finite difference scheme increases with the order. However, the use of higher order schemes is not yet recommended since they are not robust and they require a large amount of memory. Pseudospectral methods, on the other hand, are based on expanding flow variables in Fourier series. Their accuracy is therefore limited only by the total number of gridpoints used in the discretization. Finite difference schemes are more appropriate to solve problems with complex geometry but the greater accuracy[3] of pseudospectral methods makes them more suitable to investigate on fundamental problems. The pseudospectral technique has been used to investigate on the structure of turbulence in different flows, and particularly channel turbulence[4,5]. Extensive description of numerical simulations of turbulent flows can be found in refs.[6–8].

DNS was also applied to the study of turbulent dispersions. The DNS of an homogeneous and isotropic turbulent flow field has been used by Yeung and Pope[9] to calculate Lagrangian velocity autocorrelations and by Eswaran and Pope[10] to study the mixing of an inert passive scalar. Later a number of researchers[11–17] used DNS to study the behaviour of particles in turbulent flow fields. Elgobashi and Truesdell[11] simulated the experiment by Snyder and Lumley[18] of particle dispersions in decaying isotropic turbulence obtaining good agreement with data and information at a much more detailed level. The papers by Squires and Eaton[12,13] deal with the problem of preferential particle concentration. They found that particles tend to segregate in regions of low vorticity and high strain rate, and that this trend has a maximum for particles with intermediate time constant. McLaughlin[14] and Brooke et al.[15] simulated a vertical channel (non-isotropic) turbulent flow, and analyzed the deposition of particles. Interestingly, Brooke et al.[15] found that particles close to the wall are trapped by the large eddies in the wall layer. Pedinotti et al.[16] simulated horizontal open channel turbulence and found that particles of intermediate time constant tend to remain in the low speed regions in the wall layer. Finally, Soldati et al.[17] applied the DNS to investigate the fine characteristics of the transport of charged aerosols in Electrostatic Precipitators.

The fundamental advantage of DNS is that all the information relative to the motion of both particles and fluid parcels is available at every timestep. As it can be seen by the previous literature analysis, relevant information on fluid and particle behaviour and on their mutual interaction may be drawn from such complete database. This information could be studied also with methods of nonlinear dynamics, for instance to obtain characteristics of the chaotic behaviour of particle dispersions. Indeed, Lagrangian information about fluid particles has been used to investigate on particle chaotic dynamics in simple types of flow. McLaughlin[19] studied the motion of passive particles in cellular flow and found that particle sedimentation under gravitational effects may be chaotic. Particle dispersion in the same steady three-dimensional cellular flow was investigated by Wang et al.[20–22]. They found a relation between the fractal dimension of the attractor and the dispersion coefficient. In particular, a low fractal dimension is correlated with high dispersion rate. The dynamical system formed by particles settling under gravity in a two-dimensional periodic free shear flow was analyzed by Gañàn-Calvo and Lasheras[23]. The flow was obtained from the steady, inviscid and incompressible solution of the Euler equation. The result was that particles can be attracted by periodic, quasiperiodic and chaotic orbits. Furthermore, if particles are driven in chaotic orbits, the sedimentation is hampered and a long-term suspension occurs.

It appears that chaotic dynamics methods could prove to be useful in determining the characteristics of particle dispersions in turbulent flows. However, a complete study in this direction has not yet been performed.

In this paper, DNS results relative to closed channel turbulence obtained with a pseudospectral method are used to simulate a dispersion of small aerosol in air under a steady drift. The pseudospectral method is extensively described in refs.[4,5,17]. Very small aerosol particles (micrometric size), are moved in the turbulent field under the coupled effects of fluid flow field, gravity and electrostatic drift. As a first report of an ongoing work, the mechanisms dominating the dispersion of particles are described, and qualitative trends of the particle distribution are presented. Finally, an attempt to characterize the dispersion of particles in the non-homogeneous direction with the aid of its statistical properties is made.

2. Direct Numerical Simulation of Turbulence in a Closed Channel

2.1. Pseudospectral Method

The problem considered is that of a pressure driven turbulent flow of an incompressible Newtonian fluid bounded by flat walls with no-slip conditions. The coordinates x, y, and z are respectively the streamwise, the spanwise (vertical), and the cross-stream coordinates. The full time dependent Navier-Stokes equations were numerically integrated for the flow considered. The governing equations, written in dimensionless form and using the Einstein's notation where repeated indices indicate summation, are:

$$\frac{\partial u_i}{\partial x_i} = 0 \tag{1}$$

$$\frac{\partial u_i}{\partial t} = -u_j\frac{\partial u_i}{\partial x_j} + \delta_{3i} + \frac{1}{Re}\frac{\partial^2 u_i}{\partial x_j \partial x_j} - \frac{\partial p}{\partial x_i} \tag{2}$$

where u_i are the velocity components, p is the kinematic pressure (pressure divided by density), and the Kronecker symbol, δ_{3i} represents the mean kinematic pressure gradient that drives the flow. In the following, for the spatial coordinates, the notation equivalence $x_1 = x$, $x_2 = y$,$x_3 = z$ holds. Eqs. (1) and (2) were made dimensionless using the duct half spacing, h, and the shear velocity, u_τ, defined as:

$$u_\tau = \sqrt{\frac{\tau_w}{\rho_f}} \tag{3}$$

where τ_w is the shear at the wall and ρ_f is the fluid density. The Reynolds number Re is defined as $u_\tau h/\nu_f$, where ν_f is the fluid kinematic viscosity. Eq. (2) is solved by taking the curl twice. An Helmoltz type equation of the fourth order in the velocity is obtained. The solution of the flow field is calculated by a pseudospectral method. The solution is assumed periodical in the streamwise and spanwise directions, such that periodicity boundary conditions can be applied. At both walls conditions of no-slip are imposed. The solution is represented by finite Fourier expansions in the streamwise and in the spanwise (wall parallel) directions and, to deal with the finiteness of the domain, by Chebyshev polynomials in the cross-stream (wall normal) direction. Then, the velocity field is expanded in the form:

$$u(x,y,z) = \sum_{k_1}\sum_{k_2}\sum_{k_3} \hat{u}(k_1,k_2,k_3)e^{i(k_1x+k_2y)}\ cos(k_3cos^{-1}z) \tag{4}$$

where $\hat{u}(k_1,k_2,k_3)$ represents the velocity field in the phase space, k_1 and k_2 are wavenumbers in the streamwise and vertical directions and k_3 is the wavenumber in the wall-normal direction. The calculation algorithm is very accurate because of the use of the Fourier-Chebyshev expansions. However, the calculation of the products involved in the convective term (first term in the right hand side of Eq. 2) would require the calculation of a convolution integral. This is very time consuming. Therefore, through the use of the fast Fourier transform, products are calculated in the physical space and results are back transformed into wave number space. This justifies the origin of the name pseudospectral. A two level Adams-Bashforth method for the convective terms and an implicit Crank-Nicolson method for the diffusive terms provide the time advancement of the flow field.

The dimensions of the computational domain have to be chosen large enough to justify the the application of periodic boundary conditions in the streamwise and spanwise directions. The domain size is $4\pi h \cdot 2\pi h \cdot 2h$. This choice was guided by the previous measurements by Comte-Bellot[24] and Utami and Ueno[25]. These authors demonstrated, via experimental measurements, that in channel flow the streamwise spatial correlation functions become negligible at a distance of approximately 3.6 h, while in the spanwise direction they become negligible at a distance of about 1.6 h, roughly half the characteristic streamwise correlation length. A posteriori analyses of the calculations proved the dimensional choice to be suitable for the problem analyzed[17,26].

2.2. Some Features of the Structure of Turbulence at the Wall

A cross section of the simulated turbulent flow field is presented in Figure 1 (see also Picture A at the end of this paper). The picture is a single image of a movie describing the temporally and spatially varying turbulent field, a snapshot of turbulent structures in the channel. The statistical features of the flow field have been widely assessed in previous publications[17,26]. It is interesting, however, to comment on the structure of turbulence at the wall, since it determines particle transport characteristics. With the main flow perpendicular to the picture (x direction), the brightness of the vectors in the plane represents the intensity of the fluctuating streamwise component of the velocity in the cross-section. At both walls about $6 \sim 7$ vortical structures

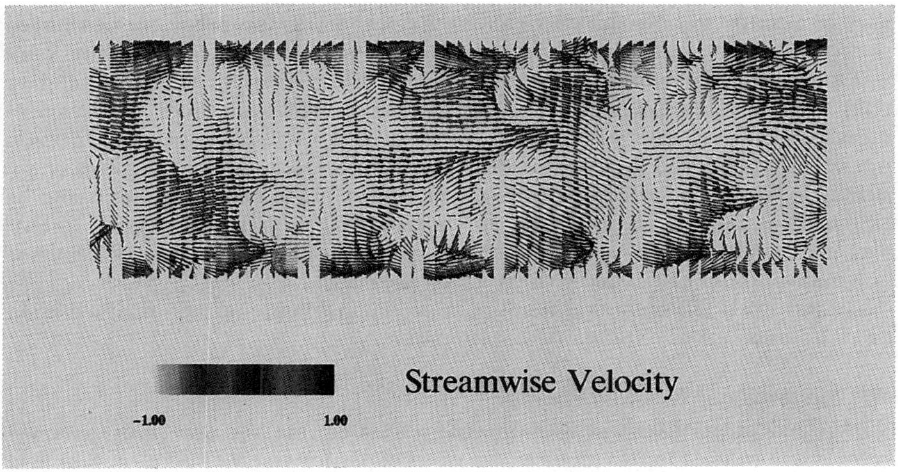

Figure 1: Instantaneous (also fluctuating) velocity vectors in a y-z plane (y is the vertical and wall parallel direction, and z is the wall normal. The picture is 90° degrees rotated clockwise). Magnitude of vectors is the modulus of the fluctuating velocity in the y-z plane. Color of vectors (dark denotes high, light denotes low) is the intensity of the streamwise (x, wall parellel horizontal) fluctuating velocity.

are present: these are cross-sections of streamwise vortices. These vortices are the *legs* of the so-called horseshoe vortices, produced by the effect of the shear at the wall. For a complete description of such structures, see Kline et al. [27], and Robinson[28], and for a new and intriguing explanation of their dynamics see Kaftori[29] and Kaftori et al.[30]. For the purposes of the present paper, however, it will be enough to assess that these structures determine the transport features of the turbulent flow. These streamwise vortical structures determine *ejections* or *bursts* of low momentum fluid from the layers close to the wall, and *insweeps* of fluid from the outer regions. This mixing greatly enhances momentum transfer across the channel and definitely affects particle transport, as it will be seen in the next sections.

3. Particle Transport in a Closed Channel with a Steady Drift

The transport of particles in boundary layer turbulence under a steady drift is relevant for many environmental and industrial problems. To name a few, dispersion and fallout of heavy pollutants in the atmosphere, aerosol transfer at the ocean atmosphere interface, electrostatic filtering.

In a previous paper Soldati et al.[17] applied DNS to the problem of aerosol transport in electrostatic precipitators (ESPs). In that work, the behaviour of charged particles in a turbulent duct flow under the action of different electrostatic fields was analyzed. The purpose was to simulate previously obtained experimental data relative to particle collection[31,32]. Results were very encouraging since perfect agreement with previous data was obtained when matching overall quantities like particle migration velocity and precipitator efficiency. However, for practical purposes of designing ESPs, the knowledge of averaged particle diffusion characteristics would be extremely useful. Moreover, a thorough understanding of particle dispersion mechanisms in boundary layer turbulence under a steady drift would be of general usefulness in a number of industrial and environmental problems.

In this work, the numerical results of that simulation are further analyzed in an effort to assess particle transport characteristics.

3.1. Simulation of Particle Dispersion

A cloud of particles was homogeneously released into the previously obtained turbulent flow field. In this problem, the effect of particles on the turbulent flow field was assumed negligible. Indeed, it is experimentally evident that for the particle size considered, the presence of the particulate phase, even at higher concentration, can be safely neglected in the computation of the flow field[29]. Particles were subject to fluid drag, gravity and electrostatic drift. Other forces, like the Basset history force, the Saffman lift force, the Magnus effect, the hydrostatic force and the added mass force are assumed, with good accuracy, to be negligible for the case treated[33]. All particles were tracked singularly by solving the following equation of motion made dimensionless using scales characteristic of the turbulent wall region, i.e. the shear velocity u_τ and the fluid kinematic viscosity ν_f:

$$\frac{d\vec{v}}{dt} = -\frac{\vec{v} - \vec{u}}{\tau_p} + \vec{g} + \frac{1}{Fr_p^2}\vec{e} \tag{5}$$

The particle and fluid velocities are $\vec{v}$ and $\vec{u}$, τ_p is the particle relaxation time, $\vec{g}$ is the gravity vector and $\vec{e}$ is the electric field unity vector. Each particle was considered rigid and pointwise. Furthermore, given the low concentration, the effects of particle-particle interactions, due to either inertial force or electrostatic repulsion force ware assumed negligible. The particle relaxation time, τ_p, measuring the particle responsiveness to a change in fluid velocity[34], can be written as:

$$\tau_p = \frac{d_p^2 \rho_p}{18\mu_f} \frac{1}{1. + 0.15 Re_p^{.697}} \quad . \tag{6}$$

where Re_p is the particle Reynolds number based on particle velocity relative to the fluid velocity, on particle diameter, d_p and on fluid kinematic viscosity, ν_f. The expression for the fluid dynamic drag force that has been used in this equation is the one proposed by Rowe and Henwood[35]. In Eq. (5), the dimensionless electric-Froude number, Fr_p, represents the square root of the ratio of the inertial force to the electrostatic force acting on the particle and it has the following form:

$$Fr_p^2 = \frac{m_p u_\tau^3}{q_p \nu_f E}. \tag{7}$$

where m_p and q_p are respectively particle mass and particle charge and E is the electrostatic field modulus. The electrostatic field is constant and uniform throughout the whole domain. The parameters of the simulation are presented in Table 1 (from ref.[17]).

$Re_{D_h} = U_m D_h / \nu_f$	6000.	Grid Size $N_x \times N_y \times N_z$	$64 \times 64 \times 65$
$Re_\tau = U_\tau h / \nu_f$	108.	$\Delta t^- = \Delta t \ (u_\tau / h)$	0.0011
U_m / u_τ	13.7	N_p	4000
U_m	$1.16 \ m \ s^{-1}$	d_p	$4.0 \ \mu m$
ν_f	$15.7 \ 10^{-6} \ m^2 \ s^{-1}$	ρ_p	$898. \ kg \ m^{-3}$
ρ_f	$1.29 \ kg \ m^{-3}$	E	$240. \ kV \ m^{-1}$
h	$0.02 \ m$	q_p	$1.602 \ 10^{-16} \ C$

Table 1: Parameters of the Simulation.

4. Results and Discussion

4.1. Particle Trajectories and Distribution

The motion of particles is determined by the coupled effects of the turbulent flow field, particle inertia and drift velocity. The effect of particle inertia, characterized by the particle relaxation time, is to cause a lag between the velocity of the particle and that of the surrounding fluid. The result is that particle r.m.s. fluctuation velocity would be less than the fluid one. However, the particles used in the present simulation have very small inertia. In Figure 2, the r.m.s. values of the three components of the fluctuating velocity for particles and fluid are presented. It is evident that particles follow the flow field fluctuations very closely. It is interesting to note that particles show r.m.s. values slightly but constantly smaller than the fluid ones in y and z directions, but larger in the streamwise direction. This effect is likely to be due to the gravitational and electrostatic forces in the two directions, causuing a redistribution of the turbulent energy.

If no drift force were applied, since particle inertia is small, particles would behave almost as a passive scalar, showing no tendency to preferential concentration[13] or segregation[16]. However, since a strong electrostatic drift is applied, owing to its lateral migration velocity, a particle tend to change its fluid neighborhood falling from one eddy to another. This effect is called crossing-trajectory effect[36]. On the contrary, a fluid parcel remains within the eddy until the eddy decays (loss of coherence). Ultimately, particles tend to lose velocity correlation faster than the fluid, and, since the particle correlation time is related to the dispersion features, particles disperse less than the fluid.

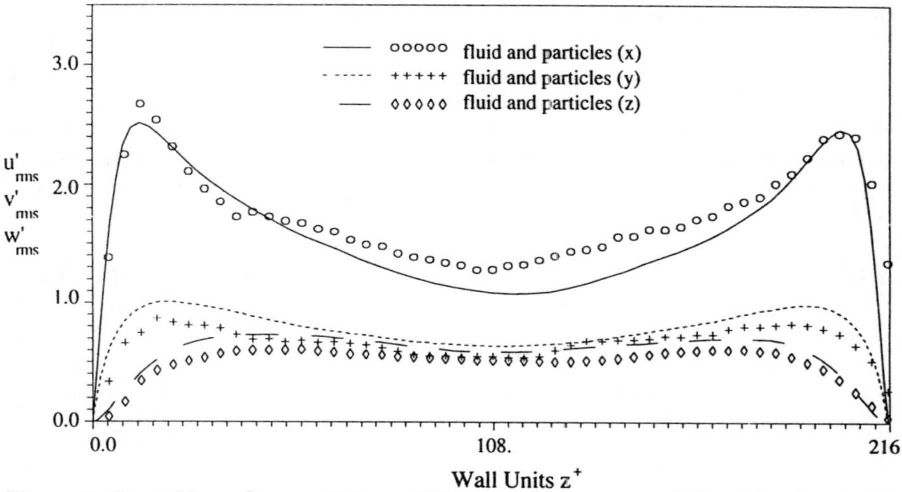

Figure 2: Root Mean Square Values of Velocity Components of Particles (symbols) and Fluid (lines) and different turbulent layers. The *Laminar sublayer*, dominated by viscous forces, $0 < z^+ < 5$. The *buffer layer*, highest turbulence intensity, $5 < z^+ < 30$. The *inertial sublayer*, outer layer with turbulence less influenced by the wall presence, $30 < z^+ < 108 =$ mid-channel.

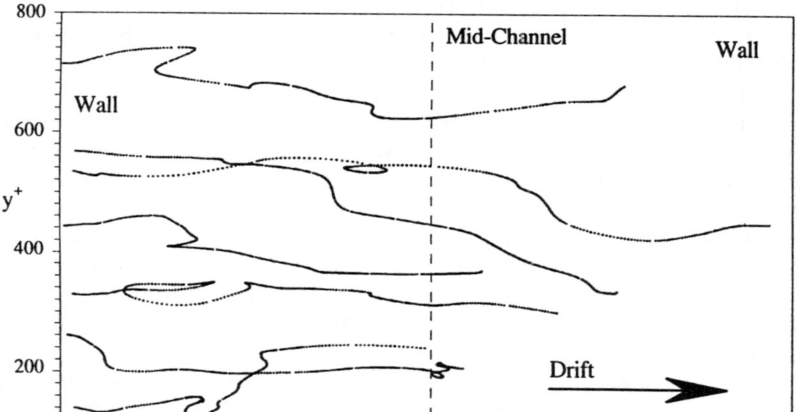

Figure 3: Trajectories of particles leaving form the laminar sublayer close to the non-collecting wall.

To appreciate the effect of turbulence on particles, in Figure 3 several trajectories of particles are shown. All the particles were initially in the laminar sublayer of the non-collecting wall. The electrostatic drift acts such to *straighten* the trajectories. However, when a particle is entrained in a streamwise vortex, the effect is quite remarkable. Few loops are present in the trajectories, assessing the relevance in turbulent transport of the wall structures. The presence of these structures is responsible also for particle spreading and diffusing. In Figure 4, a pictorial representation of the dispersion of particles enhanced by turbulence is shown. All the particles were initially in the viscous sublayer of the non-collecting wall ($z^+ < 5$). Their position in the (z^+, y^+) plane is presented every 0.33 dimensionless time units. The dispersion around the mean position is remarkable. The shape of the distribution of the same particles is shown in Figure 5 at three different times. The distribution, centered around the mean z^+ position, is initially very peaked, and it obviously broadens with time. In homogeneous turbulence, for long times, the distribution should tend to the normal (or Gaussian) form. The total number of particles sampled is not sufficient to give accurate information on the shape of the distribution curve. Firthermore, particles are constantly collected at the wall heavily influencing the statistics. However, the distribution does seem to tend toward a uniform distribution across the duct.

4.2. Lateral Dispersion of Particles

The problem of determination of the lateral dispersion of particles is addressed in this section. Usually, particle dispersion characteristics such as the mean square displacement (MSD) and the diffusion coefficients are determined on the basis of the theory initially put forth by Taylor[37]. This theory is based on the assumption of homogeneous isotropic turbulence, while the present case deals with nonisotropic boundary layer turbulence.

However, in an attempt to quantify the dispersion, it is useful to define some parameters[20–22,38–40]. The MSD in the lateral direction, (the ensemble-averaged variance of the particle position alog the z axis) gives a measure of the particle dispersion with time, and it is defind as:

$$< z(t)^2 >= \frac{1}{N(t)} \sum_{n=1}^{N(t)} (z_n(t) - < z(t) >)^2 . \tag{8}$$

where $N(t)$ is the number of realizations of a particle trajectory, that is the number of particles followed. The total number of particles varies with time because particles are collected at the wall. In Eq. (9), $< z(t) >$ is the mean displacement of the cloud of particles, that is the ensemble averaged position of particles at time t,

$$< z(t) >= \frac{1}{N(t)} \sum_{n=1}^{N(t)} z_n(t). \tag{9}$$

The MSD is a function of time. From its asymptotic analysis it is possible to derive information about the dispersion behaviour. In fact, the particle Lagrangian velocity is correlated for what is called correlation time, T_L, or integral time scale. It

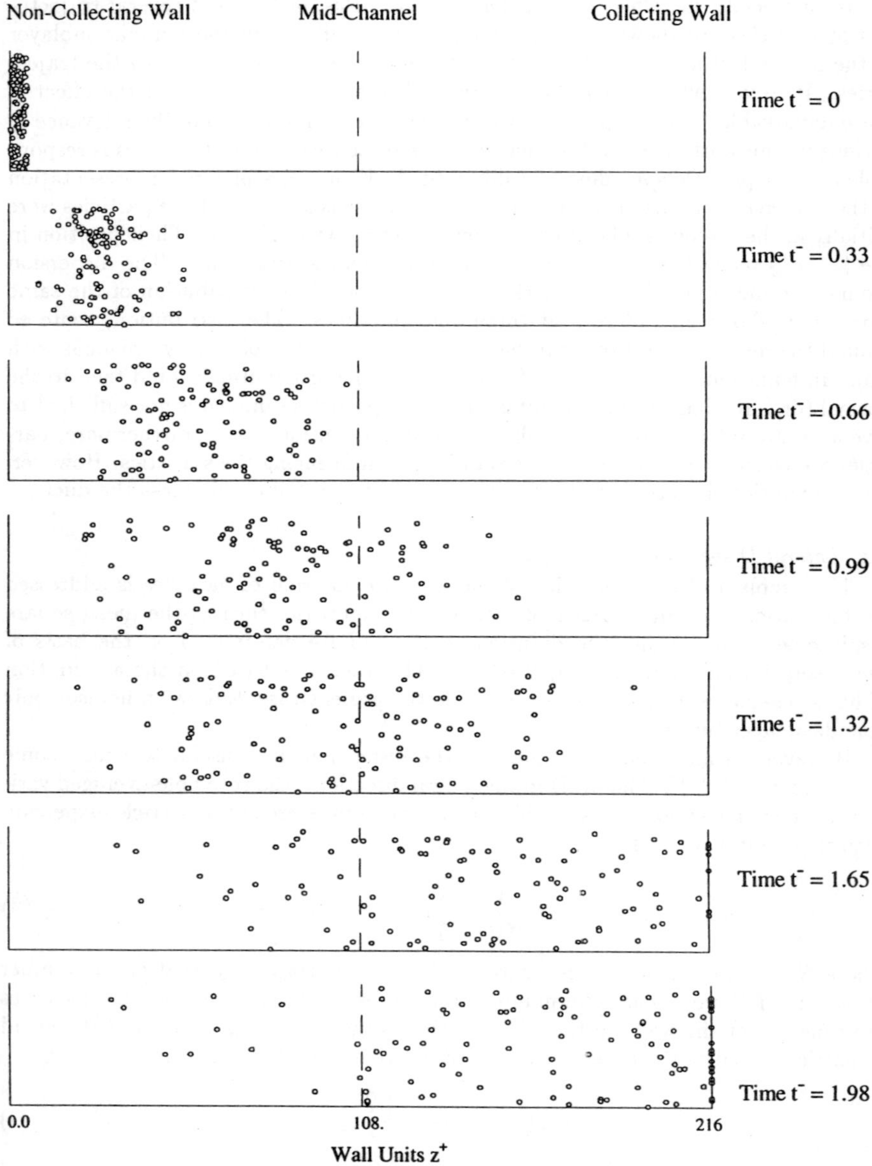

Figure 4: Lateral dispersion of particles. All particles leaving from the laminar sublayer are followed along their path. The spread around the mean position is quite remarkable.

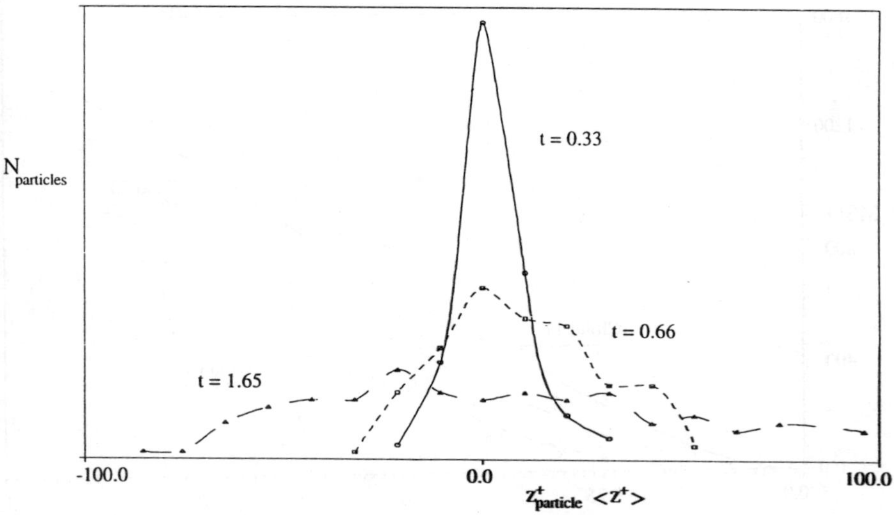

Figure 5: Particle distribution around the mean position at different times.

follows that, for $t > T_L$, the particle trajectory loses coherence and becomes very irregular. This corresponds to the asymptotical trend of the displacement probability distribution to Gaussianity. Therefore, for long times, $t >> T_L$, it is possible to define a dispersion coefficient as the rate of growth of variance of particle displacement distribution[37,38,40] in the following way:

$$D_z^2 = \lim_{t\to\infty} \frac{d < z(t)^2 >}{dt}. \tag{10}$$

To analyze the dispersion, the domain has been subdivided into N_s bins, each of thickness Δz^+. The analysis has been done for the cloud of particles that at the beginning of the simulation were in the same bin. It seems reasonable to average on particles starting with the same turbulence intensity. Therefore, since turbulence intensity shows steep gradients (see Figure 2), bins have to be chosen not too wide. Choosing $N_s = 40$, Δz^+ is about 5 and each bin comprises about 100 particles. The behaviour in time of each particle cloud is then analyzed.

In Figure 6, the MSD for cloud $n.1$, $n.10$ and $n.20$ are shown. Examining Figure 2, it is possible to appreciate the different initial turbulence intensity for the three clouds chosen. Cloud $n.1$ comprises particles initially in the laminar sublayer, cloud $n.10$ comprises particles initially in the tail of the buffer layer (high intensity of turbulence), and cloud $n.20$ comprises particles initially in the inertial sublayer (middle of the channel, low turbulence intensity). Particles of cloud $n.10$ show the largest rate of growth of the MSD, while the ones of cloud $n.1$ have the lowest. This is not

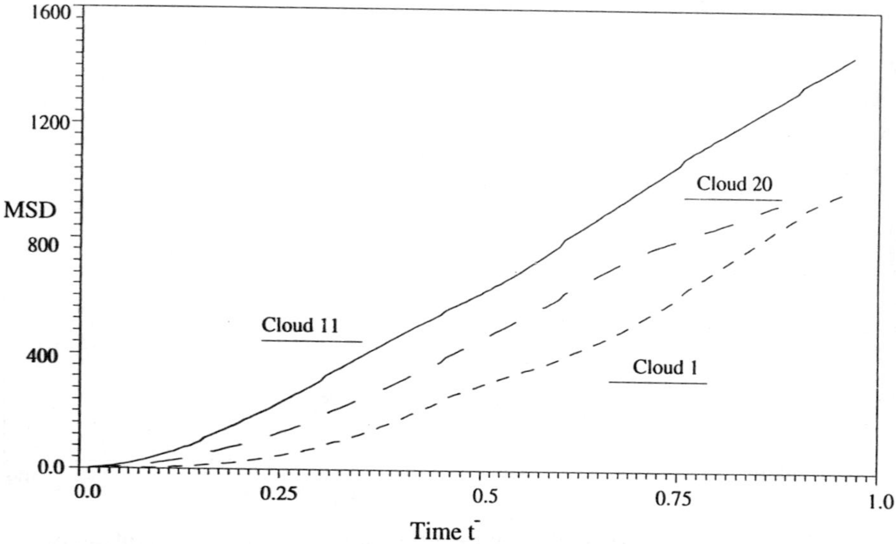

Figure 6: Mean Square Displacement of particles belonging to different initial clouds. The MSDs of Clouds $n.1$, $n.10$ and $n.20$ are shown.

surprising, since the intensity of turbulence varies remarkably in the different regions of the channel. At long-times, though, since particle distributions are very broadened around the mean displacement, the three MSDs show a similar rate of growth tending towards a linear dependence on time, thus indicating similar dispersion coefficients. However, the trends show oscillations probably due to the limited amount of particles in each layer. This procedure does not allow to draw quantitative conclusions, but only qualitative behaviour. The difficulty is that techniques valid for homogeneous turbulence do not hold for this case. However, one may observe that the profile of the turbulence intensity is rather flat in the outer region of the channel ($30 < z^+ < 186$) and the inhomogeneity of the field is less pronounced. Therefore, the MSDs of clouds $n.3b$ and $n.4b$ were calculated. Cloud $n.3b$ comprises particles with initial cross-stream position z^+ ranging from 30 to 78. Cloud $n.4b$ includes particles with initial cross-stream position z^+ ranging from 78 until the middle of the channel ($z^+ = 108$). These particles travel for the most part of their history in similar turbulence conditions. In Figure 7 the position of the center of gravity of the two particle clouds is shown. Both $< z(t) >$ grow linearly with time for $t^- < 1$. This means that the majority of the cloud is in the outer region and particles are not yet collected. MSDs can therefore be safely calculated for $t^- < 1$.

In Figure 8 the MSDs for the two clouds $n.3b$ and $n.4b$ are presented. The steadyness of the curve behaviour guarantees the appropriateness of statistics. It is possible to observe that after a certain elapsed time (larger than the particle velocity cor-

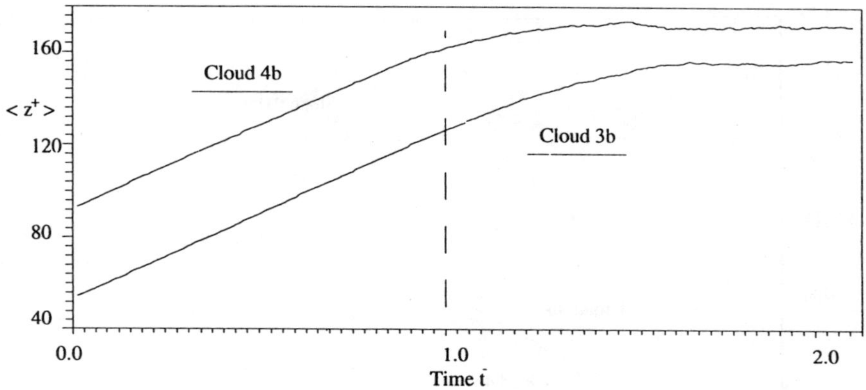

Figure 7: Position of the center of gravity of the two clouds $n.3b$ and $n.4b$.

relation time, T_L), the MSD behaviour becomes linear. Initially, particles of cloud $n.3b$ diffuse slightly more. This is in any case expected, since particles of cloud $n.3b$ experience an initial higher turbulence intensity. The particle diffusion coefficients, evaluated by means of Eq. (10), are:

$$D_z^2(3b) = 5.87 \cdot 10^{-2} \left[\frac{(z^+units)^2}{t^-}\right] = 1.96 \cdot 10^{-4} \left[\frac{m^2}{s}\right]$$

$$D_z^2(4b) = 5.67 \cdot 10^{-2} \left[\frac{(z^+units)^2}{t^-}\right] = 1.89 \cdot 10^{-4} \left[\frac{m^2}{s}\right]$$

It is possible to conclude that particles belonging to the two initial clouds therefore diffuse at the same rate. In dimensional units, the diffusion coefficients are about 2.0 cm^2/s. This value can be considered the value of the diffusion coefficient in the outer region of the channel. It actually represents an average for $30 < z^+ < 186$. This result is also confirmed by previous experimental results[32].

5. Conclusion

The purpose of this paper is to attempt a quantification of the turbulent dispersion of particles in channel turbulence. A pseudospectral method to solve the Navier Stokes equations was employed (DNS). Aerosol particles were released in the time and space varying turbulent flow field. The flow field was obtained during a previous work[26], and its detailed characteristics may be found in refs.[17,26]. The great advantage of the technique adopted consists in providing time histories for particle position and velocity. Particles released were subject to the effects of the turbulent flow field, gravity and electrostatic drift force. Since turbulence intensities vary along the non homogeneous direction, z (see Figure 2), the diffusion process is expected to be a

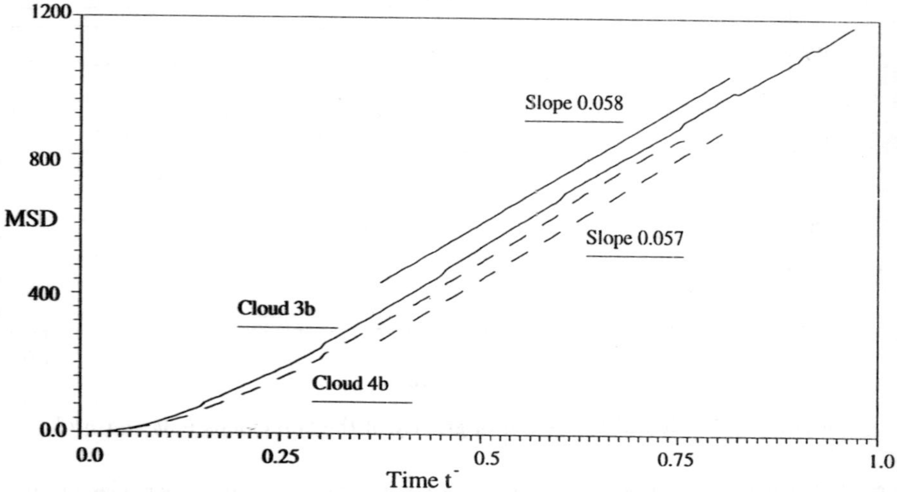

Figure 8: Mean Square Displacement of particles belonging to different initial clouds. **The MSDs of Clouds** $n.3b$ and $n.4b$ are presented.

function of the cross-stream coordinate z. In such case, the usual approach[37] may not be safely applied. However, the calculation of the MSD for particle swarms may supply interesting information.

An attempt to quantify the effect of turbulent diffusion was made evaluating the MSDs for different cloud of particles characterized by similar initial conditions of the turbulent flow field. Particles starting in the buffer region diffuse faster than particles starting in the middle of the channel. Particles initially in the wall layer have the lowest diffusion rate. Considering the non homogeneity effects in the central, outer, region of the flow field negligible, the diffusion coefficient of particle swarms was determined via their MSD[37]. Data are in good agreement with previous experimental measurements. This result further confirms the adequacy of DNS approach to turbulent transport phenomena. The determination of the *local* diffusion coefficient requires a deeper examination of the advection/diffusion equation and the use of more complex statistical means. More complete analyses of available data are under examination to elucidate better particle dispersion mechanisms.

Acknowledgements

Computational resources made available to perform this work on the CRAY YMP 4/464 at the National Center for Supercomputing Applications, Urbana-Champaign, Illinois, and visualization facilities at the San Diego Supercomputer Center, San Diego, California, are gratefully acknowledged.

References

1. A. K. M. F. Hussain, *Personal Communication* (1994).
2. P. R. Bandyopadhyay and A. K. M. F. Hussain, *Phys. Fluids* **27**, (1984) 2221.
3. S. A. Orzsag, *J. Fluid Mech.* **49**, (1971) 75.
4. J. Kim, P. Moin and R. Moser, *J. Fluid Mech.* **177** (1987) 133.
5. K. Lam and S. Banerjee, *Phys. Fluids A* **4** (1992) 306.
6. R. S. Rogallo and P. Moin, *Ann. Rev. Fluid Mech.* **16** (1984) 99.
7. W. C. Reynolds, *Whither Turbulence? Turbulence at the Crossroads.* Springer, (1990) 313.
8. M. M. Rai and P. Moin, *J. Comput. Phys.* **96** (1991) 15.
9. P. K. Yeung and S. B. Pope, *J. Fluid Mech.* **207** (1989) 531.
10. V. Eswaran and S. B. Pope, *Phys. Fluids* **31** (1988) 506.
11. S. Elgobashi and G. C. Truesdell, *J. Fluid Mech.* **242** (1992) 655.
12. K. D. Squires and J. K. Eaton, *J. Fluid Mech.* **226** (1991) 1.
13. K. D. Squires and J. K. Eaton, *Phys. Fluids A* **3** (1991) 1169.
14. J. B. McLaughlin, *Phys. Fluids A* **1** (1989) 1211.
15. J. W. Brooke, K. Kontomaris, T. J. Hanratty and J. B. McLaughlin, *Phys. Fluids A* **4** (1992) 825.
16. S. Pedinotti, G. Mariotti and S. Banerjee, *Int. J. Multiphase Flow* **18** (1992) 927.
17. A. Soldati, P. Andreussi and S. Banerjee, *AIChE J.* **39** (1993) 1910.
18. W. H. Snyder and J. L. Lumley, *J. Fluid Mech.* **48**, (1971) 41.
19. J. B. McLaughlin, *Phys. Fluids* **31** (1988) 2544.
20. L. P. Wang, T. D. Burton, and D. E. Stock, *Phys. Fluids A* **2** (1990) 1305.
21. L. P. Wang, T. D. Burton, and D. E. Stock, *Phys. Fluids A* **3** (1991) 1073.
22. L. P. Wang, M. R. Maxey, T. D. Burton, and D. E. Stock, *Phys. Fluids A* **4** (1992) 1789.
23. A. Gañàn-Calvo and J. C. Lasheras, *Phys. Fluids A* **3** (1991) 1207.
24. G. Comte-Bellot, *Thése de Doctorat*, Universitè de Grenoble, France (1963).
25. T. Utami and T. Ueno, *J. Fluid Mech.* **174** (1987) 399.
26. A. Soldati, *Ph.D. Dissertation* Università di Pisa, Italy, (1993).
27. S. J. Kline, W. C. Reynolds, F. A. Schraub and P. W. Runstadler, *J. Fluid Mech.* **30** (1967) 741.
28. S. K. Robinson, *Ann. Rev. Fluid Mech.* **23** (1991) 601.
29. D. Kaftori, *Ph.D. Dissertation* University of California, Santa Barbara, California, U.S.A. (1993).
30. D. Kaftori, G. Hetsroni and S. Banerjee, *Phys. Fluids A* (1994) In Press.
31. K. D. Kihm, *Ph.D. Dissertation* University of Stanford, Stanford, California, U.S.A. (1987).
32. S. A. Self, M. Mitchner and K. D. Kihm, *EPA/EPRI Symp. on Particulate Control Technol.*, Nashville (1988).
33. M. R. Maxey and J. J. Riley, *Phys. Fluids* **26** (1983) 883.

34. A. Soldati, P. Andreussi and S. Banerjee, *AIChE J.* (1994) In Press.
35. P. N. Rowe and G. A. Henwood, *Trans. Am. Instn. Chem. Engrs.* **39** (1961) 43.
36. M. I. Yudine, *Advances in Geophysics,* Vol. 6. Academic Press (1959) 185.
37. G. I. Taylor, *Proc. R. Soc. London Ser.* **A 20**, (1921) 196.
38. G. K. Batchelor, *J. Fluid Mech.* **3**, (1957) 67.
39. R. Chein and J. N. Chung, *AIChE J.* **34** (1988) 946.
40. D. A. Chin, *J. Env. Engng.*, ASCE **120** (1994) 256.

A color movie of the temporally varying flow field was made at the San Diego Supercomputing Center. In the next page, a color picture representing a cross section (y-z plane, where y is the vertical and wall parallel direction, and z is the wall normal) of the flow field and corresponding to Figure 1 is presented. In the picture, the magnitude of vectors is the modulus of the fluctuating velocity in the y-z plane. The color of vectors (red denotes high, blue denotes low) is the intensity of the streamwise (x, wall parellel horizontal) fluctuating velocity. The structure of turbulence at the wall (cross section of streamwise aligned vortical structures) may be appreciated.

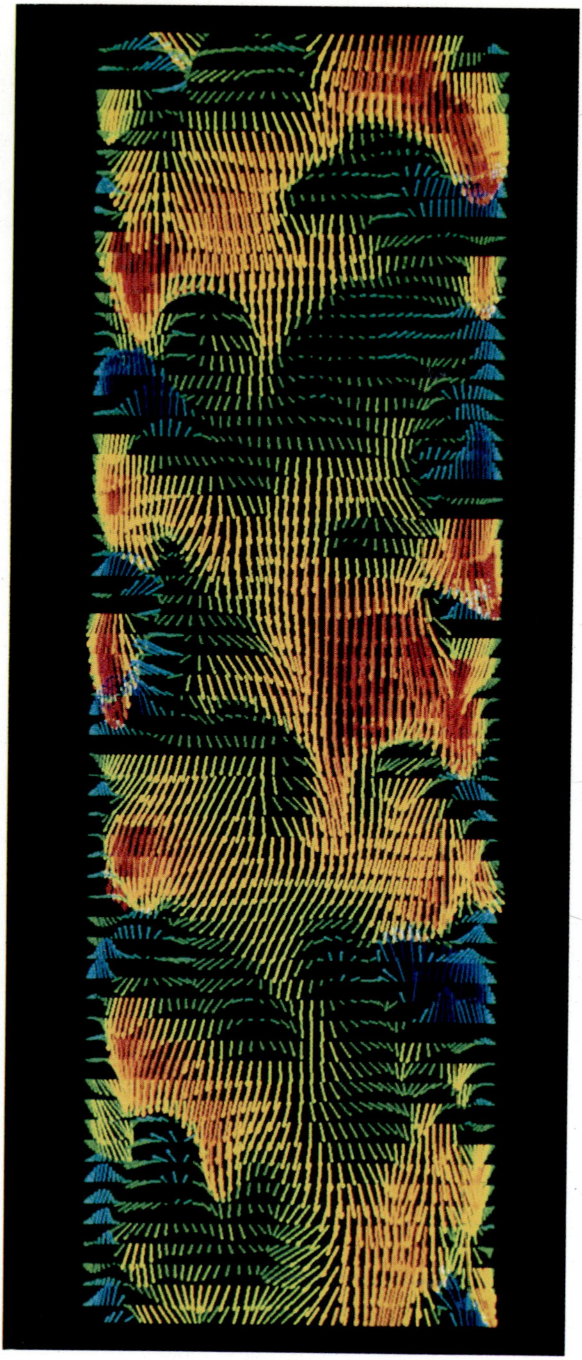

WAVELET ANALYSIS OF THE VELOCITY FLUCTUATIONS IN THE NEAR FIELD OF A COAXIAL JET

PAOLO PETAGNA
A.R.I.A., Aerodynamic Research for Industrial Applications
Via Delle Sedie 15, Livorno, 57125, ITALY

JACQUES LEWALLE
Department of Mechanical, Aerospace and Manufacturing Engineering, Syracuse University
151 E.A. Link Hall, Syracuse, NY 13244-1240, U.S.A.

and

GUIDO BURESTI
Dipartimento di Ingegneria Aerospaziale, Università di Pisa
Via Diotisalvi 2, Pisa, 56126, ITALY

ABSTRACT

After some introductive notes on the relevance of fluid dynamics aspects in jet combustion, the problem of the measurement of turbulent flow in a coaxial jet is addressed. The theory underlying the wavelet transform is then briefly reviewed, and some sample results of a wavelet analysis of the velocity traces in the near field of a coaxial jets are shown.

1. Introduction

The knowledge of the aerodynamic features of the flow field produced by a coaxial jet is relevant to many engineering applications, as, for instance, the control of mixing and combustion in burners. The requirement of an efficient reduction of the emission of pollutants by lowering their production inside the flame has brought to the attention of the research community the issue of combustion control via the control of the fluid dynamics involved in the process. Indeed, a strong connection exists between combustive and fluid dynamical phenomena in reacting flows, and the fundamental similarities between the cold flowfield and the flame for equal geometries have been largely documented in the literature. For instance, Schlieren visualizations[1] have clearly shown the presence and importance of large-scale turbulent structures in a methane-air diffusion flame, as well as the large influence of flow excitation at different frequencies. Also, Gutmark et al.[2,3], by extensive measurements in an axysimmetric combustion jet by the Planar Laser Induced Fluorescence technique, which allows the zones of the jet where the combustion process is taking place to be visualized, found that the combustion areas seem to practically coincide with those where large-scale structures are present, and that, in the initial portion of the jet, combustion takes place in the regions of high turbulence intensity surrounding the large-scale structures and in those of high vorticity. Anyway, combustion may be extinguished by excessively stretched vortical structures, as may happen in the regions connecting the jet exit to the first vortex: local extinctions of the flame may indeed occur in presence of excessive strain rates in the flow, which produce excessive gradients in the concentration of the species.

Dealing with the wide subject of the interactions between gas combustion and fluid dynamics is of course beyond the aims of the present paper; suffice it to stress here that, from the above considerations, it is apparent that the phenomena taking place in the reacting field of a jet-like burner are strongly connected with the dynamics of the cold turbulent field of a developing axisymmetric mixing layer. Developing turbulent flows can be roughly considered as characterized by three different scales: large, small and fine. Among these, only the last one (i.e. the Kolmogorov scale) may be usually regarded as isotropic and universal, whereas the other two are strongly anisotropic and vary their nature and their dynamics in different turbulent flows. In particular, in transitional jet flows, very likely to be typical of most industrial burners, the whole phenomena of transport of mass, momentum and heat (as well as those related to sound emission) are heavily influenced by the dynamics of the large scale structures and by their interactions with smaller scales. Under this perspective, turbulent flows appear ruled by energy exchanges between vortical structures of different scales, with the large ones, developed through instability mechanisms of the flow, capable of *organizing*, in a sense, the motions of the small ones and of extracting energy from them; this energy is then transferred back to smaller scales after complex processes ending with the breakdown of the structures. This scheme of *energy anticascade* and following *cascade*, widely dominated by coherent motions[4,5], carries the expectation of the possibility of modifying in a desired way the general dynamical features of a flow by suitably acting on its coherent component. However, to reach this goal, a deeper insight into the structure of the axisymmetric mixing layer formed by two coaxial naturally developing jets should be achieved.

2. Velocity measurements in the coaxial jet flow-field

2.1. On the measurement techniques in jet flows

Experimental analyses of jet flows may typically involve:

- visualizations (usually instantaneous image acquisitions phase-locked to some reference signal), which can provide a good qualitative description of the dynamical features of the flow's structures, but are poor in quantitative information, and sometimes may even be misleading.

- Instantaneous measurement of velocity at many points in 2D (e.g. Particle Image Velocimetry[6]) or in 3D (e.g. Holographic Particle Velocimetry[7]), based on the tracking of the positions of tracer particles at different instants in a plane or in the space. These measurement techniques allow a *lagrangian* approach to the flow features and, most important, a high-quality measurement of the vorticity field, but, at the present state-of-the-art, experience severe limitations on the sampling rate and on the maximum velocity that can be tracked.

- Single-point velocity measurements with Hot Wire or Laser Doppler Anemometry, which represent the classical *eulerian* approach to measurement and may as well provide multi-point information (even if with poor spatial resolution) if rakes of probes or scanning techniques are used. Single-point velocity measurements are of course unsuitable for the aim of a *quantitative visualization* of the vorticity field and their use also bears the shortcoming of a heavy resort to the Taylor's hypothesis, whose validity might be questionable in some developing flows[4]. Nonetheless, nearly any quantitative information produced up to now on the dynamics of turbulent flows has been provided by these techniques; furthermore, they have been proved useful also to study the structure of the turbulent coherent structures through phase-locked mesurement in the transitional regions of jets and mixing layers[8] (i.e. in the region of interest for the present research) where the degree of organization of the vortical structures is higher and their occurence may be regarded as *quasi-periodical.*

Besides the above mentioned distinction among different possible measurement approaches, a further discrimination has to be made about the type of developing jet flow under consideration, i.e. natural or excited. It is indeed possible, by active excitation of the flow before the jet exit, to lock the formation of a particular family of large-scale structures to its dominant frequency and to inhibit other natural instability effects of the mixing layer. This greatly helps the *eduction* (i.e. the process of identification and description) of the coherent structures allowing an easy phase-locking of the measurements with the formation of the structures. On the other hand, naturally developing axisymmetric mixing layers show turbulent structures spread over different frequencies, apparent phenomena of modulation and intermittency, and a strong sensitivity to the exit conditions: by analyising the single-point signals recorded in different regions of the field, it is possible in this case to obtain information on the *effects* of the evolution of the structure on the fluid dynamical behaviour of the flow field.

2.2. Measurements in coaxial jets

The flow field of an unexcited jet in coaxial configuration (schematically depicted in figure 1) is dominated by a complex combination of several instability mechanisms, heavily interacting with each other and depending by a large number of geometrical and fluid dynamical parameters. For this reason the extrapolation to this case of the experimental data available for the simpler case of the single axisymmetric jet is not straightforward.

Some previous works on the coaxial geometry are available in the literature. Whitelaw and co-workers[9,10,11,12,13] have performed very accurate velocity measurements on free and confined coaxial flows, both by Hot Wire and by Laser Doppler Anemometry, but focalized their attention on fully developed pipe jets: although these flows may have an application interest, they do not allow a clear analysis of the effects of the exit conditions on the mixing layer development, and, furthermore, the results are not readily related to the more common configuration with potential cores and thin boundary layers. As for this latter flow condition, in the early 70's Champagne and Wygnanski[14] first measured turbulence intensities and shear stresses by Hot Wire Anemometry with X-probes, but their results unavoidably suffer of the technological level of the time, based on analogical linearization and processing techniques. Subsequent works by Ko and co-workers[15,16,17,18,19] have widely dealt with several configurations having different values of the ratio between the inner and the outer jet exit velocities U_i/U_o. This group of papers provides a systematical analysis of the influence of the velocity ratio over the main characteristics of the flow field, but are quite limited as for the adopted measurement techniques and the characterization of the exit conditions. As a consequence, the information regarding the turbulent field features exhibits lack of completeness. Finally, recent Laser Induced Fluorescence visualizations by Dahm et al.[20] have shown that the model of the flow field of a coaxial configuration as a simple combination of the structures typical of a circular and an annular jet, as suggested by Kwan and Ko[16], does not fit direct experimental observations.

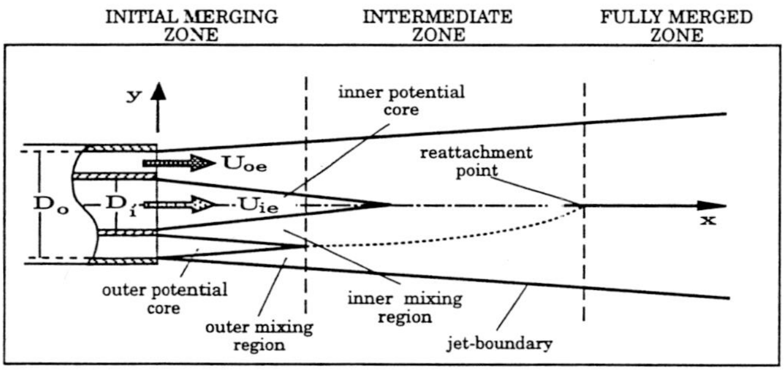

Figure 1: Flow field of a coaxial jet configuration

In order to obtain a better definition and, possibly, a deeper physical comprehension of the variation of the aerodynamic features of a coaxial jet as a function of some of their basic geometrical and fluid dynamical parameters, a long-term research was started. An experimental facility has been designed to produce coaxial jets with potential cores and thin boundary layers[21], in whose field extensive single-point measurements are performed both by Hot Wire and by Laser Doppler Anemometry. In the first part of the investigation the flow fields for a given geometrical configuration at two velocity ratios have been characterized in terms of first and second order statistical moments and of Power Spectral Density of the velocity components[21,22]. The different regions of the complex field have been thus determined and a comparison between the gross features of the flow for the two velocity ratios have been carried out. By studying the spectra of the axial and radial velocity components, also a first analysis of the characteristics of the fluctuating field have been performed and the dominating frequencies of the flow at various locations have been identified. Much more information is contained inside the signals acquired throughout the flow field and a number of different analysis techniques are currently used to gain insight into the physics of this turbulent flow, by combining the capabilities of each one and using the results obtained from one technique as suggestions to guide the application of the others. Among the techniques adopted are: analysis of higher order statistical moments; different kinds of conditional resampling and ensamble averaging; signal demodulation. A special attention is now focusing on the use of the wavelet transform analysis, the first results of whose application seem to be particularly promising.

3. The wavelet transform

The infinite wavetrain assumed in Fourier analysis cannot adequately describe the intermittency or modulation of velocity traces in large Reynolds number flows. Indeed, in the Fourier transform the information related, for instance, to a local discontinuity is distributed in the whole spectrum, because the trigonometric functions used to decompose the original signal oscillate indefinitely in both directions. Wavelet transforms provide an alternative with a *localized* waveform: the units of wavelet analysis are concentrated in short intervals, being generated by shifting and dilation of a function g (sometimes called the "mother wavelet") concentrated on one interval and satisfying the *admissibility* condition:

$$\int_{-\infty}^{\infty} g(t)\, dt = 0 \,. \tag{1}$$

Farge[23] provides a good introduction to wavelet analysis as it applies to turbulence data, while mathematical details are covered in a number of recent books, as for instance that by Daubechies[24]. Recent applications to fluid dynamics include the works by Lewis and Gharib[25], Dallard and Browand[26] and Higuchi et al.[27].

The wavelet transform of a function $x(t)$ is defined as:

$$X(\kappa, t) = \int_{-\infty}^{\infty} \kappa^{1/2} x(\tau) g^{\star}(\kappa(\tau - t))\, d\tau \,, \tag{2}$$

where $g^{\star}$ is the complex conjugate of g. In Eq. (2) the mother wavelet is shifted and dilated: the dilation parameter κ^{-1} (often referred to as "a") represents the width of the interval over which the function has significant non-zero values, i.e. the *scale* of the function, and is called *duration.* The *waveletnumber* κ is used here to stress its analogies (and its differences) with the Fourier wavenumber ω. The wavelet transform is thus a function of two variables, κ and t, and, applied to the analysis of time series, provides for each value of t the contribution to the signal from fluctuations having κ^{-1} as characteristic length. This ability to *quantify* the magnitude and temporal relationship of events of different durations is most useful for our purposes. A vivid image of the possible appication to signal analysis of the wavelet transform is provided by Liandrat and Moret-Bailly[28]: it can be compared to a microscope, with the optics given by g, the enlargement governed by κ^{-1} and the position fixed by t. Consistently with this interpretation, one has to remember that different information can be provided by different *optics*, a correct choice of the function g being thus of some relevance with respect to the objects of the analysis.

Wavelets may be either complex or real functions, the former featuring the interesting property of showing signal discontinuities (rapid evolution at each frequency) not only from the wavelet coefficient modulus, but also from its phase representation[28]. Among the complex wavelets, the most used for the analysis of turbulent signals is the Morlet wavelet, defined as:

$$g_M(t) = e^{i\omega_0 t} e^{-t^2/2} \, . \tag{3}$$

This wavelet provides a time/scale analysis particularly efficient in multiscaled signals. Furthermore, its surprising likeness with the "eddy" shape proposed by Tennekes and Lumley[29] suggests that it could be very useful for turbulence spectral dynamics study. Anyway, the conditions under which a relation exists between wavelets and eddies have not been established yet. The Morlet wavelet is actually only marginally admissible, the addition of some very small correction terms being necessary for it to be of zero average. Anyway these correction terms become negligible if the value of the parameter ω_0 is chosen between 5 and 6. In particular, from our limited experience, a choice of ω_0=5 seems to be nearly-optimal for the analysis of velocity traces in turbulent flows.

Among the real-valued wavelets, we mention here the "Mexican hat" g_2, which is in fact the second derivative of the gaussian function:

$$g_2(t) = (t^2 - 1)e^{-t^2/2} \, , \tag{4}$$

and its kin g_1

$$g_1(t) = te^{-t^2/2} \, . \tag{5}$$

In the first case, high wavelet coefficients correspond to a strong curvature of the signal at the corresponding scale of the wavelet; g_2 can then be useful when whishing to identify local maxima or minima of the signal. The g_1 wavelet, in turn, providing information on the local slope of the signal, can be used when looking for gradients.

It is well known that the Parseval theorem serves as the basis of the conventional power spectra. Its extension to the wavelet formalism, giving the total energy of the function $x(t)$, is summarized by the relation:

$$\int_{-\infty}^{\infty} \mid x(t) \mid^2 dt = \frac{1}{C(g)} \int_0^{\infty} d\kappa \int_{-\infty}^{\infty} dt \mid X(\kappa, t) \mid^2 = \int_0^{\infty} 2E(\kappa) \, d\kappa \, , \tag{6}$$

by which the expression $\mid X(\kappa, t) \mid^2 / C(g)$ can be interpreted as the *local* energy spectral density. The mean spectrum $E(\kappa)$ is of course qualitatively similar to the Fourier spectrum. In Eq. (6), $C(g)$ is the norm of the wavelet g, defined as:

$$C(g) = \int_{-\infty}^{\infty} \frac{\mid \hat{g}(\omega) \mid^2}{\omega} \, d\omega \, , \tag{7}$$

$\hat{g}(\omega)$ being the Fourier transform of $g(t)$.

Rebuilding of the original signal is also possible from its wavelet coefficients, by use of the inverse wavelet transform formula, whose general formulation is written:

$$x(t) = \frac{1}{C(g)} \int_0^\infty d\kappa \int_{-\infty}^\infty \kappa^{1/2} X(\kappa, \tau) g(\kappa(t-\tau))\, d\tau \ , \tag{8}$$

In the following, only of the g_2 wavelet will be used. As previously stated, this function belongs to the family of the even-indexed Gaussian wavelets (i.e. those obtained as even derivatives of the Gaussian function): for these wavelets the inverse transform formula is greatly simplified. Indeed, if we now indicate with $x_n(\kappa, t)$ the Gaussian wavelet transform of the function $x(t)$ for the Gaussian wavelet of index n, it is possible to show[30] that a fast inversion formula exists for even values of n:

$$x(t) = \frac{(-1)^n \pi^{-1/2} 2^{-n+\frac{1}{4}}}{\Gamma(n)} \int_{-\infty}^\infty \kappa^{-1/2} x_{2n}(\kappa/\sqrt{2}, t)\, d\kappa =$$

$$= \int_0^\infty X_n(\kappa, t)\, d\kappa \ . \tag{9}$$

In analogy with the Parseval relation for the energy spectrum, Eq (9) yelds an *amplitude* spectrum, i.e. the contribution from point t and scale κ to the signal's local amplitude. For this reason X_n is called "amplitude transform".

It is noteworthy that the primary concepts in the wavelet analysis are those of time and duration, whereas frequency is clearly a secondary concept applicable if some repetition of individual events occurs. In contrast to Fourier's familiar formulae, frequency and duration are not reciprocal of one another in non-periodic data. However, reference to "frequency" for nearly periodic signals is a convenient figure of speech. In the case of Gaussian wavelets, for a sine wave, the relation between the "frequency" f and the dominant wavelet duration κ^{-1} is:

$$f/\kappa = \frac{\sqrt{n+0.5}}{2\pi} \ , \tag{10}$$

where n is the index of the Gaussian wavelet.

4. Processing and sample results

4.1. Generalities

In the following some sample results obtained by application of the wavelet transform to velocity traces will be presented. The time-series have been recorded with Hot Wire Anemometry in the flow field of the coaxial jet produced by the experimental facility described in Buresti et al.[21]. The configuration considered has an inner to outer diameter ratio D_i/D_o of 0.485, with 76.2 mm inner diameter and a thickness of the inner duct at the outlet t=5 mm. The tests were performed at a velocity ratio $U_i/U_o = 0.67$, at a Reynolds number (based on the inner diameter and velocity) $Re_{Di} = 1.04 \times 10^5$. The isocontours of mean axial velocity for this configuration are shown in figure 2, from which the extention of the potential cores and of the mixing layers may be inferred. All the velocity traces herein used have been acquired along the line $y/D_i = 0.6$ at axial stations between $x/D_i = 1.0$ and $x/D_i = 5.0$, i.e. inside the inner mixing layer. These points have been chosen here to show some interesting potentialities of the wavelet transform. A systematical scan of the near-exit region by the processing techniques presented in this paper is currently in progress.

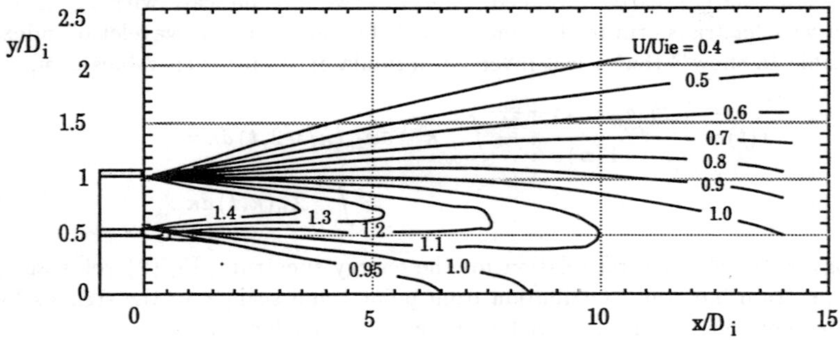

Figure 2: Isocontours of mean axial velocity for $D_i/D_o = 0.485$ and $U_i/U_o = 0.67$

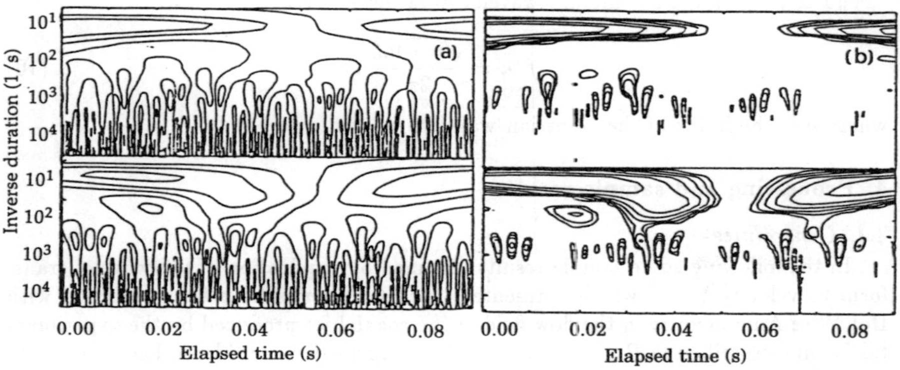

Figure 3: Example of g_2 wavelet map (a) and energy map (b) of the axial velocity trace at $x/D_i = 3.75$

4.2. Wavelet and energy maps

The typical result of the wavelet transform of a time-series is a wavelet map showing the wavelet coefficient of each scale at any instant. An example of such a map is shown in figure 3a for the velocity signal acquired at $x/D_i = 3.75$. As mentioned in relation with Eq (6), the two-dimensional array of numbers corresponding to this map is easily transformed into an energy map (figure 3b). Observation of this kind of map permits to isolate singular sequences of events related to particular scales, and also to focus on individual events in each sequence. For instance, in figure 3b the presence of an intermittent organization at waveletnumbers of order 10^3 is clearly detectable, corresponding to a quite broad peak visible in the Fourier spectrum (see figure 5) between 100 and 200 Hz. Unfortunately, observations of this kind, almost the only one produced at the moment by people working in this area, can provide suggestions to the researcher, but are strongly qualitative and, hence, poorly able to provide information otherwise hardly achievable by other, more traditional, analysis techniques. Anyway, a lot of quantitative data are contained in the wavelet and energy maps, whose further processing can elicit valuable information: in the following paragraphs some possible processing techniques will be introduced.

4.3. Event eduction and structure maps

The spectral/temporal resolution of the signal in the wavelet map lends itself to the eduction of individual events. The criteria for eduction are obviously at discretion of the investigator: having matched the shape of a wavelet (e.g. g_2 identifies local maxima or minima) to the signal, events can consist of local maxima of the wavelet map or energy map, or of particular sequences or simultaneous occurrences of several features identified by thresholding in a search algorithm.

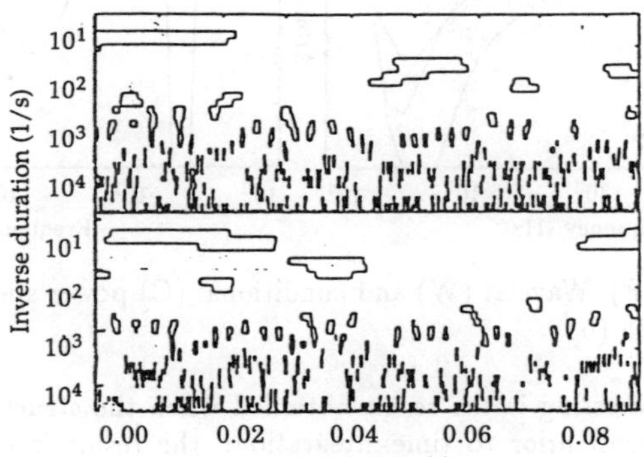

Figure 4: Structure map for the velocity trace of figure 3

In particular, the algorithm of scanning for local maxima of the energy map is detailed in Higuchi et al.[27]. The resulting structure map is shown in figure 4: after the identification of local maxima of energy and the application of an acceptability criterion for the identified maximum, a local normalization is performed and all values smaller than a selected threshold value (in this case 70% of the local maximum) are discarded from the plot. Regardless of the screening criterion, the presence or the absence of a structure meeting it a given time and duration results in a "black-and-white" map (structure map), over which classical signal analysis algorithms are easily implemented (e.g. the quantification of the intermittency at the scales where energy is concentrated), and that can be used for conditional statistics, as explained below.

4.4. Conditional spectra

According to Eq (6), time integration of the energy map yelds the mean power spectrum. Except in cases where a periodic component is strongly dominant in the signal, Fourier and wavelet spectra convey roughly the same information, the latter being usually smoother and showing lower high-frequency content, as the Fourier approach requires substantial high-frequency corrections to non-sinusoidal low frequencies. The wavelet spectrum must be of course shifted to the left by the scale factor given in Eq (10) to make its peaks coincide with those of the Fourier spectrum.

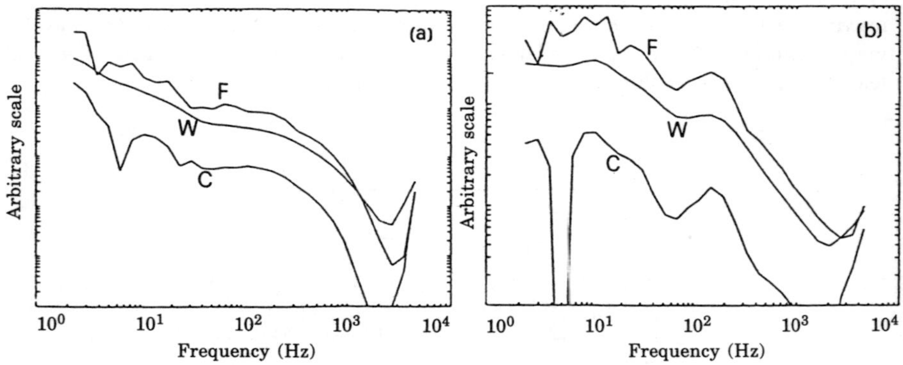

Figure 5: Fourier (F), Wavelet (W) and conditional (C) power spectra at $x/D_i = 2.5$ (a) and $x/D_i = 3.75$ (b)

Much more interesting is the result obtained when the structure map is used to mask the energy map prior to time-integration: the result is called a *conditional spectrum*. In figure 5 Fourier, wavelet and conditional spectra for two axial stations along the line $y/D_i = 0.6$ are shown. The high threshold adopted (75% of the local maximum value of energy) accounts for the order-of-magnitude difference between the mean wavelet spectrum and the conditional spectrum. The ability of the latter to *isolate* the scales at which a considerable level of coherency exists is evident in figure 5b, relative to the station $x/D_i = 3.75$, where, after the end of the outer potential core, the activity of the turbulent structures of the inner mixing layer considerably rises.

Conditional spectra can be interpreted as a measure of the coherence of the flow independently of any underlying periodicity. In figure 5, higher coherence, as measured by the ratio of the conditional spectrum to the wavelet spectrum, is seen in the region corresponding to shear-layer structures (around 180 Hz)

4.5. Correlation maps

Another processing method is suggested by the anectodal observation of simultaneous occurrence or mutual exclusion of events at different scales on the wavelet maps. For any pair of maps, the wavelet transform or energy map at any duration can be correlated with other durations on the companion map. Specifically, given the maps $X(\kappa,t)$ and $Y(\kappa,t)$ at N_t points in time and N_κ different durations, we can calculate various statistics: specifically the mean

$$\bar{X}(\kappa) = \frac{\sum_t X(\kappa,t)}{N_t}\,, \tag{11}$$

the spectral variance

$$\sigma_X^2(\kappa) = \frac{\sum_t (X-\bar{X})^2}{N_t - 1}\,, \tag{12}$$

and the spectral correlation

$$\rho_{X,Y}(\kappa_X,\kappa_Y) = \frac{\sum_t (X-\bar{X})(Y-\bar{Y})}{N_t^2 \sigma_X(\kappa_X)\sigma_Y(\kappa_Y)}\,. \tag{13}$$

Provided use is made of the "amplitude transform", we have

$$\sum_{\kappa_X,\kappa_Y} \rho_{X,Y}(\kappa_X,\kappa_Y) = \frac{\sum_t (X(t)Y(t)))}{N_t}\,, \tag{14}$$

so that the correlation map is directly related to the distribution of correlations among pairs of scales κ_X and κ_Y. Thus, spectral information about the transport terms can be obtained from these maps.

As an example, we report in figure 6 the evolution of the correlation between the axial energy and the radial velocity as the measurement point moves from the axial station $x/D_i = 3.0$ to $x/D_i = 4.5$. Note that positive/negative correlations are indicative of outward/inward radial transport of energy. The dramatic change in the shape of the correlation map after $x/D_i = 3.5$ is impressive: the end of the outer potential core brings the large-scale structures from the inner side of the outer (stronger) mixing layer inside the region of the innner mixing layer, causing the appearance of an island of inward transport at durations typical of the signature of these structures. Proceeding downstream, the negative correlation island tends to reduce its size and to lower its peak value, while a quite strong island of positive correlation is again formed at slightly lower durations (i.e. higher "frequencies"), a trend that goes on moving further downstream after the station $x/D_i = 4.5$: as the end of the inner potential core is approached, the characteristics structures of the inner mixing layer's jet-column mode rise their level of organization, whilst those from the inner side of the outer shear layer start to collapse and loose importance.

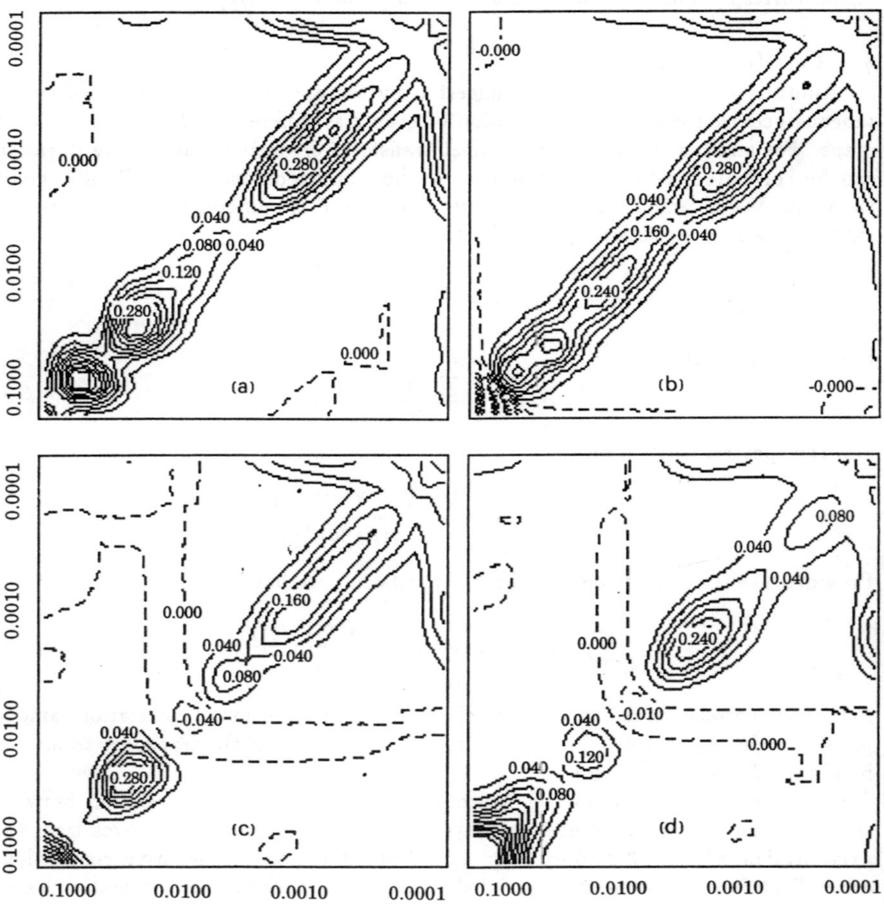

Figure 6: Correlation maps between axial energy and radial velocity at $x/D_i = 3.0$ (a), $x/D_i = 3.5$ (b), $x/D_i = 4.25$ (c), $x/D_i = 4.5$ (d)

5. Conclusions

To elicit really usable (i.e. quantitative) information from the traces of the velocity components in the three-dimensional turbulent field of a coaxial jet configuration simultaneous and coordinate resort to different signal analysis techniques is needed. Among these, wavelet analysis offers new flexibility in the identification of flow structures and spectral/temporal discrimination, but the simple application of the wavelet transform to the time-series to be analized bears most of all qualitative information. Based on the wavelet maps of one or several signals, algorithms can be devised to isolate dynamical features of potential interest. Needless to say, a number of possible algorithms, different from those presented here, can be tested and can provide new insight into the *structural* behaviour of turbulent flows: processing of wavelet-transformed signals seems to be a very promising way to manage them, and much work is still needed before standard procedures for the signal treatment may be established.

References

1. A. W. Strawa and B. J. Cantwell, *Phys. Fluids A* **28, 8** (1985) 2015.
2. E. Gutmark, T. P. Parr, D. P. Hanson-Parr and K. C. Schadow, *Comb. Flame* **75** (1989) 229.
3. E. Gutmark, T. P. Parr, D. P. Hanson-Parr and K. C. Schadow, *Exp. Fluids* **10** (1990) 147.
4. A. K. M. F. Hussain, *Phys. Fluids A* **26, 10** (1983) 2816.
5. A. K. M. F. Hussain, *J. Fluid Mech.* **173** (1986) 305.
6. R. J. Adrian, *Int. J. Heat Fluid Flow* **7** (1986) 127.
7. H. Meng and A. K. M. F. Hussain, *Fluid Dyn. Res.* **8** (1991) 33.
8. K. B. M. Q. Zaman and A. K. M. F. Hussain, *J. Fluid Mech.* **138** (1984) 325.
9. D. Durão and J. H. Whitelaw, *J. Fluids Eng.* **95** (1973) 467.
10. M. M. Ribeiro and J. H. Whitelaw, *J. Fluids Eng.* **98** (1976) 284.
11. M. A. Habib and J. H. Whitelaw, *J. Fluids Eng.* **101** (1979) 521.
12. M. A. Habib and J. H. Whitelaw, *J. Fluids Eng.* **102** (1980) 47.
13. M. M. Ribeiro and J. H. Whitelaw, *J. Fluid Mech.* **96** (1980) 769.
14. F. H Champagne and I. J. Wygnanski, *Int. J. Heat Mass Transf.* **14** (1971) 1445.
15. N. W. M. Ko and A. S. H. Kwan, *J. Fluid Mech.* **73** (1976) 305.
16. A. S. H. Kwan and N. W. M. Ko, *J. Fluid Mech.* **82** (1977) 273.
17. N. W. M. Ko and H. Au, *J. Fluids Eng.* **103** (1981) 335.
18. N. W. M. Ko and H. Au, *J. Sound Vibr.* **100** (1985) 211.
19. H. Au and N. W. M. Ko, *J. Sound Vibr.* **116** (1987) 427.
20. W. J. A. Dahm, C. E. Frieler and G. Tryggvason, *J. Fluid Mech.* **241** (1992) 241.
21. G. Buresti, P. Petagna and A. Talamelli, *Exp. Therm. Fluid Sci.* (1994) (in

press).
22. G. Buresti, P. Petagna and A. Talamelli, in *Proc. of the XII Nat. Congr. AIDAA* (Como, Italy, Jul.1983) (in press).
23. M. Farge, *Ann. Rev. Fluid Mech.* **24** (1992) 395.
24. I. Daubechies, *Ten Lectures on Wavelets* (S.I.A.M., 1992).
25. C. G. Lewis and M. Gharib, *Phys. Fluids A* **4** (1992) 104.
26. T. Dallard and F. K. Browand, *J. Fluid Mech.* **247** (1993) 339.
27. H. Higuchi, J. Lewalle and P. Crane, *Phys. Fluids* **6** (1994) 297.
28. J. Liandrat and F. Moret-Bailly, *Eur. J. Mech., B/Fluids* **9** (1990) 1.
29. H. Tennekes and J. L. Lumley, *A First Course in Turbulence* (M.I.T. Press Cambridge, Massachussets, 1972).
30. J. Lewalle, *Acta Mech.* (1994) in press.

USE OF DIFFUSIONAL TECHNIQUES IN GAS-LIQUID TWO-PHASE FLOW

ALESSANDRO PAGLIANTI
Dipartimento di Ingegneria Chimica, Chimica Industriale e Scienza dei Materiali
University of Pisa, Via Diotisalvi 2, Pisa, 56126, Italy

and

PAOLO ANDREUSSI
Dipartimento di Scienze e Tecnologie Chimiche, University of Udine, Via Cotonificio 108
Udine, 33100 Italy

ABSTRACT

The first regime maps were based on visual identification of phase distribution. More recently, other maps based on the analysis of pressure drops have been proposed. All these maps have the same defect: for many situations they are inapplicable or seem to be too subjective. The goal of this work is to propose an objective method to distinguish between flow regimes and to show as with diffusional techniques it is possible to identify subregimes that previous methods cannot evidence.

1. Introduction

Gas-liquid two-phase flow occurs in a wide range of industrial plants. It is easily found in the petroleum industry, chemical plants, in all equipment in which changes of phase occur, in the transport of geothermal fluids, in nuclear reactors and in many other applications. When gas and liquid flow together in a pipe many configurations, called flow regimes, can be encountered. These flow patterns differ from each other in the spatial distribution of the interface and the configuration that settles in the pipe depends on the superficial velocities of the phases, the geometry of the pipelines, diameter and inclination, and on the physical properties of the working fluids.

Fig. 1 schematically reproduces the possible flow patterns that can be met in a horizontal pipe, superimposed on the Mandhane et al.[1] map that is usually used to predict flow regime.

The hydrodynamics of the flow, and therefore the pressure drop and heat and mass transfer, depend strongly on the flow regime that occurs in the pipeline.

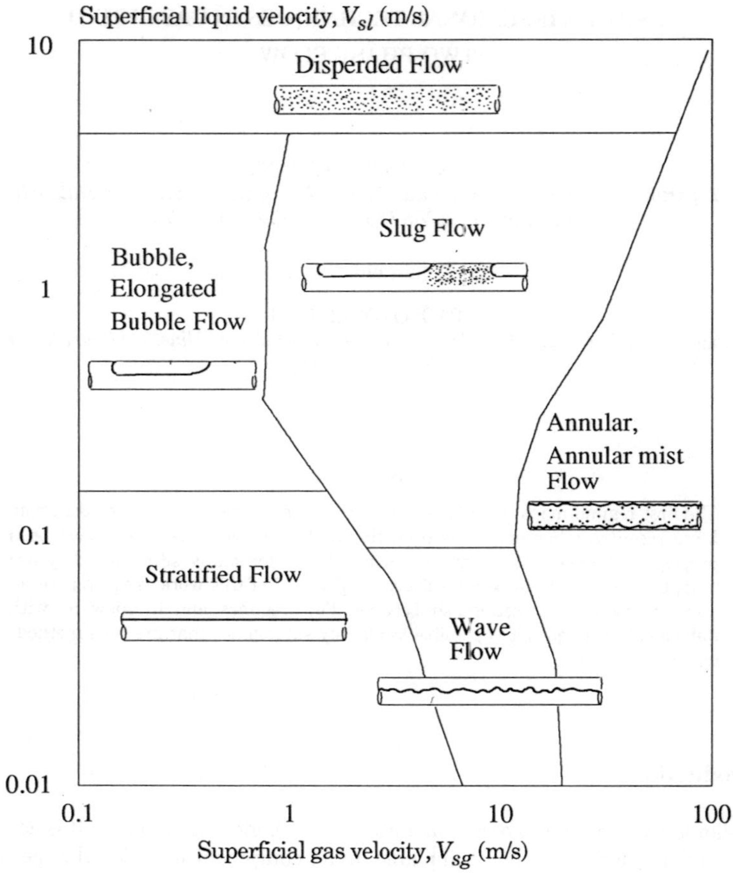

Figure 1: Mandhane et al.[1] map and schematic diagram of flow patterns in horizontal pipe flow

For instance, in horizontal or near-horizontal pipes, at low gas and liquid flow rates, a stratified configuration is established with the liquid flowing in the bottom of the pipe and the gas flowing above the liquid. Even in this simple configuration two different subregimes have been identified: when the interface is almost flat and undisturbed the subregime is called Stratified Smooth, whereas if it is covered by a wave structure it is called Stratified Wavy. The Stratified Wavy regime can be characterized by two-dimensional regular waves, by three-dimensional wavelets, and by large disturbance waves.

At higher gas velocities, part of the liquid is entrained by the gas in the form of droplets and Annular flow conditions are established in the pipe. In this flow pattern the gas-liquid interface is covered by low-amplitude capillary waves and by large disturbance waves that cause most of the liquid entrainment. As for stratified flow patterns, also in Annular flow the interfacial structure determines the transport processes between the gas and liquid phases. The Annular flow pattern is highly nonsymmetrical in the cross section at lower gas velocities in horizontal pipes and becomes more and more symmetrical as the gas velocity or the inclination increases.

At large liquid flow rates, a situation in which the liquid represents the continuous phase and the gas presents various degrees of dispersion in the liquid phase is typically encountered. At high liquid velocities, the gas flows in the form of dispersed spherical bubbles (Dispersed Bubble flow). This flow pattern can be observed for the full range of pipe inclinations, but at different inclinations the gas distribution in the cross section and also the slip between the phases vary appreciably. At lower liquid velocities, dispersed bubbles take on an ellipsoidal shape, tend to pulsate, and bubble coalescence is greatly enhanced. In these cases, the stable flow pattern is identified as Plug or Slug flow. These flow regimes are encountered in all pipe inclinations, but obviously both the flow structure and the transition between various regimes and subregimes are affected by pipe inclination.

Each flow pattern displays a particular hydrodynamics that implies great differences in pressure drop, heat and mass transfer and in all the characteristics of the flow, such as liquid holdup. Usually the studies of the transition are based on analysis of the pressure drop time traces[2]; in this work signals of liquid holdup coming from a gamma densitometer and from a capacitance probe have been analyzed.

2. Diffusional Techniques

The diffusional method, described by, among others, Fujisaka[3,4], Suzuki[5,6] and Takayoshi and Mori[7], can be adopted in the study of multiphase flow. In particular, as pointed out by Giona et al.[8,9], this approach can be applied to analysis of the transitions between flow patterns and presents interesting characteristics that make it preferable to the usual R/S analysis. In this work, time series $\{x_i\}$ have been modified to obtain normalized series $\{\xi_i\}$ in the following way:

$$\xi_i = \frac{(x_i - \langle x \rangle)}{\sigma_x} \tag{1}$$

where

$$\langle x \rangle = \frac{1}{N} \cdot \sum_{i=1}^{N} x_i \tag{2}$$

and

$$\sigma_x^2 = \frac{1}{N} \cdot \sum_{i=1}^{N} \left(x_i - \langle x \rangle\right)^2 \tag{3}$$

where N is the length of the time series. From the series $\{\xi_i\}$ it is possible to create a new series representing a diffusional process as

$$z_i = z_{i-1} + \xi_i \tag{4}$$

and from it, according to the general statistical theory of diffusive phenomena, compute the fundamental quantity associated with the diffusion process as

$$R_z^2(n) = \left\langle (z_{k+n} - z_k)^2 \right\rangle = \frac{1}{N_{av}} \cdot \sum_{k=1}^{N_{av}} (z_{k+n} - z_k)^2 \tag{5}$$

where N_{av} is the number of averaging points and n is an index of the time. As shown by Giona et al.[8], it is possible to rescale $R_z^2(n)$ as

$$R_z^2(n) \approx n^{\beta_1} \;,\; n \le n_c \qquad R_z^2(n) \approx n^{\beta_2} \;,\; n \gg n_c \tag{6}$$

where n_c is crossover time and the coefficients β_i depend on the type of regime.

3. Flow Pattern Detection

In the study of the transition of two-phase flow many empirical maps (Mandhane et al.[1], Spedding and Nguyen[10], Baker[11]) and maps based on physical models after the work by Taitel and Dukler[12] have been published. In this section the results obtained from diffusional techniques have been compared with published data.

The first works on the two-phase flow transitions have been performed identifying the flow pattern by visual observation of the phenomena; clearly a similar approach is inapplicable for many situations and appears to be too subjective. More recently, several authors, in an attempt to more objectively classify flow regimes, have developed other methods. Weisman et al.[2] analyzed the pressure drop signals, Jones and Zuber[13] used the power spectral density function and the probability density function analyzing mean void fraction fluctuations. The approach suggested in this work allows identifying the flow regime from the analysis of two typical parameters, β_1 and β_2. In the following table a list of β_i for the main two-phase flow regimes is reported:

Flow Regime	β_1	β_2
Stratified	≈ 2	$\gg 1$
Intermittent	$\gg 1$	$\ll 1$ (≈ 0)
Bubble	≈ 1	≈ 1
Annular	$\gg 1$	< 1

Table 1: Typical two-phase flow regime parameters

The choice of this method offers many advantages, the largest being the objectivity of the model and the possibility of writing a simple code for the on-line analysis. The usual maps published generally identify the transition between two different flow regimes as a line neglecting the existence of regions with no definite characteristics. On the contrary, the method proposed in this work, by defining typical values of the parameters β_1 and β_2 for each regime, allows identifying regions that display intermediate characteristics between two different flow patterns.

The experimental transitions identified in this work have been obtained by analyzing liquid holdup signals recorded with a gamma densitometer. The experimental loop used is assembled at the Institute for Energy Technology (IFE) at Kjeller, Norway. It works at atmospheric conditions and consists of an acrylic pipe 14 m long with an internal diameter of 0.032 meters; the working fluids are oil and air.

The comparison between the results obtained with the proposed analysis and the transitions proposed by Weisman et al.[2] and Taitel and Dukler[12] have been plotted in Fig. 2.

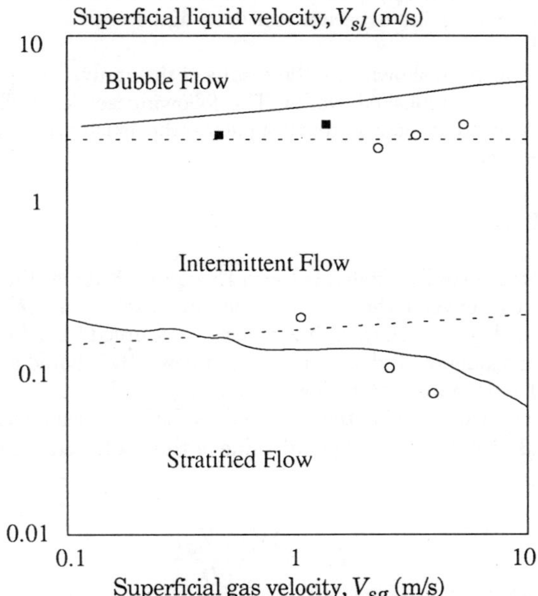

Figure 2: Comparison between published flow pattern maps (Weisman et al. [2], – – – and Taitel and Dukler [12], ——) and experimental data (fully developed Intermittent Flow,O-fully developed Dispersed Bubble Flow,■)

Concerning the model proposed by Taitel and Dukler[12] some observations seem to be necessary.

- The transition between Dispersed Bubble and Intermittent flow has been identified when the turbulent fluctuations are strong enough to overcome the buoyant forces that tend to keep the gas in the upper part of the pipe. This approach seems to identify the fully developed Dispersed Bubble flow leaving the intermediate region with no definite characteristics in the Intermittent flow region

- The transition between Intermittent and Stratified flow regimes occurs when a wave of infinitesimal amplitude formed from a flat sheet of liquid grows until it blocks the whole pipe section. This choice seems to identify fully developed Intermittent flow leaving the intermediate region in the Stratified flow region.

As for the transition between Dispersed Bubble and Intermittent flow, our experimental results, which allow identifying the fully developed Dispersed Bubble flow, agree with the model proposed by Taitel and Dukler[12], whereas our experimental results that identify fully developed Intermittent flow agree with the experimental map of Weisman et al.[2].

As for the transition between Stratified and Intermittent flow regimes, the same observations can be made; the experimental data agree with the model proposed by Taitel and Dukler[12].

In this section it has been shown how the results of the analysis proposed in this work agree with the previous published works. The following section will show how this method in the Intermittent flow pattern suggests the existence of subregimes that cannot otherwise be evidenced.

4. Intermittent Flow

Intermittent flow is usually subdivided into Elongated Bubble (Plug flow) and Slug flow. The differences between these subregimes are exclusively related to the void fraction in the liquid slug. In both cases it is generally assumed that the flow pattern can be represented by a sequence of identical units composed by a liquid slug, which can be aerated (Slug flow) or deaerated (Plug flow), and a long bubble.

As shown in the previous section the proposed method is based on the analysis of the coefficient β_1. Fig. 3 plots the experimental value of this coefficient against the mixture Froude number, defined as

$$Fr_m = \frac{\left(V_{sl} + V_{sg}\right)}{\sqrt{gD}} \tag{7}$$

where V_{sl} and V_{sg} are respectively the superficial velocity of the liquid and gas phases.

This figure shows the existence of a critical value of the Froude number (Fr_m=6); below it β_1 remains constant and equal to about 1.9, and above it there is a monotone decreasing trend. This behavior suggests the existence of two different subregimes with different characteristics of coherency. As shown by Giona et al.[8], a periodic signal is characterized by a scaling of $R_z^2(n)$ having β_1=2, whereas if stochastic fluctuations are superimposed on the regular oscillations the scaling of $R_z^2(n)$ leads to β_1<2. This observation allows us to conclude that if $\beta_1 \approx 2$ the flow pattern is characterized by a sequence of regular units, whereas if β_1<2 the flow pattern presents much more irregular units.

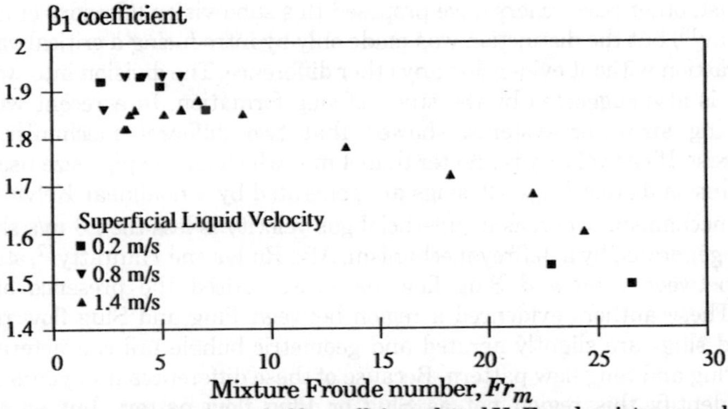

Figure 3: Experimental trend of β_1 coefficient plotted Vs Froude mixture number

This different behavior cannot be attributed to the transition between Plug and Slug regime. Indeed, all the time traces analyzed are referred to a fully developed Slug flow, as shown by Fig 4. The results obtained analyzing time series according to the method presented in Par. 2, suggest to subdivide Slug flow region in two different subregimes.

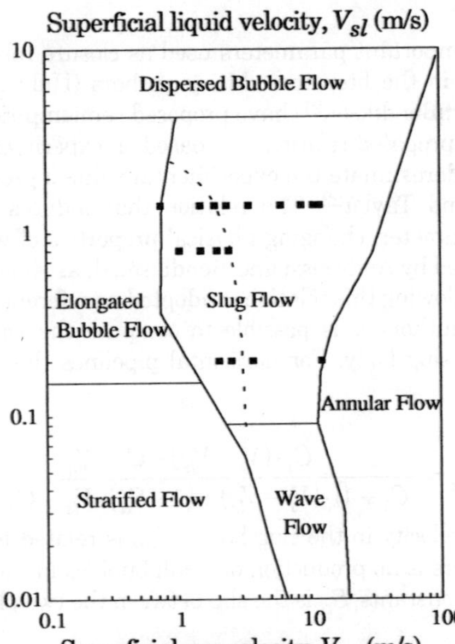

Figure 4: Comparison between Mandhane et al.[1] Flow map (——), proposed transition (- - -) and experimental data

In the past, other researchers have proposed this subdivision (Gregory et al.[14], Kokal and Stanislav[15]) but the distinction was made only by introducing a critical value on the Slug void fraction without evidencing any other difference. The division into two different subregimes is also suggested by the study of slug formation. In a recent work Fan et al.[16], studying air-water systems, showed that two different mechanisms of slug formation exist. If gas velocity is greater than 4 m/s, which for the pipe size used by these authors corresponds to a $Fr_m \approx 4.2$, slugs are generated by a nonlinear Kelvin-Helmhotz instability mechanism, whereas if superficial gas velocity is less then 3 m/s slugs evolve from waves generated by a Jeffrey mechanism. Also Ruder and Hanratty[17], studying the transition between Plug and Slug flow patterns, noticed the presence of different behaviors. These authors evidenced a region between Plug and Slug flow patterns in which liquid slugs are slightly aerated and geometric bubble tail characteristics differ both from Plug and Slug flow pattern. Because of these differences it appears to be more correct to identify this region not as Slug or Plug flow pattern but as a different subregime, Elongated bubble flow with dispersed bubbles (in the following, *EB+DB*).

The next section will show how a distinction of Slug flow pattern into two different subregimes, *EB+DB* and fully developed Slug flow (in the following, *SL*), allows a satisfactory description of the experimental trends of the void fraction.

5. Void Fraction

This is one of most important parameters used as closure relation in all the mechanistic models available in the literature. Many authors (Hubbard[18], Gregory et al.[19], Malnes[20], Andreussi and Bendiksen[21]) have proposed semiempirical relations to be used for this purpose. All the proposed relations are based on experimental data performed in small diameter that underestimate the experimental value if great diameters are used, as shown by Jepson and Taylor[22]. The relation that induces smaller errors in the computation of this parameter, changing physical properties of working fluids and size diameter, is that proposed by Andreussi and Bendiksen[21], as shown by Paglianti et al.[23]. For this reason in the following this relation is adopted as reference.

According to these authors it is possible to compute the void fraction by a mass balance written for the slug body. For horizontal pipelines this approach leads to the relation

$$\alpha_s = \frac{C_1 \cdot (V_t - V_{ls}) - C_1 \cdot V'_{mf}}{(C_1 + 1) \cdot (V_t - V_{ls}) + (1 - S_d) \cdot V_{ls} + C_2 \cdot V_{go}} \tag{8}$$

where V_{ls} is the liquid velocity in the slug body, V'_{mf} is related to the limiting mixture velocity below which there is no production of small bubbles in the slug body, C_1 and C_2 are two proportionality constants, S_d is the slip between the two phases expressed as

$$S_d = \frac{V_{gs}}{V_{ls}} \tag{9}$$

where V_{gs} and V_{ls} are respectively the gas and the liquid velocity in the slug body. Finally V_{go} is the rising velocity expressed in the form proposed by Harmathy[24].

$$V_{go} = 1.53 \cdot \left[\frac{g \cdot \sigma}{\rho_l^2} \cdot (\rho_l - \rho_g) \right]^{0.25} \tag{10}$$

The bubble velocity can be computed with the relation proposed by Bendiksen[25],

$$V_t = C_o \cdot V_{ls} + V_o \tag{11}$$

where the values of the constants C_o and V_o depend from the value of the Froude number. In fact he shows that for air-water systems there exists a limit value of the Froude number, Fr_{crit}, which we have assumed as the transition between *EB+DB* and *SL*, equal to 3.5, below which C_o and V_o can be computed as

$$C_o = 1.05 \quad V_o = 0.542 \cdot \sqrt{g \cdot D} \tag{12}$$

whereas for Froude numbers over that limit those parameters have to be computed as

$$C_o = 1.2 \qquad V_o = 0 \tag{13}$$

Using Eq. (11) relation (8) can be rewritten as

$$\alpha_s = \frac{V_{ls} - V_{mf}^*}{V_{ls} \cdot \left[\frac{(C_1 + 1)}{C_1} + \frac{(1 - S_d)}{C_1 \cdot (C_o - 1)} \right] + V_o \cdot \frac{(C_1 + 1)}{C_1 \cdot (C_o - 1)} + V_{go} \cdot \frac{C_2}{C_1 \cdot (C_o - 1)}} \tag{14}$$

where V^*_{mf} indicates the transition mixture velocity between Plug and Slug flow pattern and is related to V'_{mf}:

$$V'_{mf} = V_{mf}^* \cdot (C_o - 1) + V_o \tag{15}$$

where C_o and V_o are obtained from Eq. (11) and V^*_{mf} can be computed as proposed by Barnea[26].

$$V_{mf}^* = \left\{ \left[0.725 \cdot \left(\frac{\sigma}{\rho_l} \right)^{\frac{3}{5}} \cdot \frac{1}{d_c} \right]^{\frac{5}{2}} \cdot \frac{D}{2 \cdot f_m} \right\}^{\frac{1}{3}} \tag{16}$$

where d_c is the bubble diameter at the transitions between Dispersed Bubble and Intermittent flow pattern. This is a function of liquid velocity, of the physical properties of the working fluids and of the friction factor f_m. The value of d_c to be used in Eq. (16), as suggested by Barnea[26], is the smallest between d_{cd} and d_{cb}, where d_{cd} is the critical bubble diameter above which the bubble is deformed and d_{cb} is the bubble diameter below which the migration to the upper part of the pipe is prevented.

$$d_{cd} = 2 \cdot \sqrt{\frac{0.4 \cdot \sigma}{(\rho_l - \rho_g) \cdot g}} \tag{17}$$

$$d_{cb} = \frac{3}{8} \cdot \left[\frac{\rho_l}{(\rho_l - \rho_g)} \right] \cdot \frac{f_m \cdot V_m^2}{g} \tag{18}$$

In order to simplify the model, Andreussi and Bendiksen[21] proposed using the value of C_o and V_o obtained with Fr_m greater than Fr_{crit} in the whole range of the mixture velocities. In this work, as suggested by Diffusional analysis, Slug flow regime has been subdivided in two different subregimes. This assumption implies that, against to what has been published in literature, two different equations, to compute slug void fraction in Intermittent flow region, have to be used.

If Fr_m is below Fr_{crit}, as suggested by Malnes[20] and Bendiksen[25], $S_d = C_o$. According to these assumptions, therefore in *EB+DB* range, relation (14) can be rewritten as

$$\alpha_s = \frac{V_{ls} - \frac{1}{(C_o - 1)} \cdot (V'_{mf} - V_o)}{V_{ls} + \frac{1}{(C_o - 1)} \cdot C_4 \cdot V_{go} + \frac{1}{(C_o - 1)} \cdot C_3 \cdot V_o} \tag{19}$$

where $C_3 = (C_1 + 1)/C_1$, $C_4 = C_2/C_1$ and V_{ls} in horizontal flow can be computed as

$$V_{ls} = \frac{V_m}{1 + \alpha_s \cdot (S_d - 1)} \tag{20}$$

whereas if Fr_m is over Fr_{crit}, the slip ratio is equal to 1 and the rising velocity V_o is equal to 0. In this case, that is in *SL* range, relation (14) assumes the form

$$\alpha_s = \frac{V_m - \frac{1}{(C_o - 1)} \cdot V'_{mf}}{C_3 \cdot V_m + \frac{1}{(C_o - 1)} \cdot C_4 \cdot V_{go}} \tag{21}$$

To solve the model it is necessary to define the constants C_3 and C_4. Constant $1/C_3$ represents the maximum value of the void fraction obtained for high mixture velocities and for this reason it is a function of physical properties of the working fluids and of the size diameter. Constant C_4 can be obtained by imposing the equality between Eq. (19) and Eq. (21) for $Fr_m = Fr_{crit}$. Fig. 5 shows the comparison between our experimental data of void fraction, the proposed relations and the relation of Andreussi and Bendiksen[21]. The good agreement between the data and the proposed relations is significant, but it is more important to underscore the presence of a discontinuity in the experimental trend for $Fr_m = Fr_{crit} = 6$. The simplified correlation proposed by Andreussi and Bendiksen[21] cuts the region with $Fr_m < Fr_{crit}$, that is, the *EB+DB* subregime. While this assumption can be acceptable if diameter is small, if pipe size increases large errors can be induced.

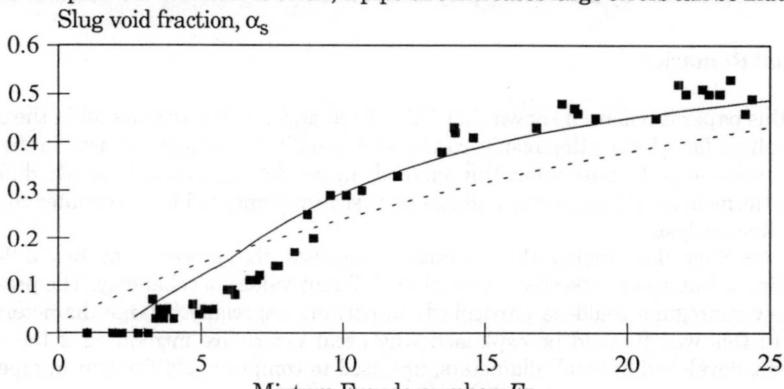

Figure 5: Slug Void fraction plotted Vs Mixture Froude Number (Andreussi and Bendiksen[21] - - -, proposed relations—— and our experimental data, ■)

As shown by Jepson and Taylor[22], at constant superficial velocity the liquid holdup decreases increasing the pipe diameter, so the available relations fail when they are used to predict this parameter in pipelines with large diameters. According to our analysis this error is due to having neglected the presence of the subregime *EB+DB*. Fig. 6 shows the comparison between the experimental data of Jepson and Taylor[22], the relation of Andreussi and Bendiksen[21] and the relations suggested in this paper.

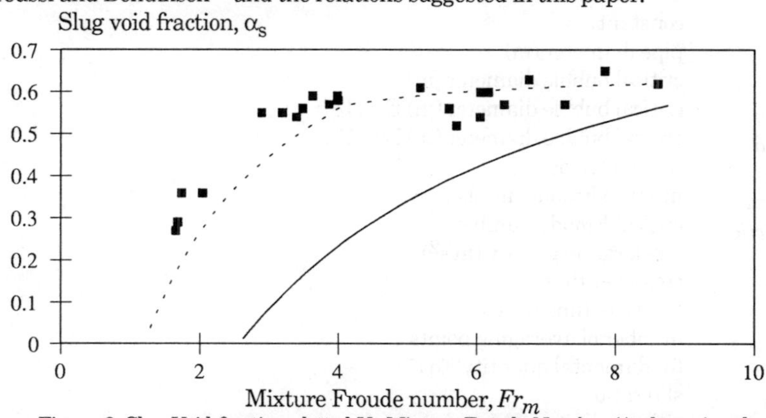

Figure 6: Slug Void fraction plotted Vs Mixture Froude Number (Andreussi and Bendiksen[21] ——, proposed relations - - -, and experimental data of Jepson and Taylor[22], ■)

The values of the constants V_o and C_o used in relations (19) and (21) have been obtained by analyzing the experimental data of slug translational velocity according to the way suggested by Bendiksen[25]. The discontinuity in the experimental trend is clear and is well predicted by the relation proposed in this work.

6. Final Remarks

In this paper it has been shown that Diffusional analysis is a useful tool in the study of two-phase flow. Interesting results can be achieved if this technique is used in the two-phase transitions. In particular, this method allows distinguishing between different flow patterns in an objective way and can be easily implemented in a computer code for the on-line analysis.

In the Slug flow region this technique suggests the presence of two different subregimes that seem to be characterized by different values of coherency. The presence of a new subregime could be particularly important, especially if large diameters are used. In this way it could be explained why great errors are introduced if the usual relations, developed in small diameters, are used to compute void fraction in pipelines with large diameters.

Acknowledgments

We would like to thank you Prof. K. Bendiksen, O.J. Nydal, M. Giona and A. Soldati for useful discussions and precious suggestions.

Nomenclature

C	constant
D	pipe diameter (m)
d_c	critical bubble diameter (m)
d_{cb}	critical bubble diameter (m) Eq. (18)
d_{cd}	critical bubble diameter (m) Eq. (17)
f_m	friction factor
Fr_m	mixture Froude number
Fr_{crit}	critical Froude number
g	acceleration gravity (m/s^2)
n_c	crossover time
N	length of time series
N_{av}	number of averaging points
$R^2{}_z$	fundamental quantity Eq.(5)
S_d	slip ratio
V_{go}	rising velocity (m/s)
V_{gs}	gas velocity in slug body (m/s)
V_{ls}	liquid velocity in slug body (m/s)

V_m	mixture velocity (m/s)
V_{mf}	velocity of aeration onset (m/s)
V_{sg}	superficial gas velocity (m/s)
V_{sl}	superficial liquid velocity (m/s)
V_t	translational velocity (m/s)
α_s	void fraction
β_i	flow regime coefficient Eq.(6)
x	time series
ξ	normalized series
z	diffusional series
ρ_g	gas density (Kg/m^3)
ρ_l	liquid density (Kg/m^3)
σ	surface tension (N/m)
σ_x	variance
$< >$	mean value

References

1. J.M. Mandhane, G.A. Gregory and K. Aziz, *Int. J. Multiphase Flow* **1** (1974) 537.
2. J. Weisman, D. Ducan, J. Gibson and T. Crawford, *Int. J. Multiphase Flow* **5** (1979) 437.
3. H. Fujisaka, *Prog. Theor. Phys.* **71** (1984) 513.
4. H. Fujisaka, in *Caos and Statistical Methods,* ed. Y.Kuramoto (Springer, Berlin, 1984) 117.
5. M.Suzuki, *Physica A* **117** (1983) 103.
6. M.Suzuki, in *Caos and Statistical Methods*, ed. Y.Kuramoto (Springer, Berlin, 1984) 260.
7. K.Takayoshi and H. Mori, *Prog. Theor. Phys.* **68** (1982) 439.
8. M.Giona, A.Paglianti and A.Soldati, in press on *Fractals.*
9. M.Giona, A.Paglianti and A.Soldati, *Fractals* **2** (1994) 265.
10. P.L.Spedding and V.T.Nguyen, *Chem. Eng. Sci.* **35** (1980) 779.
11. O.Baker, *Oil and Gas J.* **53** (1954) 185.
12. Y.Taitel and A.E.Dukler, *AIChE J.* **22** (1976) 47.
13. O.C.Jones and N.Zuber, *Int. J. Multiphase Flow* **2** (1975) 273.
14. G.A.Gregory, M.K.Nicholson and K.Aziz, in *27th Can. Chem. Eng. Conference* (1977).
15. S.L.Kokal and J.F.Stanislav, *Chem. Eng. Sci.* **44** (1989) 665.
16. Z.Fan, F.Lusseyran and T.J.Hanratty, *AIChE J.* **39** (1993) 1741.
17. Z.Ruder and T.J.Hanratty, *Int. J. Multiphase Flow* **2** (1990) 233.
18. M.G.Hubbard, *Ph. D. Dissertation* University of Houston, Texas, (1965).
19. G.A.Gregory, M.K.Nicholson and K.Aziz, *Int. J. Multiphase Flow* **4** (1978) 33.
20. D.Malnes, *Report IFE/KR/E/-83/002* V. Inst. for Energy technology, Kjeller,

Norway (1982).
21. P.Andreussi and K.Bendiksen, *Int. J. Multiphase Flow* **6** (1989) 937.
22. W.P.Jepson and R.E.Taylor, *Int. J. Multiphase Flow* **19** (1993) 411.
23. A.Paglianti, G.Trotta, P.Andreussi and O.J.Nydal, in *MultiPhase Production*, ed.A. Wilson (MEP, London, 1993) 193.
24. M.G.Harmathy, *AIChE J.* **6** (1960) 281.
25. K.H.Bendiksen, *Int. J. Multiphase Flow* **10** (1984) 467.
26. D.Barnea, *Int. J. Multiphase Flow* **13** (1987) 1.

IDENTIFICATION OF MULTIPHASE FLOWS BY MEANS OF FLUCTUATION ANALYSIS

MASSIMILIANO GIONA
Dipartimento di Ingegneria Chimica, Universitá di Cagliari
Piazza d'Armi, 09123 Cagliari, ITALY

ALESSANDRO PAGLIANTI
Dipartimento di Ingegneria, Chimica Industriale e Scienza dei Materiali, Univ. di Pisa
Via Diotisalvi 2, 56100 Pisa, ITALY

and

ALFREDO SOLDATI
Dipartimento di Scienze e Tecnologie Chimiche
Via Cotonificio 108, 33100 Udine, ITALY

ABSTRACT

After reviewing the various methods used to identify multiphase flow, we outline how diffusional analysis provides an objective identification of multiphase flow regimes and how the method of relative diffusion furnishes the basis for an efficient filtering procedure. The choice of the experimental time series (pressure drop signal vs density signal) is discussed briefly.

1. Introduction

The last five years have seen a growing interest in the quantitative characterization of multiphase flows by means of time-series analysis associated with on-line measurements of a characteristic observable of the flow (pressure drop, density fluctuations, liquid hold-up)[1,2,3,4,5,6,7,8,9,10]. The underlying concept in all these articles is that the signature of the flow conditions is carried by the statistical properties of fluctuations of some characteristic observable and hence that the detailed analysis of these fluctuations can provide both qualitative (e.g. flow-pattern identification) and quantitative (e.g. fundamental frequency of oscillations in intermittent flow) information on flow conditions. The area of application of these methods is extremely wide indeed since it ranges from bubble-column dynamics[9] to fluidized-bed operations[1,3,5,6,7,8] and multiphase flow regimes[2,4,10]. The methods employed also cover a broad spectrum of analytical tools. Some authors make use of methods typical of chaotic time series analysis[1,4,5,6,7] (Grassberger-Procaccia analysis of correlation dimension[11], estimation of Kolmogorov entropy[12]) and others employ mathematical tools of stochastic fluctuation analysis[2,3,4,8,9] (rescaled-range analysis[13] - known as R/S analysis -, or modelling of flow by means of fractional Brownian motion[14,15]).

Reviewing the different approaches, a *philosophical* distinction can be drawn between the approach based on the claim[1,6] that multiphase flow and fluidized-bed dynamics have a deterministic explanation, i.e. are ultimately chaotic systems, and the approach based on the assumption that the evolution of the system can be described as a stochastic process whose statistics depends on the flow conditions[2]. On this important question we wholly agree with the analysis of Kennel and Isabelle[16] and their warning that some purely stochastic processes (e.g. colored noise) may exhibit the same features as chaotic systems in the scaling of the correlation dimension and that therefore the stochastic/noisy nature should be ascertained by means of methods other than the behaviour of the correlation dimension (e.g. through the analysis of the local predictability properties).

An alternative approach to the analysis of fluid dynamic time series making use of diffusional methods has been proposed by Giona et al.[17,18], showing that the application of this approach provides a clear, unambigous characterization of two-phase flow regimes. Indeed, diffusional analysis is a typical tool used to obtain information on the correlation and statistical properties of a time series. The physical motivation for its application is ultimately the equivalence between diffusion processes and stochastic fluctuations[19,20].

The expression *diffusional analysis* encompasses several distinct methods casting light on different statistical aspects of the time series under consideration. In this article we analyze all these methods, focusing on the identification of two-phase flow regimes and on the separation of the deterministic and stochastic components of the signal. A brief discussion on the choice of the observable (pressure drop, density signal) is added in order to facilitate the application of these methods to other fluid dynamic problems.

2. Experimental conditions

The experimental results presented in this paper were obtained from the experimental loop of the Institutt for Energiteknikk of Kjeller, Norway. The apparatus consisted of an inclinable bench supporting a 14-meter pipeline with an inner diameter of 0.031 meters. The working fluids are air, flowing in all the cases close to atmospheric conditions, and light oil.

The test section consisted of a γ-densitometer for measuring the density of the mixture flowing in the tube and a DP cell for measuring the pressure drops. The signals of the γ-densitomer and those of the DP cell were both recorded at a sampling time $\Delta t = 1.0$ ms. The pressure drops were measured with a DP cell made by Fisher and Porter, ensuring an accuracy of 2%. The distance between the pressure taps was 1.0 m and they were positioned symmetrically with respect to the position of the γ-densitometer. The fluid dynamic conditions are represented in terms of the gas and liquid superficial velocities j_g, j_l; these parameters were obtained by dividing the volumetric flowrate by the section of the pipeline.

3. Methods of diffusional analysis

Let $\{x_i\}$, $(i = 1, .., N)$, be the experimental time series of an observable (density, pressure drop) sampled at constant time intervals Δt. In order to analyze normalized data we will consider the dimensionless time series $\{\xi_i\}$, with $\xi_i = (x_i - <x>)/\sigma_x$, having zero mean and unit variance. $<x>$ is the average of $\{x_i\}$, and σ_x^2 its variance. In the analysis of multiplicative processes it is convenient to define a normalized time series $\{\zeta_i\}$, with $\zeta_i = (x_i - x_{min})/(x_{max} - x_{min})$.

3.1. Linearly driven diffusion processes

The simplest way to generate a diffusion process starting from an arbitrary deterministic, chaotic or stochastic time series $\{\xi_i\}$ is to consider a random walk $\{z_i\}$ on the real line linearly driven by the signal $\{\xi_i\}$ under consideration. If z_i is the position at time i of the walk, the dynamics of the process is described by the equations

$$z_{i+1} = z_i + \xi_{i+1} \, , \tag{1}$$

with the initial condition $z_0 = 0$ (which means that the random walk starts from the origin). The definition of the random walk represented by Eq. (1) can be regarded as a filtering of the signal with a linear filter having a pole at $\lambda = 1$. As an example, figures 1 and 2 show an experimental time series $\{\xi_i\}$ associated with density fluctuations and the corresponding diffusion process $\{z_i\}$ defined by Eq. (1). Information about the correlation properties of the random walk is provided by mean square displacement $R_z^2(n)$, defined as usual as

$$R_z^2(n) = <(z_{k+n} - z_k)^2> = \frac{1}{N_{av}} \sum_{k=1}^{N_{av}} (z_{k+n} - z_k)^2 \, , \tag{2}$$

where N_{av} is the number of averaging points. If the signal $\{\xi_i\}$ is a realization of fractional Brownian motion, characterized by the Hurst exponent[13] H, then

$$R_z^2(n) \sim n^{2H} \, , \tag{3}$$

and in particular, if the driving signal is a regular Brownian motion process, $H = 1/2$, $R_z^2(n) \sim n$.

The opposite case is represented by deterministic oscillating signals. In this case the mean square displacement exhibits a crossover behaviour: at short time scales $n \leq n_c$ then $R_z^2(n) \sim n^2$, as it should given since the deterministic biasing of the signal; at long time scales the behaviour of $R_z^2(n)$ is oscillating with an average slope equal to zero. A similar situation arises in the analysis of signal $\{\xi_i\}$ which can be regarded as a superposition of deterministic oscillating modes and random fluctuations. In multiphase flow this situation occurs in connection with intermittent (slug) flow. In this case, the mean square displacement $R_z^2(n)$, is characterized by a crossover behaviour (see e.g. figure 4)

$$R_z^2(n) \sim \begin{cases} n^{\beta_1} & n \leq n_c \\ n^{\beta_2} & n >> n_c \end{cases} \, . \tag{4}$$

The crossover instant n_c is related to the fundamental period of oscillations T_c by the relation $T_c = 4 n_c \Delta t$, where Δt is the sampling time.

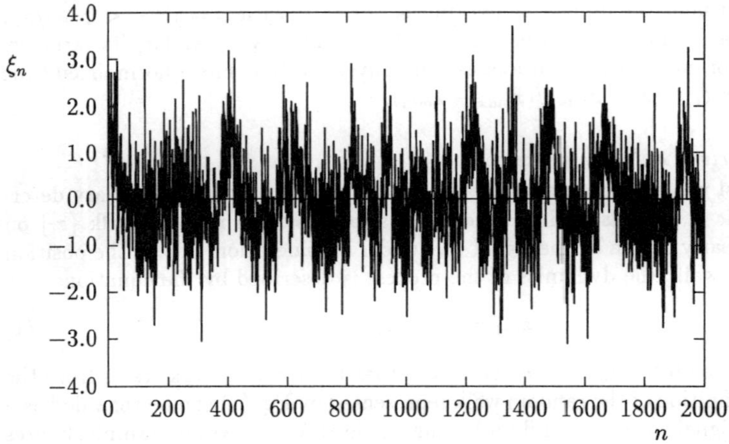

Figure 1: Normalized experimental time series $\{\xi_n\}$ vs n associated with density fluctuations. The liquid superficial velocity is $j_l = 2.48$ [m/s], the gas superficial velocity is $j_g = 3.86$ [m/s].

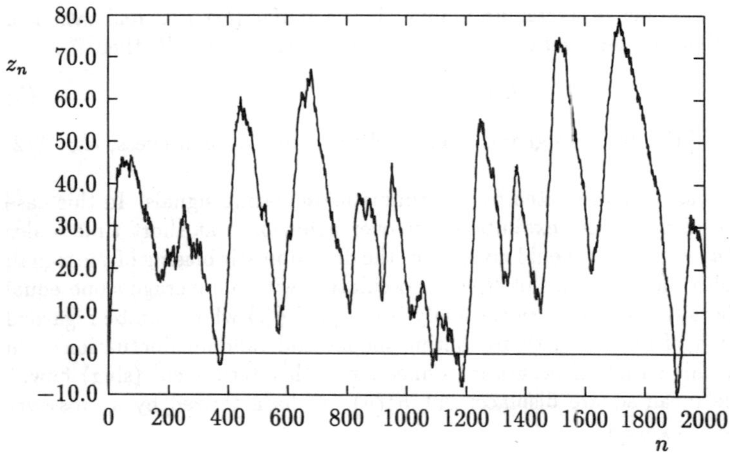

Figure 2: $\{z_n\}$ vs n for the time series of figure 1.

To conclude, in general the mean square displacement is characterized by the two exponents β_1, β_2 (and by the crossover instant n_c in intermittent flow). The dependence of these exponents on the superficial gas and liquid velocity j_g, j_l makes the identification of flow regime possible.

3.2. Identification of multiphase flow

The identification of multiphase flow by means of diffusional methods has been extensively treated by Giona et al.[17,18], Paglianti and Andreussi[21]. In this section we are concerned solely to point out how diffusional analysis has a direct physical motivation based on the nature of the signal.

A schematic review of the main results of diffusional analysis in terms of the exponent β_1 and β_2 can be found in Giona et al.[18] and in Paglianti and Andeussi[21]. *Bubble flow* is characterized by random fluctuation in the local density deriving from the Brownian motion of the gas bubble inside the main liquid flow. Consequently $\beta_1 = \beta_2 = 1.0$, as it should given the Brownian nature of the signal. The opposite case is *Stratified flow* which is a fairly regular regime with oscillations at the gas-liquid interface: in this case diffusional analysis furnishes exponents β_1 close to 2 and β_2 definitely greater than 1 deriving from the strong biasing effects in the motion. *Intermittent flow*, as discussed in the previous section, exhibits a typical crossover behaviour in $R_z^2(n)$. This is due to the presence of deterministic oscillations associated with the fairly regular motion of the slugs in the pipe, and of random fluctuations, with an exponent β_1 much larger than 1 and close to 2 in the presence of high coherent fluctuations.

From the above observations, it appears clear that each result obtained by applying diffusional analysis has its physical counterpart in terms of the fluid dynamics of the flow and of the statistical nature of the resulting signal which makes this kind of analysis attractive as a way to achieve an objective characterization of the flow regimes.

A similar claim cannot be made for R/S analysis, as discussed by Giona et al.[17]. In dealing with slug flow in particular, the R/S scaling does not furnish well defined results susceptible of a simple physical interpretation.

Some fluid dynamic regimes may be difficult to identify by means of the scaling of $R_z^2(n)$. For instance, Stratified flow with highly pronounced wavy oscillations and *Annular flow* have similar statistical properties in terms of $R_z^2(n)$ scaling. However, in similar situations it is possible to make use of other methods (e.g. relative diffusion analysis, see sections 3.4 and 4) which may help in discriminating between these flow conditions.

A final comment should be made on the physical meaning of the asymptotic exponent β_2. In our analysis, the expression ***asymptotic scaling*** means long time scales compared with the period of oscillation of the signal $\{\xi_i\}$. In practice, it means time scales up to $t =$ 1-2 s, i.e. $n =$ 1000-2000. The results obtained for β_2 refer to these characteristic time scales.

3.3. Multiplicative processes

Another way of characterizing fluctuations is by means of a multiplicative dynamics driven by the time series under consideration. This method was proposed by Fujisaka[22] in order to analyze deviations from Gaussianity in stochastic and chaotic fluctuations. The multiplicative process $\{w_i\}$ is defined as

$$w_i = w_{i-1}\zeta_i \ , \tag{5}$$

where ζ_i is the experimental time series normalized between 0 and 1.

The generalized moments of $\{w_n\}$, $< w_n^q >$, $q \in (-\infty, \infty)$ scale exponentially with time n

$$< w_n^q > \sim \exp(q\mu_q n) \tag{6}$$

and μ_q depends on the statistical properties of the driving signal $\{\zeta_i\}$. In particular, for a Gaussian process, μ_q scales linearly with q

$$\mu_q = \mu_o + Dq \ , \tag{7}$$

where D is the diffusion coefficient associated with the Gaussian process $\{\zeta_i\}$. A deviation from linear behaviour in μ_q is an indicator of the non-Gaussian nature of the signal.

3.4. Relative diffusion analysis

Relative diffusion analysis was introduced by Suzuki[23,24] to study the statistical properties of chaotic dynamical systems and turbulence. Given a time series $\{\xi_i\}$ and a positive number $\varepsilon > 0$ (arbitrarily small), the mean square displacement in relative diffusion $R_r^2(n)$ is defined as

$$R_r^2(n) = \frac{1}{\mathcal{N}(\varepsilon)} \sum_{i=1}^{N} \sum_{j=i+1}^{N} (\xi_{i+n} - \xi_{j+n}\eta(\varepsilon - |\xi_i - \xi_j|))^2 \ , \tag{8}$$

where $\eta(\varepsilon)$ is the step function, $\eta(x) = 1$ if $x \geq 0$, $\eta(x) = 0$ if $x < 0$, and

$$\mathcal{N}(\varepsilon) = \sum_{i=1}^{N} \sum_{j=i+1}^{N} \eta(\varepsilon - |\xi_i - \xi_j|) \ . \tag{9}$$

Physically, the quantity $R_r^2(n)$ represents the mean square deviation in time of two points of the time series initially lying ε-close to each other.

In the case of the chaotic time series $\{\xi_i\}$, relative diffusion give as direct estimate of the maximal Ljapunov exponent λ,

$$R_r^2(n) \simeq \varepsilon^2 \exp(2n\lambda) \ . \tag{10}$$

On the opposite edge, for pure white noise

$$R_r^2(n) = 2 \ , \qquad \forall n \neq 0. \tag{11}$$

These results can be extended to a general signal constituted by the superposition of a deterministic continuous signal $\{s_i\}$ and of a stochastic component $\{r_i\}$,

$$\xi_i = f_s s_i + f_n r_i \,. \tag{12}$$

In Eq. (12) we have assumed that $\{s_i\}$ and $\{r_i\}$ are normalized to unit variance and are uncorrelated to each other, so that $f_s = \sigma_s/[\sigma_s^2 + \sigma_n^2]^{1/2}$, $f_n = \sigma_n/[\sigma_s^2 + \sigma_n^2]^{1/2}$, where σ_s^2 and σ_n^2 are the variances of the deterministic and stochastic components. The parameters f_s^2, f_n^2, represent the fraction of the power spectrum associated with the deterministic and the random component, $f_s^2 + f_n^2 = 1$. In this case, the relative mean square displacement displays the following behaviour[25]

$$R_r^2(n) = 2f_s^2\psi(n) + 2f_n^2 + mf_n^2f_s^2 \tag{13}$$

where $0 \leq m \leq 2$ is a constant (in practice $m \simeq 0.2$) and $\psi(n)$ an oscillating function, $\max(\psi) \sim O(1)$, $\psi(1) \simeq \psi(0) \simeq 0$, which implies that the deterministic component is continuous in time and $2\lambda\Delta t << 1$. Therefore,

$$R_r^2(1) = \lim_{n\to 0+} R_r^2(n) \simeq f_n^2(2 + mf_s^2) == f_n^2(2 + m(1 - f_n^2)) = \alpha \,, \tag{14}$$

and from this equation it is possible to estimate the fraction of the power spectrum associated with stochastic oscillations. Eq. (14) can be used both for the identification of the fluid dynamic regime and for filtering purposes.

4. On the choice of the experimental time series

Most of the analysis developed by our group[17,18,21,25] refers to experimental observations of density fluctuations measured by means of optical methods (γ-densitometric measurements).

Examination of to the literature shows that the overwhelming majority of applications of fluctuation analysis methods consider pressure-drop signals[1,3,8,5,6,7,8,9]. It is thus worth discussing this choice briefly.

Density signals measured by means of a γ-densitometer are *local* signals representing the pointwise fluid density averaged across the section of the tube. Pressure-drop measurements are intrinsically *non-local*, in that they are measured across a given length L of the tube and therefore represent volume-averaged quantities. The length L introduces a level of arbitrariness, which may make flow-regime identification more complex. In principle, the length L should be as short as possible in order to measure local pressure gradients. On the other hand, it can be also argued that the main statistical features of a given fluid dynamic regime should be independent of the choice of the observable analyzed (density, pressure drops) as long as the observable is representative of the flow conditions and the measurement is accurate. For example, figures 3 and 4 show the temporal behaviour of $R_z^2(n)$ evaluated on the basis of density and pressure-drop measurement in the two limiting cases of Stratified and Intermittent flow. As can be observed from these figures, the results obtained are

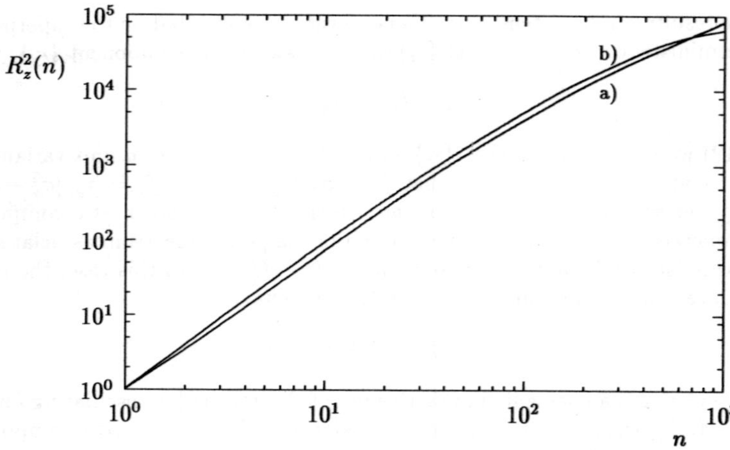

Figure 3: Mean square displacement $R_z^2(n)$ in Stratified flow: $j_g = 2.52$ [m/s], $j_l = 0.10$ [m/s]. a) γ-densitometric signal; b) pressure-drop signal.

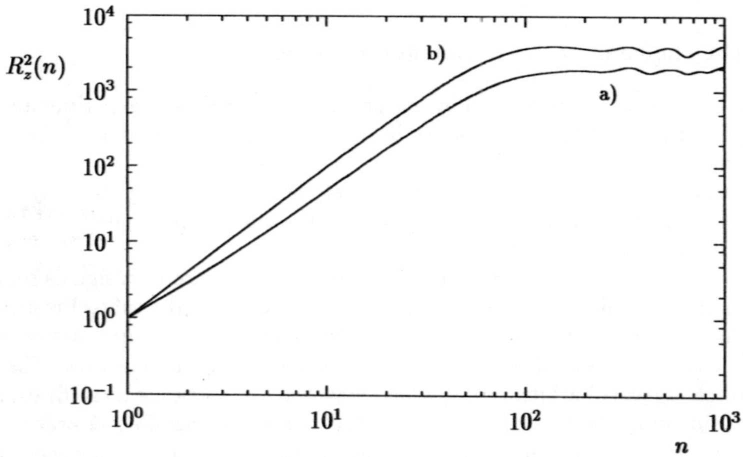

Figure 4: Mean square displacement $R_z^2(n)$ in Intermittent flow; $j_g = 2.07$ [m/s], $j_l = 2.07$ [m/s]. a) γ-densitometric signal; b) pressure-drop signal.

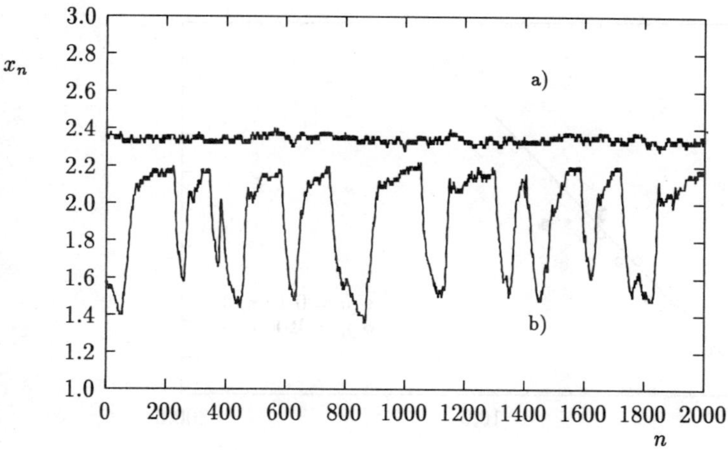

Figure 5: Non-normalized pressure-drop signals $\{x_n\}$ vs n: a) Bubble flow $j_g = 0.55$ [m/s], $j_l = 2.89$ [m/s]; b) Intermittent flow, $j_g = 2.07$ [m/s], $j_l = 2.07$ [m/s].

practically identical. The main problem with the diffusion analysis of pressure-drop signals is related to Bubble flow. Figure 5 shows the experimental time-series $\{x_i\}$, without normalization, of pressure-drop fluctuations in Bubble and Intermittent flow. In Bubble flow, pressure drops are almost constant (this is due to the facts that volume averages cancel out the fluctuations) and therefore the slight oscillation in the pressure signal derives mostly from experimental noise.

For this reason, we prefer to consider density measurements, although it has been experimentally proved that diffusion analysis can be applied in both cases.

5. Fluctuation intensity and filtering of experimental signals

Relative diffusion analysis provides a simple way to identify the intensity of stochastic fluctuations present in a given time series. This information can be used both to characterize the signal and to filter out the stochastic component.

5.1. Fluctuation intensity

By applying relative diffusion analysis making use of Eq. (14) it is possible to obtain an estimate of the corresponding intensity of stochastic fluctuations present in a given time series. Figures 6 shows the behaviour of α vs the fluid dynamic conditions $j_T = j_l + j_g$ for relatively small liquid velocities $j_l \leq 1$ [m/s]. The quantity α is a monotonically increasing function of the fraction f_n^2 of stochastic fluctuations present in the time series. In this region of operating conditions, the fluctuation intensity is monotonically increasing with j_T (i.e. with the Froude number). This behaviour

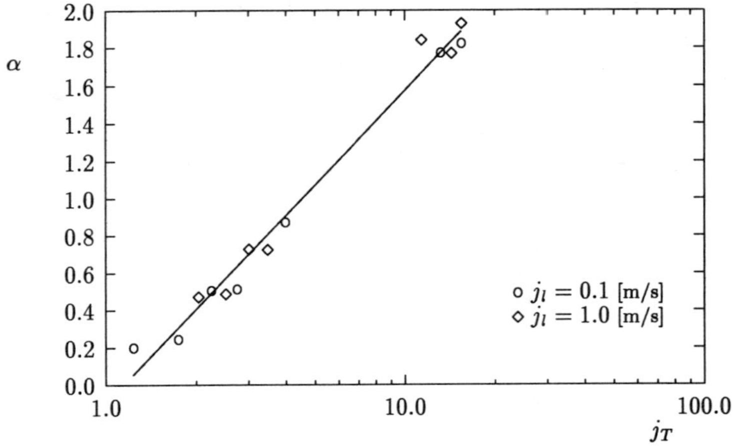

Figure 6: α (see Eq. (14)) vs $j_T = j_g + j_l$ at low liquid velocity.

encompasses the region in the Mandhane map associated with the transition from Stratified flow to Wavy flow and from Wavy flow to Annular flow.

Although there is a continuous increase of α with j_T, the result expressed by figure 6 indicates that it is possible to recognize annular flow from wavy motion by looking at the level of stochastic fluctuations present in the density signal.

The behaviour of α vs j_T or vs j_g is no longer monotonic in the slug region as depicted in figure 7. In particular, there exists a value of the superficial gas velocity (or rather an interval of superficial gas velocities) in which the intensity of fluctuations attains a local minimum. This region corresponds quantitatively to the zone in which the exponent β_1 associated with the scaling of $R_z^2(n)$ attains its maximum value, and the exponent β_2 its minimum, as discussed by Paglianti and Andreussi[21]. In this region, slug oscillations have the maximal degree of coherency. The presence of this *coherency zone* is associated with the existence of two subregimes within the Slug flow, as proposed by Paglianti and Andreussi[21].

5.2. Filtering of fluid dynamic data

Another interesting aspect related to the estimate of fluctuation intensity is the possibility of filtering out the stochastic component (which can be associated with intrinsic fluctuations or with experimental noise) from the deterministic part of the signal. This can be done by applying the method suggested by Sauer[26] or by applying Wiener filters and low-pass filters starting from the knowlegde of f_n. The filtering of experimental fluid dynamic signals has some important practical perspectives. It can be used as a preprocessing of the signal in order to apply chaotic methods of analysis to the deterministic component or, more practically, it can be coupled to

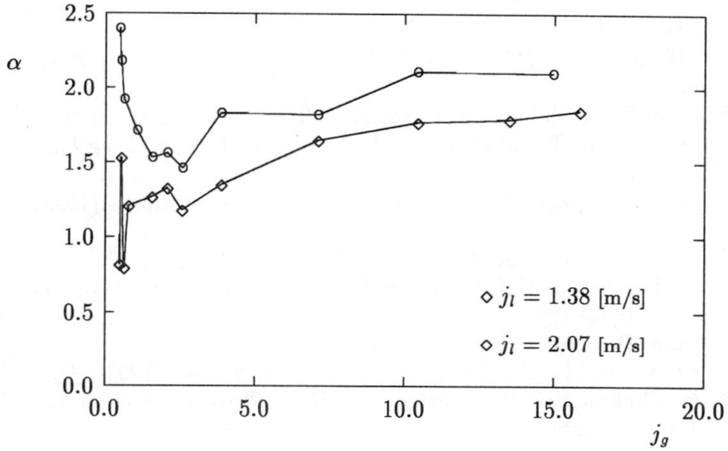

Figure 7: α vs j_g in Intermittent flow.

experimental measuring devices in order to eliminate noise levels. The development of this filtering method is discussed by Giona et al.[25].

6. Concluding remarks

In this article we have analyzed the various methods related to diffusional analysis in all the aspects that are useful in the study of multiphase flow.

In particular, we have extended relative diffusion analysis to estimate the intensity of stocastic fluctuations. The non-monotonic behaviour of the fluctuation intensity in Slug flow confirms the hypothesis of the presence of two distinct subregimes with different coherency properties. We have also briefly outlined the possibility of filtering fluid dynamic signals. The development of this method could have a wide range of application in all the chemical processes involving estimation and the on-line control of fluid dynamic conditions within the process units.

Acknowledgments

The authors are grateful to A. Grisanti for useful discussions and for the joint research developed in connection with the reconstruction and filtering of noisy chaotic time series.

References

1. C. Stuart Daw, W. F. Lawkins, D. J. Downing and N. E. Clapp Jr., *Phys.*

Rev. **A 41** (1990) 1179.
2. G. Sæther, K. Bendiksen, J. Müller and E. Frøland, *Int.J. Multiphase Flow*, **16** (1990) 1106.
3. L.T. Fan, D. Neogi, M. Yashima and R. Nassar, *AIChE J.* **36** (1990) 1529.
4. F. Franca, R. Acikgoz, R. T. Lahey and A. Clausse, *Int. J. Multiphase Flow* **17** (1991) 545.
5. D. P. Skrycke, K. Nguyen and C. S. Daw, *Fluidized Bed Combustion* **1** (1993) 155.
6. C. M. van den Bleek and J. C. Schouten, *Chem. Eng. J.* **53** (1993) 75.
7. M. L. M. van der Stappen, J. C. Schouten and C. M. van den Bleek, *Fluidized Bed Combustion* **1** (1993) 129.
8. L.T. Fan, Y. Kang, D. Neogi and Yashima, *AIChE J.* **39** (1993) 513.
9. J. Drahos, F. Bradka and M. Puncochar, *Chem. Engng. Sci.* **47** (1993) 4069.
10. M. Bernicot, H. Dhulesia and P. Deheuvels, in *Multi-Phase Production*, ed. A.P. Burns (Elsevier, London, 1993).
11. P. Grassberger and I. Procaccia, *Physica* **D 9** (1983) 189.
12. P. Grassberger and I. Procaccia, *Phys. Rev.* **A 28** (1983) 2591.
13. J. Feder, *Fractals* (Plenum Press, New York, 1988).
14. B. B. Mandelbrot and J. W. Van Ness, *SIAM Rev.* **10** (1968) 422.
15. B. B. Mandelbrot and J. R. Wallis, *Water Resour. Res.* **5** (1969) 228.
16. M. B. Kennel and S. Isabelle, *Phys. Rev. A* **46** (1992) 3111.
17. M. Giona, A. Paglianti and A. Soldati, *Fractals* (1994) (in press).
18. M. Giona, A. Paglianti and A. Soldati, *Fractals* **2** (1994) 265.
19. N. Wax, (Ed.) *Noise and Stochastic Processes* (Dover, New York, 1954).
20. A. Lasota and M. C. Mackey, *Chaos, Fractals and Noise* (springer-Verlag, Berlin, 1994).
21. A. Paglianti and P. Andreussi, this book.
22. H. Fujisaka, in *Chaos and Statistical Methods*, ed. Y. Kuramoto (Springer-Verlag, Berlin, 1984) 117.
23. M. Suzuki, *Physica* **A 117** (1983) 103.
24. M. Suzuki, in *Chaos and Statistical Methods*, ed. Y. Kuramoto (Springer-Verlag, Berlin, 1984) 260.
25. M. Giona, A. Grisanti, A. Paglianti and A. Soldati (unpublished results).
26. T. Sauer, *Physica* **D 58** (1992) 193.

Chaos in Chemical Process Units

Albeit intellectually stimulating, deterministic chaos was regarded until a few years ago for more as a "divertissement", than as a subject the knowledge or the mastery of which, could be considered a necessary or useful part of an engineer's training or skills.

Now, however, the pace of developments has accelerated sharply with the realization that what had been considered as an exception seems to be the rule and that disordered systems play an important role in both natural and applied sciences. The result is that all engineers, and chemical engineers in particular, must develop a new attitude and develop the capacity to deal with all the pathological effects of nonlinear dynamic systems, from the multiplicity of steady states to permanent oscillations and the onset of chaos.

As a certain amount of valuable research work has been carried out in this field over the last few years in Italy, we attempted to gather together almost all the researchers working in the chemical engineering area and provide them with a forum where research lines and results could be compared and opportunities provided to enhance interaction and cooperation.

The selected papers published here mainly cover the chaotic behaviour of chemical reactors and complex fluid flow and, as a whole, provide a clear and sharply focused illustration of the objects and level of research work carried out in Italy.

DETERMINISTIC CHAOS IN CHEMICAL ENGINEERING: STATE OF THE ART AND TRENDS

LAURA PELLEGRINI, GIUSEPPE BIARDI
Dipartimento di Chimica Industriale e Ingegneria Chimica "G. Natta"
Politecnico di Milano - P.le Leonardo da Vinci, 32
Milano, 20133, Italy

and

CRISTINA TABLINO POSSIO
Dipartimento di Matematica "F. Enriques" Università di Milano - via Saldini, 50
Milano, 20133, Italy

ABSTRACT

The paper reviews the most significant works on deterministic chaos in Chemical Engineering. The majority of the papers in this field concerns chemical reacting systems; among them a further subdivision can be made between systems where chaos is induced by the configuration of the reacting systems (an apart mention must be devoted to the systems under feedback control) and systems where the reaction scheme itself gives rise to chaos (the Belousov Zhabotinskii reaction is a typical example). Another fundamental distinction is between dynamical systems whose model is represented by Ordinary Differential Equations and systems that can be modelled by means of Partial Differential Equations, i.e. infinite dimensional systems. The basis of chaos characterisation and bifurcation analysis are then illustrated. In the end, the authors enlight the most promising applications of chaos theory.

1. Introduction

It is well known that nonlinearity is responsible for a variety of interesting patterns of behaviour.
With his work of 1879 "Sur les courbes descrites par une equation differentielle" Poincaré was the first to recognise that nonlinear systems may exhibit instability and multiplicity problems.
In the field of Chemical Engineering we must wait till the 1950's with the pioneering papers by Aris and Amundson (1958)[6] to find significative studies on nonlinear dynamics, though we can cite an abundance of literature on steady state multiplicity, instability and limit cycles, ranging from the 1950's to the early 1980's.
Chaos is perhaps one of the most interesting and stimulating aspects of nonlinear dynamics and its "discovery" in chemical engineering goes back to the early 80s.
In fact, if works on fluid dynamics are considered apart, the first example of a strange attractor is that attributed to Jorgensen and Aris (1984)[63] for two exothermic consecutive reactions in a CSTR.
Chemical systems coupling transfer processes with complex reactions can exhibit a

great variety of complex dynamic behaviour, among which, is chaos.
The chemical reactors, in particular, are certain to be a favourite subject for research studies.
Besides the fundamental books in the strictly mechanistic field (Guckenheimer and Holmes (1983)[40], Bergé (1984)[15]), which constitutes, however, the basis for nonlinear dynamic studies in any field, specific books of interest for chemical engineers have recently been published (Marek and Schreiber (1991)[78], Scott (1993)[123]).
Apart from these, a number of more general books, that have an applicative scope, can be found (Applied Chaos (1992)[72], XAOC (1990)[17], etc.); many of them are the proceedings of congresses on chaos, that are now beginning to be held everywhere in the world.
Before writing anything on chaos it is "a must" to ask the following question that is not merely rhetorical:
What is the use?
Often researchers that deal with chaos are in charge of only one academic exercise.
The first spontaneous answer is that one cannot and must not discuss what one doesn't know and this field still presents a lot of unknown factors.
Moreover, many examples where researchers may significantly influence the industrial reality can be cited (Villermaux (1993)[143]), but we will will come back to this topic in the paragraph dealing with it.
The main point is to be able to evaluate what one is studying critically and to foresee the possible applications even if in a long term perspective.
In the following sections, after a brief review of chaos literature for topics in chemical process engineering, a classification according to the type of mathematical models and the numerical methods employed to solve them is reported.
Then, the main lines of the Floquet theory and of the more general bifurcation analysis are expected to be the basic tools in chaos studies: an outline of future developments in their application is given.
At the end, trends on chaos science in chemical engineering are dealt with.

2. Examples of chaos in Chemical Engineering

Most of the cases of chaotic behaviour in the chemical process industry are relevant to chemically reacting systems. So, we dwell particularly upon these examples, that can generally be grouped into the following major classes:

1. Generic chemical systems where chaos does not arise for a specific chemistry but for the type of arrangement of the equipment itself. In this category we can include chaos examples in non reacting systems, provided that their nonlinearity is sufficient, when coupled to another outbreaking mechanism, to start strange patterns of behaviour (Boe and Chang (1989)[16]);

2. The Belousov-Zhabotinskii reaction, which has been perhaps the most studied

subject both from theoretical and experimental viewpoints;

3. Heterogeneous catalytic oxidation reactions. Most of the literature deals with CO and ammonia oxidations;
4. Gas phase reactions, including thermal explosions;
5. Electrochemical reactions, mainly iron and nickel electrodissolution;
6. Biochemical reactions, where oscillatory patterns of behaviour are very common;
7. Polymerisation reactors.

A through review of literature involving each of the previous points would take too space and would go beyond the scope of this paper. We will simply make a selection to indicate those papers that, in our opinion, deserve attention.

2.1. Generic chemical systems

This category groups examples of chaotic behaviours belonging to different areas. We can, however, recognise two fundamental lines: studies on "forced systems" and studies on "controlled systems".

Forcing a chemical system is based on the idea of improving the steady state process operation.

When forced oscillations are applied to a system of coupled ordinary differential equations (Mc Karnin et al. (1988)[82]), the response can be periodic with either the same period as the forcing or an integer multiple of the forcing period. The response can also be quasi-periodic or chaotic, and more than one kind of response is possible, depending on the initial conditions. When low amplitude forced oscillations are introduced into a naturally oscillating system there are regions of subharmonic resonance (synchronisation) at forcing frequencies which are rational multiples of the natural frequency of the system. Outside of these regions, the behaviour is quasi-periodic (Kuramoto (1984)[75]). With large amplitude forcing the perturbing influence is prevailing and the original nonlinear behaviour takes a back seat. For intermediate amplitudes the interaction between the original and imposed time-dependencies differs quantitatively from system to system (but many universal qualitative features can be identified).

The earliest studies on forced systems followed the Brusselator model for behaviour (Tomita and Kai (1979)[136], Kai and Tomita (1979)[64]) by varying the concentration of a feed species. Additional works on this matter are by Aronson et al. (1986)[7], Schreiber et al. (1988)[120].

Mankin and Hudson (1984)[77], Kevrekidis et al. (1984[67], 1986[68 69 70]), Caneba and Crossey (1987)[18], considered a single irreversible exothermic reaction in a CSTR with the coolant temperature as forcing.

Analogous studies have been performed on forced biological systems (Guevara et al.

(1981)[41], Guevara and Glass (1982)[42], Tomita and Daido (1980)[137]).
Controlled systems play an important role in a review on chaos.
From one standpoint, we can see the controller as an external forcing that compels the system to operate, in some cases, rather far from its open loop conditions. If the constraints imposed by the control action are too rigid, the system runs away from the set point but without being free to fall again on the open loop condition (that, however, is no longer the singular point of the closed loop model); so chaotic behaviour can arise.
Chang and Chen (1984)[19] codify the influence of each type of feedback control mode on the dynamic behaviour of nonlinear systems; the possibility of control induced chaos is theoretically proved.
The chaos occurrence in a PI controlled CSTR with a single exothermic reaction is analysed by Pellegrini and Biardi (1990)[100]; similar results have been obtained by Pellegrini et al. (1992[101], 1993[102]), Tablino Possio and Pellegrini (1994)[131] in the analysis of a feedback controlled tubular reactor.
A separate case is represented by Boe and Chang (1989)[16] that consider a classical liquid level control system where the inlet flow rate into the tank is the manipulated variable. In this case, as in the other few examples we can find in literature (Inamdar et al. (1991)[55], Pellegrini et al. (1994)[103]) chaos is induced by the presence of a time delay.

2.2. The Belousov-Zhabotinskii reaction

The Belousov-Zhabotinskii reaction is known for its long train of oscillations in closed vessels. The simple model called "Oregonator" has proven successful at reproducing and predicting the observed patterns of behaviour in a wide range of experimental situations. Since 1980's closed vessels (batch reactors) have been replaced by open systems (flow reactors) for studies on BZ reaction.
The first observation of chaos in BZ reaction is attributed to Schmitz et al. (1977)[119] and since then many research groups have devoted themselves to the analysis of BZ dynamics.
Hudson and co-workers (Hudson et al. (1979[46], 1981[47]), Hudson and Mankin (1981)[48], Baier et al. (1989)[8]) studied the chaotic behaviour of the system as well as the Texas group (Turner et al. (1981)[138], Roux and Swinney (1981)[109], Simoyi et al. (1982)[126], Roux et al. (1982[110], 1983[112]), Swinney and Roux (1984)[129]). The differences between these and the Hudson studies are in the range of residence times considered; the Texas work concentrated on long residence time analysis.
Other groups of experiments have been carried out by the Bordeaux school, starting from the pioneering studies by De Kepper et al. (1976)[24] up to those by Vidal et al. (1979[140], 1982[141]), Roux et al. (1980)[108], Roux (1983)[111], Roux and Rossi (1984)[113].
Other investigations were carried out by Maselko and Swinney (Maselko (1980)[79 80], Swinney and Maselko (1985)[130], Maselko and Swinney (1986)[81]).

2.3. Heterogeneous catalytic oxidation reactions

Heterogeneous catalytic systems have been widely studied in nonlinear dynamics. The oxidation of CO, either by O_2 or NO, has been studied from ultra-high-vacuum condition (10^{-8} torr) to atmospheric pressure.
There are two reasons for the success of this reaction: its kinetics seems particularly simple because there is only one overall reaction and because of the few possible intermediate absorbed states; moreover, the atmospheric reaction can be applied to catalytic converters in car exhausts.
Ertl and co-workers (see, among others, Ertl (1980[30], 1990[31]), Imbihl et al. (1985[51], 1986[52], 1988[53] [54]), Eiswirth et al. (1985[27], 1988[28], 1989[29])) studied the catalysed $CO + O_2$ reaction under ultra-high-vacuum conditions on single Pt crystals.
The $CO+O_2$ reaction over Pt at atmospheric pressure has been intensively studied by different researchers in various ways; we remember the papers by Onken and Wicke (1986)[93], Wicke and Onken (1986[144], 1988[145]), Schuth and Wicke (1989)[121] [122].
For studies on the chaotic patterns of behaviour during ammonia oxidation you can refer to the study by Sheintuch and Schmidt (1988)[125].

2.4. Gas-phase reactions

The gas-phase systems may exhibit significant departures from isothermal operation and temperature plays the role of "autocatalyst".
Since in gas-phase reactions real chemical "autocatalysts" are often present especially in oxidation processes involving oxygen, the nonlinear response of such systems may be extremely complex.
However, compared with heterogeneous catalysis reactions, this topic has received much less attention.
Most of the papers are relevant to bifurcation studies rather than to chaotic behaviour analysis. For a detailed survey refer, among others, to Gray and Scott (1985[37], 1990[38]) and Griffiths and Scott (1987)[39].

2.5. Electrochemical reactions

Anodic electrochemical reactions often exhibit irregular oscillations. Many of these electrochemical studies involve the electrodissolution of metals. Oscillatory and chaotic behaviour has been observed by several investigators during the corrosion of iron in acid solution (Bartlett (1945)[9], Franck and FitzHugh (1961)[34], Rius and Lizarbe (1962)[107], Podesta et al. (1979)[105], Russel and Newman (1986)[114], Orazem and Miller (1986)[94], Diem and Hudson (1987)[25]).
Copper will also undergo current oscillations under potentiostatic conditions in acid medium (Cooper et al. (1980)[22], Kawczynski et al. (1984)[65], Lee et al. (1985)[76], Bas-

set and Hudson (1987[10], 1988[11], 1989[12 13], 1990[14]), Albahadily et al. (1989)[4], Schell and Albahadily (1989)[115] as well as cobalt (Franck and Meunier (1953)[35], Jaeger et al. (1985)[60], Hudson et al. (1989)[49]).
The precision with which electrical voltages and currents can be controlled and measured makes electrochemical reactions the ideal field to experiment and prove nonlinear principles.

2.6. Biochemical reactions

Oscillatory patterns of behaviour are quite common in biochemical systems (heart beat and respiration, population and epidemic periodic fluctuations, etc.)
Ghosh and Chance (1964)[36], for the first time, observed oscillations in biochemical systems during their experiments on glycolysis. Subsequent studies on glycolysis were carried out by Herzel et al. (1987)[45] and Tomita and Daido (1980)[137] to name a few, while studies on enzymatic oxidase-peroxidase reactions were done mainly by Olsen (1977[89], 1979[90], 1983[91], 1985[92]).
Another class of papers on biochemical systems concerns the analysis of cycling input conditions on their dynamic behaviour (Abulesz and Lyberatos (1987[1], 1989[2]), Parulekar and Wei (1988)[97], Pedersen et al. (1987)[99], Stephens and Lyberatos (1988)[127], Schmidt et al. (1989)[118], Parulekar and Lee (1993)[98]).

2.7. Polymerisation reactors

The dynamic behaviour of solution polymerisation reactors has been the main point of a series of investigations by Ray and co-workers (Jaisinghani and Ray (1977)[61], Schmidt and Ray (1981)[116], Hamer et al. (1981)[43], Schmidt et al. (1984)[117], Teymour and Ray (1989[132], 1991[133], 1992[134 135]).
In their most recent studies Teymour and Ray have analysed the behaviour of CSTR for the free radical polymerisation of vinyl acetate, discovering the emergence of chaos via period-doubling and hysteresis.

3. Mathematical models and numerical methods

In the previous section, dynamic systems exhibiting chaotic patterns of behaviour in the field of chemical engineering have been reported, following a chemical classification.
Now, we review such a classification, or more precisely we build up a parallel one, by means of a different key, i. e. the mathematical model representing these systems, the aim being to emphasise their analytical structures and the numerical methods that have been successfully applied in order to perform their investigation.
Most quoted chemical systems can be modelled by means of a few autonomous ordi-

nary differential equations.
Naturally, the characteristic of low dimension is a favourable circumstance for the corresponding analysis: the geometrical features of the strange attractor, like its structure or the stretching and folding of trajectories around it, can be properly appreciated.
Preliminary studies of the equilibrium stability or the determination of the Hopf locus may, at least in some cases, be performed analytically: the extensive analysis performed by Uppal et al. (1974)[139] about the dynamic behaviour of a CSTR is rightly famous.
In addition, the computational cost related to numerical simulation is not so expensive owing to the few equations to be handled.
In general, explicit integration schemes can be successfully applied, though in some cases the stiff nature of the system may require the use of implicit (or semimplicit) integration schemes, properly formulated to take into account the characteristic eigenvalues spectrum of the phenomenon under study.
In any case, special attention must be given to the choice of the numerical method adopted for the simulation of the system, since some dynamic features are quite delicate and it is essential to avoid that the algorithm itself introduces more complicated and less ordered structures.
In particular, such sensitivity is more amplified near bifurcation conditions, so that careful schemes like the Villadsen-Michelsen one (1978)[142] or the Fatunla one (1980)[33] are to be preferred.
Another important class of dynamic systems is given by the external T−periodic forced ones modelled by means of systems of non autonomous differential equations

$$\frac{dy}{dt} = F(y,t) \quad y \in R^n,\ t \in R, \tag{1}$$

$$F(y,t+T) = F(y,t) \quad T > 0 \tag{2}$$

Such a different intrinsic nature can be handled in the following way: because of the periodicity of the function $F(y,T)$ with respect to the time variable t, it is possible to rewrite the system in the form

$$\begin{cases} \dfrac{dy}{dt} & = F(y,s) \\ \\ \dfrac{ds}{dt} & = 1 \end{cases} \tag{3}$$

where the new dependent variable s is defined on a circle of perimeter equal to T (i.e. s is a real number modulo T). In such a way the phase space dimension is increased by one, but now the system is in an autonomous form.
This fact also explains why for a periodically forced system with only two degrees of freedom, chaotic patterns of behaviour are also possible. If no steady state exists, the possible nature of other related dynamic patterns of behaviour are the periodic and

the quasi-periodic ones; the frequency characteristic of the forcing is always present and the periodic trajectory will have periods that are integer multiples of the forcing period (Kevrekidis et al. (1986)[68 69]).

As largely outlined in the previous section, many examples of chaotic motions are reported in chemical engineering literature for lumped parameters systems. The same doesn't hold for distributed parameters systems.

First we have to mention the static and Hopf bifurcation analysis performed by Heinemann and Poore (1981)[44] and by Jensen and Ray (1982)[62] in the case of a first order irreversible reaction in a tubular reactor with axial diffusion.

For a more detailed discussion you can refer to the paragraph on bifurcation analysis.

Now, by considering the analysis that more properly can be attributed to the field of chaos characterisation, mention is to be made to Nandapurkar et al. (1986)[88]. The authors analysed a one dimensional reaction-diffusion system related to the Brusselator model ascribed to Prigogine and Lefever (1968)[106].

After calculating the Hopf bifurcation points analytically, the numerical simulations are performed by means of an high-order finite difference scheme and the Crank-Nicholson approach, so obtaining a fully discretised system.

These numerical simulations reveal two of the classical roads to chaos predicted by the Floquet theory: the period-doubling and the quasi-periodicity scenarios, apparently caused by diffusion.

Both of them are characterised by means of the computation of the quoted Floquet multipliers related to the ODEs system in which the original PDEs one can be converted by applying the method of lines. In the case of a toroidal nature of the attractor, this is also highlighted by showing the closed invariant curve obtained with the first return map of successive solution extrema.

The period-doubling scenario is also found in the case of a tubular reactor with axial diffusion under a feedback PI controller (Pellegrini et al. (1992)[101]).

Numerical investigations are accomplished by applying the method of lines and by integrating the resulting ODEs system by means of a semi-implicit Runge-Kutta method (Villadsen and Michelsen (1978)[142]) taking into account the stiff nature of the system to be handled.

Such period doubling degeneration to chaos is also present, together with hysteresis phenomena, when a countercurrent coolant is considered in the model (Pellegrini et al. (1993)[102]). Once more, the Floquet multipliers and power spectra can prove the universality features proper of such a road to chaos.

In addition, the peculiar phenomenon of a Type III intermittency transition to chaos, as far as we know the first example in chemical engineering, appears in this last model whose characterisation requires the use of suitable numerical and statistical methods in order to recognise the phenomenon from the theory (Tablino Possio and Pellegrini (1994)[131]).

We can really cite only the few forementioned papers as PDEs models if we don't consider all the studies on fluid dynamics as peculiar of chemical engineering (e.g. Knobloch et al. (1986)[73], Moon et al. (1982[86], 1983[87]), Hyman et al. (1986)[50])

Lastly, we can consider those systems that take into account the presence of a time delay, due to the system configuration itself or due to measurement and actuator lags. Few papers exist on the analysis of the possible destabilising effect of time delay for chemical reacting systems (Inamdar et al. (1991)[55], Boe and Chang (1989)[16], Pellegrini et al. (1994)[103]).

A system with time delay is infinite dimensional and its simulation usually requires a remarkable numerical and computational effort, as in the case of each infinite dimensional system quoted above. In addition, such a system may be intrinsically "stiff", owing to the different magnitude of the intrinsic time constant of the system and the time delay.

There are two ways of approaching the problem; the first, followed by Inamdar et al. to study a CSTR with delayed recycle, determines the value of the dependent variables at the retarded time using an interpolation/extrapolation algorithm.

The second way, followed by Boe and Chang to study the dynamic behaviour of delayed systems under a feedback controller, consists of substituting the delay operator with a plug flow system (hyperbolic PDE) that represents the physical device responsible for the delay.

The reduction to an ODEs system can be obtained by means of spatial subdivisions that are, in some way, the equivalent of the knots in the interpolation algorithm.

Boe and Chang recognise a class of resonant systems for which there exist critical values of the delay time corresponding to Hopf bifurcation points with two distinct pairs of purely imaginary eigenvalues. The unfoldings of such singularities are analysed by means of a center manifold projection and normal form analysis; in fact, in their neighbourhood limit cycles and invariant tori are usually found. The authors show, in the case of cascaded liquid level control systems, the presence of a torus and its probable degeneration into a chaotic attractor.

Pellegrini et al., after testing both techniques without significative changes, have analysed some representative examples of retarded chemical reacting systems, with the aim of pointing out the occurrence of possible destabilisations, for physically reasonable values of time delay τ_D: numerical computations have shown cases where high τ_D values can stabilise a limit cycle on a steady state or make a strange attractor periodic. On the contrary, in other cases, a τ_D increase may destabilise a steady state to a periodic orbit or a periodic orbit on chaotic behaviour.

Once more, it must be outlined that the computational effort spent to analyse distributed parameters systems is not negligible respect to the corresponding one required in the case of low dimensional lumped parameters models.

First, the convergence of the finite dimensional system to the infinite dimensional one must be assured in the numerical simulation.

In addition, the analysis may become a ponderous task in the case of the application of the Floquet theory (see Bergè et al. (1984)[15]), whose importance will be explained in the next section. Such an application requires the integration, during one period, of the complete linearised system, together with the nonlinear original one, i.e.

$$\begin{cases} \dfrac{dy}{dt} & = F(y) \quad y \in R^n \\ \\ \dfrac{dH}{dt} & = J \cdot H \end{cases} \tag{4}$$

where J is the Jacobian matrix evaluated along the orbit.
Thus the integration of $n(n+1)$ ODEs and, in addition, the computation of the eigenvalues of sensitivity matrix H are required.
Similarly, the computation of Lyapunov characteristic exponents (integration of the same $n \cdot (n+1)$ ODEs, see Wolf et al. (1985)[146]), or more precisely of the largest one (only $n+n$ ODEs), results in a more expensive target owing to the requirement of reaching an appropriate approximation of the corresponding asymptotical value.
Besides its undoubted effectiveness as a tool for chaos detection, the largest Lyapunov characteristic exponent can be used to calculate the parametric sensitivity of batch reactors (Strozzi and Zaldivar (1994)[128]).
The advantage is the possibility that Lyapunov exponents can be computed from measured experimental data by reconstructing the phase space which should allow on-line sensitivity calculations in chemical plants in the future.

4. Bifurcation analysis

Because real processes are often described by models characterised by variable parameters, possible qualitative change in the asymptotical behaviour obtained by varying such parameters is of great interest in process modelling: the purpose being to reach, or to avoid, some specific invariant states showing prefixed properties.
In such a way, the application of the bifurcation theory constitutes a proper tool to obtain foresight concerning model dynamic behaviour. Such foresight can be gone into in more depth if higher codimension bifurcations are taken into account.
The simplest "qualitative changes" in dynamics occur in a close neighbourhood of equilibrium point or periodic solutions which are known as local bifurcations. On the contrary, global bifurcations include such transformations of the orbit structure which occur simultaneously at many different locations of the phase space.
Local bifurcations are usually classified according to increasing codimension, that is according to the increasing number of related conditions to be satisfied in order to induce structural instability and to produce the desired bifurcation as a generic case.
Let's consider a generic dynamic system

$$\frac{dy}{dt} = F(y,p) \quad y \in R^n,\ p \in R^m, \tag{5}$$

the equilibrium manifold Σ, defined by the algebraic condition $F(y,p) = 0$, may contain the submanifold of codimension one constituted by all the nonhyperbolic equilibrium points (i.e. almost one Jacobian eigenvalue showing zero real part). These

points can be subdivided into two different groups of generical ones and critical ones: the latter belong to a submanifold of codimension two on Σ, and so on.
Such a stratification in submanifolds of higher and higher codimension gives the so called bifurcation structure proper of the equilibrium manifold in consideration.
Consequently, we can start by considering codimension one bifurcation points:
- simple bifurcation points, for which a simple eigenvalue is zero, all the others are stable (i.e. the real parts are negative) and the zero eigenvalue crosses the imaginary axis transversally. The best known examples are the tangent (or saddle-node) bifurcation, for which two equilibria, one stable and the other unstable, collide and disappear and the transcritical bifurcation in which two different equilibrium branches intersect and a change of stability occurs.
- Hopf bifurcation points, for which there is a pair of purely imaginary eigenvalues, all the others are stable and the purely imaginary pair crosses the imaginary axis transversally. Such a bifurcation marks the transition from a steady state solution to an isolated periodic orbit, called limit cycle.
The temptation to extend the previous classification to the case of periodic orbits is very strong if we consider that the Poincaré section of a periodic orbit is a fixed point. However, via the Floquet theory, it is possible to discover how the analogy between an equilibrium point of a continuous dynamic system and the fixed point of a Poincarè section is not complete: among the bifurcations of an equilibrium point, no analogy exists with the loss of linear stability resulting from an eigenvalue equal to -1, i.e. for the period doubling cascade that represents a common observed case of degeneracy to chaos (Bergé (1984)[15]).
Moreover, the loss of stability owing to the crossing by two complex conjugate eigenvalues giving rise to an invariant torus exhibits some specific features.
The second level of possible stratification for the equilibrium manifold Σ is represented by codimension two bifurcations.
In fact, moving along the codimension one submanifold, some other eigenvalues may approach the imaginary axis and change the dimension of the center manifold (or some of the nondegenerancy conditions of the bifurcation may be violated) giving rise to higher order bifurcations:
- Bogdanov-Takens bifurcations, i.e. two zero eigenvalues (with geometric multiplicity one).
- Gavrilov-Guckenheimer bifurcation, i.e. one zero eigenvalue plus a pair of pure imaginary eigenvalues.
These bifurcation curves are common curves of the two surfaces corresponding to the tangent and Hopf singularities respectively. However they are the first two examples of the more general case of degenerate Hopf bifurcation of type I, i.e. more than one zero eigenvalue or more than one pair of purely imaginary eigenvalues.
- Cusp bifurcation, i.e. zero eigenvalue with normal form second order coefficient vanishing. Three equilibrium points exist inside the cusp region, two stable and one unstable; only one equilibrium point exists outside.
- Degenerate Hopf bifurcation of type III, i.e. Hopf bifurcation with zero first Lyapunov value. Such a value provides information about the direction of limit cycle

appearance and its stability: its vanishing describes the change of supercritical bifurcation into a subcritical one or vice versa.
The particular case given by the degenerate Hopf bifurcation of type II involves a degenerancy in the Taylor series expansion of the real part of critical eigenvalues. This type of bifurcation means that the system is non generic respect to the fixed parameters: another parameter must be added to the free ones, without a corresponding increase of the codimension order.
Now, among codimension two bifurcation of a periodic orbit, we can cite the cases of simple intersection between the curves corresponding to two different codimension one bifurcations (critical Floquet multipliers equal to $+1$, $\exp(\pm i\omega)$ or to -1, $\exp(\pm i\omega)$ or to -1 and $+1$) or the case of degeneration of these bifurcations related to the cubic order term in the relevant normal form.
More interesting cases are the strong resonance bifurcations, among which we can quote the strong resonance $1:1$ (two critical Floquet multipliers equal to $+1$) and the strong resonance $1:2$ (two critical Floquet multipliers equal to -1).
According to the dimension of the dynamic system under study and its complexity, only in a few cases can a quite complete bifurcation analysis be accomplished analytically.
So, for the numerical analysis of these bifurcations we are confronted with the following tasks:
- to detect a bifurcation point while following a branch of invariant sets;
- to accurately locate the bifurcation point by a defining equation;
- to create a good initial approximation for starting a branch of the new invariant sets.

A basic tool is, of course, the continuation technique adopted to perform the parametric analysis (see for example the pseudo-arclength continuation technique (Kubicek (1976)[74]).
In any case, the theoretical basis remains the continuation strategy, related primarily to the graph of adjacency of singularities (Afrajmovich et al. (1994)[3]) that, giving a hierarchical structure, allows it to run across also complex stratifications.
In recent years, several software packages have been developed to study the behaviour of dynamic systems, some of them primarily devoted to perform the bifurcation analysis (e.g. AUTO (1986)[26]; LINLBF (1990)[71]).
The effective power of bifurcation analysis is not only confined to the faculty of foreseing and of analysing "regular" patterns of behaviour of dynamic systems. In fact, degenerate Hopf bifurcations reveal themselves to be in general the source of interesting dynamic patterns of behaviour as the presence of multiple limit cycles, the transition to invariant tori or the appearance of a period doubling cascade.
In this sense, the bifurcation theory is no longer only one of the preliminary tools for the study of deterministic chaos.
Some works about degenerate Hopf bifurcations exist in chemical engineering literature related to lumped parameters systems and approach the analysis by different ways (Farr and Aris (1987)[32], Planeaux and Jensen (1986)[104], Moiola et al. (1990[83] [84], 1991[85]).

Farr and Aris systematically deal with degenerate Hopf bifurcations in a CSTR with a consecutive reaction $A \rightarrow B \rightarrow C$ by using the techniques proper of the singularity theory and giving some global diagrams in the neighbourhood of some degenerate bifurcations. Such techniques are shown very powerful but the global amount of the involved algebra is not negligible.

Planeaux and Jensen analysed the dynamic behaviour of a CSTR with extraneous thermal capacitance by applying the normal form theory (see Guckenheimer and Holmes (1983)[40]). Such a method consists of reducing the system at the singular point into a low dimension polynomial expansion that leaves its essential dynamic features unchanged.

The authors identified bifurcation curves in the parameter space and described the possible dynamic behaviour in the unfoldings of these singularities like bifurcations of periodic solutions to invariant tori, isolas of periodic branches, bifurcation diagrams with four Hopf points and multiple stable periodic orbits.

Also in this case, the global amount of analytical calculations is far from negligible for general engineering systems.

Lastly, Moiola et al. translated defining conditions of some static and dynamic bifurcations to the frequency domain counterpart by transforming the ODEs system into an equivalent feedback formulation and by doing a second-order harmonic balance. Their analysis recognises in the quoted models degenerate Hopf bifurcations of codimension two and three.

On the other hand, an application of static and Hopf bifurcation theory in the case of distributed parameters systems can be found in the above quoted papers by Heinemann and Poore (1981)[44] and by Jensen and Ray (1982)[62].

Heinemann and Poore deal with numerical bifurcation techniques that can be developed to investigate the multiplicity and stability of steady states and the oscillatory dynamics of tubular reactors.

These techniques, that can be successfully used for systems of parabolic partial differential equations, consist of the application of both collocation and finite difference methods coupled to Keller's modification of the Euler-Newton continuation method (Keller (1977)[66]) to solve the steady state problem: the Hopf points are detected by using a QR algorithm to compute the Jacobian eigenvalue at each point of the steady state branch and then a bisection or secant method to accurately locate the critical value of the bifurcation parameter. The subsequent stability analysis of periodic oscillations is performed by means of the Floquet theory application.

On the same subject, Jensen and Ray give a detailed analysis of different collocation methods for the solution of the steady state problem, showing how the highest accuracy is reached with the orthogonal spline collocation method. The eigenvalues of the linearised operator are calculated by both a collocation and Galerkin procedure. Moreover, different direct or indirect strategies to compute bifurcation points are analysed in depth.

This application of the static and Hopf bifurcation theory for distributed parameter systems allows the authors to point out the influence of the Péclet and Lewis numbers on multiplicity and oscillatory behaviour: the former introduces an additional set of

Hopf and static bifurcation points with respect to the two asymptotical situations given by CSTRs and plug-flow reactors, the latter reveals if periodic oscillations are possible or not in fixed bed reactors and for growing Péclet numbers these oscillations are possible only for short reactors, even in the case of empty ones.

5. Chaos: trends in Chemical Engineering

Current chaos theory has provided in many cases explanations for a variety of phenomena once considered too complex to admit simple mechanistic description. At the same time it has set bounds on the expectations on the kind of things that can and cannot be predicted: now, we know for instance that weather prediction is unattainable.

However, we should also recognise that many of the uses of chaos, no matter how elegant and intellectually satisfying they might be, are just explanations of the fact. So, the question is posed again as to whether chaos studies are just an academic exercise with no practical purposes. Some points go against this thesis.

Chaos theory provides a clean explanation for the complexity of some results. In many cases we might even be able to use theory to anticipate how things will change if parameters are changed. This is the reason for the abundance of studies involving the bifurcation theory, aiming at identifying safety operation zones in the parameter space.

This request, i.e. to work outside regions characterised by periodic or chaotic patterns of behaviour, sometimes doesn't correspond to economic optimum conditions for the chemical processes. This is proved in many examples (Seider et al. (1990)[124]): exothermic reactions (allyl chloride process, benzene to maleic anhydride, propylene oxide to propylene glycol), isothermal reactors (aerobic fermentations, emulsion polymerisation reactions, ethylene hydrogenation), thermally coupled distillation towers, heterogeneous azeotropic distillation towers, phase splitting for product recovery, supercritical extraction.

These examples support the contention that industrial practitioners often install conservative designs in part, to avoid difficult operating regimes. Some are sceptical that processes can operate more economically near or within these regimes and would improperly classify such process designs as pathological.

To improve the characterisation of these operating regimes, design engineers could apply the tools for nonlinear analysis such as algorithms for homotopy continuation, bifurcation analysis and dynamic simulation.

Another application where the chaos theory can be used to produce beneficial consequences is fluid mixing. The possibility to achieve laminar mixing of viscous fluids in a regime of "deterministic chaos" produced by intermittent change of initial conditions, as shown by Aref (1984)[5] with his famous simulation of two intermittent vortices is exciting. These ideas were extensively developed by Ottino (1992)[96].

Such chaotic processes constitute initial examples where mixing efficiency is not directly related to energy dissipation but results from a fundamental kinematic property.

This phenomenon might also play a significant role in static mixers. High efficiency mixing devices based on these entirely new principles might appear in the near future. It should be clear, however, that the use of the chaos theory in chemical engineering is not limited to the invention of new processes, but, as already discussed, mainly in the revision of the old ones.

Chaos might have been used implicitly in many applications and it is undoubtedly already present in many applications (for example the frequency of bubble detachment in a distillation plate, regimes in fluidized beds, interfacial waves in two phase flows , etc.).

It is also interesting to consider the idea that chaos gives inbuilt flexibility and makes it comparatively easy to effect changes in outputs by making slight alterations in operating conditions. The OGY method is based on this (Ott et al. (1990)[95]).

Namely, by choosing a suitable unstable low-periodic orbit embedded in a chaotic attractor, modification to the system's dynamic features can be made and stabilisation on such an orbit can be achieved.

Controlling chaotic nonlinear dynamic systems is a promising new field of research because of its possible engineering applications. Besides the quoted OGY method, some other successful techniques that can be found in literature are the so called entrainment and migration controls (Jackson (1990[56], 1991[57]), Jackson and Hubler (1990)[58], Jackson and Kodogeorgiou (1991)[59]) and linear and nonlinear feedback controls (Chui and Chen (1989[20], 1992[21]), deFigueiredo and Chen (1993)[23]).

We hope in these few pages to have made the chaos theory a less abstract concept for the reader, giving at the same time a significance to its study not only by theoricians but also, and mainly, by chemical engineering researchers.

References

1. E. M. Abulesz and G. Lyberatos, *Biotechnol. Bioengng.* **29** (1987) 1059.
2. E. M. Abulesz and G. Lyberatos, *Biotechnol. Bioengng.* **34** (1989) 741.
3. V. S. Afrajmovich, V. I. Arnold, Yu. S. Il'yashenko and L. P. Sil'nikov, in *Dynamical Systems - Encyclopaedia of Mathematical Sciences - Vol. 5*, ed. V. I. Arnold (Springer, New York, 1994).
4. F. N. Albahadily, J. Ringland and M. Schell, *J. Chem. Phys.* **90** (1989) 813.
5. H. Aref, *J. Fluid. Mech.* **143** (1984) 1.
6. R. Aris and N. R. Amundson, *Chem. Engn. Sci.* **7** (1958) 121.
7. D. Aronson, R. McGehee, I. Kevrekidis and R. Aris, *Phys. Rev.* **A33** (1986) 2190.
8. G. Baier, K. Wegmann and J. L. Hudson, *Phys. Lett.* **A 141** (1989) 340.
9. J. H. Bartlett, *Trans. Electrochem. Soc.* **87** (1945) 521.
10. M. R. Bassett and J. L. Hudson, *Chem. Engn. Commun.* **60** (1987) 145.
11. M. R. Bassett and J. L. Hudson, *J. Phys. Chem.* **92** (1988) 6963.
12. M. R. Bassett and J. L. Hudson, *J. Phys. Chem.* **93** (1989) 2731.
13. M. R. Bassett and J. L. Hudson, *Physica D* **35** (1989) 289.

14. M. R. Bassett and J. L. Hudson, in *Spatial inhomogeneties and transient behaviour in chemical kinetics*, ed. P. Gray, G. Nicolis, F. Baras, P. Borckmans and S. K. Scott (Manchester University Press, Manchester, 1990).
15. P. Bergè, Y. Pomeau and C. Vidal, *Order within chaos* (Herman and Wiley, Paris, 1984).
16. E. Boe and H. C. Chang, *Chem. Engn. Sci.* **44** (1989) 1281.
17. *Chaos/Xaoc - Soviet-American perspectives on nonlinear sciences*, ed. D. K. Campbell (AIP, New York, 1990).
18. G. T. Caneba and M. J. Crossey, *Chem. Engn. Commun.* **51** (1987) 1.
19. H. C. Chang and L. H. Chen, *Chem. Engn. Sci.*, **39** (1984) 1127.
20. C. K. Chui and G. Chen, *Lynear systems and optimal control* (Springer Verlag, New York, 1989).
21. C. K. Chui and G. Chen, *Signal processing and systems theory: selected topics* (Springer Verlag, New York, 1992).
22. J. Cooper, R. Muller and C. Tobias, *J. Electrochem. Soc.* **127** (1980) 1733.
23. R. J. P. deFigueiredo and G. Chen, *Nonlynear feedback control systems: an operator theoretic approach* (Academic Press, New York, 1993).
24. P. De Kepper, A. Rossi and A. Pacault, *CR Acad. Sci.*, **C283** (1976) 371.
25. C. B. Diem and J. L. Hudson, *AIChE J.*, **33** (1987) 218.
26. E. Doedel and J. P. Kernevez, *AUTO: Software for continuation and bifurcation problems in ordinary differential equations* (California Institute of Technology, Pasadena, 1986).
27. M. Eiswirth, R. Schwankner and G. Ertl, *Z. Phys. Chem.* **144** (1985) 59.
28. M. Eiswirth, K. Krischer and G. Ertl, *Surf. Sci.* **202** (1988) 565.
29. M. Eiswirth, P. Moller and G. Ertl, *Surf. Sci.* **208** (1989) 13.
30. G. Ertl, *Pure Appl. Chem.* **52** (1980) 2051.
31. G. Ertl, in *Spatial inhomogeneties and transient behaviour in chemical kinetics*, ed. P. Gray, G. Nicolis, F. Baras, P. Borckmans and S. K. Scott (Manchester University Press, Manchester, 1990).
32. W. W. Farr and R. Aris, *Can. math. Soc. Conf. Proc.* **8** (1987) 397.
33. S. O. Fatunla, *Math. Comput.* **34** (1980) 373.
34. U. F. Franck and R. FitzHugh, *Z. Electrokhem.* **65** (1961) 156.
35. U. F. Franck and L. Meunier, *Z. Naturforsch.* **8b** (1953) 396.
36. A. Ghosh and B. Chance, *Biochem. Biophys. Res. Commun.* **16** (1964) 174.
37. P. Gray and S. K. Scott, in *Oscillations and travelling waves in chemical systems*, ed. R. J. Field and M. J. Burger (Wiley, New York, 1985).
38. P. Gray and S. K. Scott, *Chemical oscillations and instabilities: non-linear chemical kinetics* (Oxford University Press, Oxford, 1990).
39. J. F. Griffiths and S. K. Scott, *Prog. Energy Combust. Sci.* **13** (1987) 161.
40. J. Guckenheimer and P. Holmes, *Nonlinear oscillations, dynamical systems and bifurcations of vector fields* (Springer, New York, 1983).
41. M. R. Guevara, L. Glass and A. Shrier, *Science* **214** (1981) 1350.
42. M. R. Guevara and L. Glass, *J. Math. Biol.* **14** (1982) 1.
43. J. W. Hamer, T. A. Akramov and W. H. Ray, *Chem. Engn. Sci.* **36** (1981)

1897.
44. R. F. Heinemann and A. B. Poore, *Chem. Engn. Sci.* **36** (1981) 1411.
45. H. Herzel, W. Ebeling and Th. Schulmeister, *Z. Naturforsch.* **42** (1987) 136.
46. J. L. Hudson, M. Hart and D. Marinko, *J. Chem. Phys.* **71** (1979) 1601.
47. J. L. Hudson, J. Mankin, J. McCullough and P. Lamba, in *Nonlinear phenomena in chemical dynamics.*, ed. C. Vidal and A. Pacault (Springer, Berlin, 1981).
48. J. L. Hudson and J. C. Mankin, *J. Chem. Phys.* **74** (1981) 6171.
49. J. L. Hudson, J. C. Bell and N. I. Jaeger, *Ber. Bunsenges. Phys. Chem.* **92** (1989) 1383.
50. J. M. Hyman, B. Nicolaenko and S. Zaleski, *Physica* **23D** (1986) 265.
51. R. Imbihl, M. P. Cox, G. Erlt, H. Muller and W. Brenig, *J. Chem. Phys.* **83** (1985) 1578.
52. R. Imbihl, M. P. Cox and G. Erlt, *J. Chem. Phys.* **84** (1986) 3519.
53. R. Imbihl, S. Ladas and G. Erlt, *Surf. Sci.* **206** (1988) L903.
54. R. Imbihl, M. Sander and G. Erlt, *Surf. Sci.* **204** (1988) L701.
55. S. R. Inamdar, V. Ravi Kumar and B. D. Kulkarni, *Chem. Engn. Sci.* **46** (1991) 901.
56. E. A. Jackson, *Phys. Lett.* **A 151** (1990) 478.
57. E. A. Jackson, *Physica* **D 50** (1991) 341.
58. E. A. Jackson and A. Hubler, *Physica* **D 44** (1990) 407.
59. E. A. Jackson and A. Kodogeorgiou, *Physica* **D 54** (1991) 253.
60. N. I. Jaeger, P. J. Plath and Ngo Quoc Quyen, in *Temporal order*, ed. Resing and Jaeger, (Springer, Berlin, 1985).
61. R. Jaisinghani and W. H. Ray, *Chem. Engn. Sci.* **32** (1977) 811.
62. K. F. Jensen and W. H. Ray, *Chem. Engn. Sci.* **37** (1982) 199.
63. D.V. Jorgensen and R. Aris, *Chem. Engn. Sci.* **38** (1984) 44.
64. T. Kai and K. Tomita, *Prog. Theor. Phys.* **61** (1979) 54.
65. A. L. Kawczynski, M. Przasnyski and B. Baranowski, *J. Electroanalyt. Chem.* **179** (1984) 285.
66. H. B. Keller, *Applications of bifurcation theory*, ed. P. H. Rabinowitz (Academic Press, New York, 1977).
67. I. G. Kevrekidis, L. D. Schmidt and R. Aris, *Chem. Engn. Commun.* **30** (1984) 323.
68. I. G. Kevrekidis, L. D. Schmidt and R. Aris, *Chem. Engn. Sci.* **41** (1986) 1263.
69. I. G. Kevrekidis, R. Aris and L. D. Schmidt, *Chem. Engn. Sci.* **41** (1986) 1549.
70. I. G. Kevrekidis, L. D. Schmidt and R. Aris, *Physica* **23D** (1986) 391.
71. A. I. Khibnik, in *Continuation and bifurcation numerical techniques and applications*, ed. D. Roose, B. de Dier and A. Spence (Kluwer, Dordrecht 1990).
72. *Applied chaos*, ed. J. Hyun Kim and J. Stringer (Wiley, New York, 1992).
73. E. Knobloch, D. R. Moore and T. Toomre, *J. Fluid. Mech.* **166** (1986) 409.
74. M. Kubicek, *ACM Trans. Math. Software* **2** (1976) 98.

75. Y. Kuramoto, *Chemical oscillations, waves and turbulence* (Springer, Berlin, 1984).
76. H. P. Lee, K. Nobe and A. J. Pearlstein, *J. Electrochem. Soc.* **132** (1985) 1031.
77. J. C. Mankin and J. L. Hudson, *Chem. Engn. Sci.* **39** (1984) 1807.
78. M. Marek and I. Schreiber, *Chaotic behaviour of deterministic dissipative systems* (Academia, Praha, 1991).
79. J. Maselko, *Chem. Phys.* **51** (1980) 473.
80. J. Maselko, *Chem. Phys. Lett.* **73** (1980) 194.
81. J. Maselko and H. L. Swinney, *J. Chem. Phys.* **85** (1986) 6430.
82. M. A. McKarnin, L. D. Schmidt and R. Aris, *Chem. Engn. Sci.* **43** (1988) 2833.
83. J. Moiola, M. C. Colantonio, A. Desages and J. Romagnoli, *Chem. Engn. Sci.* **45** (1990) 297.
84. J. Moiola, A. Desages and J. Romagnoli, *Chem. Engn. Sci.* **45** (1990) 1863.
85. J. Moiola, A. Desages and J. Romagnoli, *Chem. Engn. Sci.* **46** (1991) 1475.
86. H. T. Moon, P. Huerre and L. G. Redekopp, *Phys. Rev. Lett.* **49** (1982) 458.
87. H. T. Moon, P. Huerre and L. G. Redekopp, *Physica* **7D** (1983) 135.
88. P. J. Nandapurkar, V. Hlavacek and P. Von Rompay, *Chem. Engn. Sci.* **41** (1986) 2747.
89. L. F. Olsen and H. Degn, *Nature* **527** (1977) 177.
90. L. F. Olsen, *Z. Naturforsch.* **34a** (1979) 1544.
91. L. F. Olsen, *Phys. Lett.* **94A** (1983) 454.
92. L. F. Olsen, *Z. Naturforsch.* **40** (1985) 1283.
93. H. U. Onken and E. Wicke, *Ber. Bunsenges. Phys. Chem.* **90** (1986) 976.
94. M. E. Orazem and M. G. Miller, *J. Electrochem. Soc.* **134** (1986) 392.
95. E. Ott, C. Grebogi and J. A. Yorke, *Phys. Rev. Lett.* **64** (1990) 1196.
96. J. M. Ottino, in *Applied chaos*, ed. J. Hyun Kim and J. Stringer (Wiley, New York, 1992).
97. S. J. Parulekar and D. Wei, in *Proceedings of the 1988 American Control Conference* (Atlanta, GA, 1988).
98. S. J. Parulekar and J. Lee, *Chem. Engn. Sci.* **48** (1993) 3007.
99. H. Pedersen, G. H. Cho, R. Hamilton and C. K. Chin, *Ann. N. Y. Acad. Sci.*, **506** (1987) 163.
100. L. Pellegrini and G. Biardi, *Computers Chem. Engn.*, **11** (1990) 1237.
101. L. Pellegrini, S. Albertoni and G. Biardi, *Chem. Engn. Sci.* **47** (1992) 2463.
102. L. Pellegrini, C. Tablino Possio, S. Albertoni and G. Biardi, *Chaos, Solitons & Fractals* **3** (1993) 537.
103. L. Pellegrini, G. Biardi, C. Tablino Possio and S. Albertoni, in *Proceedings of the First Italian Conference on Chaos and Fractals in Chemical Engineering*, ed. G. Biardi and M. Giona (World Sci., Singapore, 1994).
104. J. B. Planeaux and K. F. Jensen, *Chem. Engn. Sci.* **41** (1986) 1497.
105. J. J. Podesta, R. C. V. Piatti and A. J. Arvia, *J. Electrochem. Soc.* **126** (1979) 1363.

106. I. Prigogine and R. Lefever, *J. Chem. Phys.* **48** (1968) 1695.
107. A. Rius and R. Lizarbe, *Electrochim. Acta* **7** (1962) 513.
108. J. C. Roux, A. Rossi, S. Bachelart and C. Vidal, *Phys. Lett.* **A 77** (1980) 391.
109. J. C. Roux and H. L. Swinney, in *Nonlinear phenomena in chemical dynamics*, ed. C. Vidal and A. Pacault (Springer, Berlin, 1981).
110. J. C. Roux, J. S. Turner, W. D. McCormick and H. L. Swinney, in *Nonlinear problems: present and future*, ed. A. R. Bishop, D. K. Campbell and B. Nicolaenko (North-Holland, Amsterdam, 1982).
111. J. C. Roux, *Physica* **7D** (1983) 57.
112. J. C. Roux, R. H. Simoyi and H. L. Swinney, *Physica* **8D** (1983) 257.
113. J. C. Roux and A. Rossi, in *Non-equilibrium dynamics in chemical systems*, ed. C. Vidal and A. Pacault (Springer, Berlin, 1984).
114. P. P. Russell and J. Newman, *J. Electrochem. Soc.* **133** (1986) 2093.
115. M. Schell and F. N. Albahadily, *J. Chem. Phys.* **90** (1989) 822.
116. A. D. Schmidt and W. H. Ray, *Chem. Engn. Sci.* **36** (1981) 1401.
117. A. D. Schmidt, A. B. Clinch and W. H. Ray, *Chem. Engn. Sci.* **39** (1984) 419.
118. A. J. Schmidt, J. M. Lee and G. Ann, *Biotechnol. Bioengng.* **33** (1989) 1437.
119. R. A. Schmitz, K. R. Graziani and J. L. Hudson, *J. Chem. Phys.* **67** (1977) 3040.
120. I. Schreiber, M. Dolnik, P. Choc and M. Marek, *Phys. Lett.* **A 128** (1988) 66.
121. F. Schuth and E. Wicke, *Ber. Bunsenges. Phys. Chem.* **93** (1989) 191.
122. F. Schuth and E. Wicke, *Ber. Bunsenges. Phys. Chem.* **93** (1989) 491.
123. S. K. Scott, *Chemical chaos* (Clarendon Press, Oxford, 1993).
124. W. D. Seider, D. D. Brengel, A. M. Provost and S. Widagdo, *Ind. Eng. Chem. Res.* **29** (1990) 805.
125. M. Sheintuch and J. Schmidt, *J. Phys. Chem.* **92** (1988) 3404.
126. R. H. Simoyi, A. Wolf and H. L. Swinney, *Phys. Rev. Lett.* **49** (1982) 245.
127. M. L. Stephens and G. Lyberatos, *Biotechnol. Bioengng.* **31** (1988) 464.
128. F. Strozzi and J. M. Zaldivar, in press on *Chem. Engn. Sci.* (1994).
129. H. L. Swinney and J. C. Roux, in *Non-equilibrium dynamics in chemical systems*, ed. C. Vidal and A. Pacault (Springer, Berlin, 1984).
130. H. L. Swinney and J. Maselko, *Phys. Rev. Lett.* **55** (1985) 2366.
131. C. Tablino Possio and L. Pellegrini, *Chem. Engn. Sci.* **49** (1994) 131.
132. F. Teymour and W. H. Ray, *Chem. Engn. Sci.* **44** (1989) 1967.
133. F. Teymour and W. H. Ray, *Chaos, Solitons & Fractals* **1** (1991) 295.
134. F. Teymour and W. H. Ray, *Chem. Engn. Sci.* **47** (1992) 4121.
135. F. Teymour and W. H. Ray, *Chem. Engn. Sci.* **47** (1992) 4133.
136. K. Tomita and T. Kai, *J. Stat. Phys.* **21** (1979) 65.
137. K. Tomita and H. Daido, *Phys. Lett.* **A 79** (1980) 133.
138. J. S. Turner, J. C. Roux, W. D. McCormick and H. L. Swinney, *Phys. Lett.* **A 85** (1981) 9.
139. A. Uppal, H. Aris and A. B. Poore, *Chem. Engn. Sci.* **29** (1974) 967.
140. C. Vidal, J. C. Roux, A. Rossi and S. Bachelart, *CR Acad. Sci.* **C289** (1979)

73.
141. C. Vidal, S. Bachelart and A. Rossi, *J. Phys.* **43** (1982) 7.
142. J. Villadsen and M. L. Michelsen, *Solution of Differential Equation Models by Polynomial Approximation* (Prentice-Hall, Englewood Clifs, New Jersey, 1978).
143. J. Villermaux, *Chem. Engn. Sci.* **48** (1993) 2525.
144. E. Wicke and H. U. Onken, *Chem. Engn. Sci.* **41** (1986) 1681.
145. E. Wicke and H. U. Onken, *Chem. Engn. Sci.* **43** (1988) 2289.
146. A. Wolf, J. B. Swift, H. L. Swinney and J. A. Vastano, *Physica* **16D** (1985) 285.

STABILITY AND CHAOTIC BEHAVIOUR OF A CONTROLLED IMPERFECTLY MIXED CSTR: Order and non-ideality

OMBRETTA PALADINO, MARCO RATTO and PAOLO COSTA
ISTIC - Facoltà di Ingegneria - Università di Genova
Via Opera Pia 15, 16145 GENOVA, Italy

ABSTRACT

The dynamic behaviour of a controlled non-ideal continuous stirred tank reactor is investigated here in terms of multiplicity of steady states, stability of limit cycles and transition to chaos. The bypass coefficient in the non-ideal Cholette (1983) CSTR model plays the role of order parameter since the stability region widens as the bypass increases; on the other hand the chaotic regions found for the perfectly mixed CSTR tend to vanish, but similar ones appear for different values of the control gains also at very low bypass flow rates. So, any lack of knowledge on the exact efficiency of mixing results in a serious uncertainty in dynamic behaviour expectations.

1. Introduction

It is well known that feedback control systems may induce instability and oscillations[1,2]: a single ideal CSTR under a proportional, PI or PID control action shows a complex dynamic behaviour with the presence of limit cycles and degeneration to chaos for certain values of the control parameters.

A detailed analysis of complex chaotic behaviour in a single controlled perfectly mixed CSTR has been presented in Pellegrini and Biardi[3] as a function of the control gains associated with PI control. They found numerically that this simple reactor model exhibits all the features of the classical chaotic transitions (period dubling, intermittency, positive maximal Lijapunov exponent, fractal structure of the limit attractor). Furthermore Giona and Paladino[4] extended the study to a CSTR with n-order reaction and different control possibilities (inlet temperature or coolant flowrate as manipulated variables and control delay). They analyzed the stability regions of the desired set point in the control parameter space in order to well individuate the range of control gains to be avoided, since they correspond to windows of chaos. In this case the only bifurcation parameters are the control gains and their values can be perfectly fixed during plant operation in order to avoid chaotic behaviour.

The extention to a system of two backmixed PI controlled ideal CSTRs has been subsequently investigated in Giona et al.[5] and Paladino[6]. This system shows complex dynamic behaviour with limit cycles and windows of chaos by varying the backmixing ratio, assumed as bifurcation parameter. If the two coupled CSTRs have equal volume $V/2$ at very high backmixing ratios the system can approximate imperfect mixing in a single reactor of volume V. Also in this case backmixing ratio can be set during plant operation, but since chaos is very sensitive with process parameters (the dynamic behaviour can drastically change in the neighbourhood of critical values: e.g. chaotic regions disappear and similar ones appear elsewhere), some uncertainty on the real value assumed by the backmixing ratio during operation, can affect our plant design based on previous analysis and the real dynamic behaviour of the two reactors can be very different from the expected one.

In this work the non-ideal CSTR model of Lo and Cholette[7] is taken into account. We present a complete stability analysis of the non-ideal PI controlled CSTR, performing a first order exothermal reaction (results can be easily extended to the general n-order reaction). The bypass coefficient, assumed as bifurcation parameter, represents the grade of non-ideality of the CSTR model and it is connected, in a certain way, to mixing efficiency. Attention is focused on how Hopf locus moves at different bypass values in the range of operation efficiencies. Already for very low bypass flow rates the windows of chaos found in the instability region for the ideal CSTR model disappear but similar windows are present for different values of the control gains. This important fact is due to sensitivity of stochastic phenomena, and makes that the dynamic behaviour of the reactor depends on the choice of the model itself.

2. Stability study of a PI controlled ideal CSTR

The equations expressing mass and energy balance for a single non-isothermal ideal CSTR performing a first order reaction $A \rightarrow B$ read:

$$V\frac{dc_A}{dt} = Q(c_{A0} - c_A) - k_0\,exp\left(-\frac{E}{RT}\right)c_A V$$

$$V\rho c_p\frac{dT}{dt} = Q\rho c_p(T_0 - T) - \Delta H_r k_0\,exp\left(-\frac{E}{RT}\right)c_A V - UA(T - T_e) \tag{2.1}$$

This autonomous system of ordinary differential equations can be written in adimensional form by means of the following manipulations (notation adopted by Pellegrini and Biardi[3]):

$$\xi = \frac{c_{A0} - c_A}{c_{A0}} \qquad \tau = \frac{V}{Q} \qquad t = \frac{\bar{t}}{\tau} \qquad N = \frac{UA}{Q\rho c_p} \qquad \Delta T_a = \frac{-\Delta H_r c_{A0}}{\rho c_p} \qquad \vartheta = \frac{T - T_{ref}}{\Delta T_a}$$

The system becomes:

$$\frac{d\xi}{dt} = -\xi + k_0\tau exp\left(-\frac{E}{R\left(T_{ref} + \Delta T_a\vartheta\right)}\right)(1-\xi)$$

$$\frac{d\vartheta}{dt} = \vartheta_0 + N\vartheta_e - (1+N)\vartheta + k_0\tau exp\left(-\frac{E}{R\left(T_{ref} + \Delta T_a\vartheta\right)}\right)(1-\xi) \tag{2.2}$$

It is known that this open-loop system cannot exhibit chaotic behaviour but different operating conditions may occur with three possible steady states. When the outlet temperature is taken as the controlled variable and the inlet temperature as the manipulable one, a proportional-integral (PI) controller acts in the following way:

$$\vartheta_0 = \vartheta_{0s} - k_p(\vartheta - \vartheta_s) - k_1\tau\int_0^t(\vartheta - \vartheta_s)d\tilde{t} \tag{2.3}$$

and by positing

$$\varphi = \int_0^t(\vartheta - \vartheta_s)d\tilde{t} \tag{2.4}$$

system (2.1) is transformed into:

$$\begin{aligned} \frac{d\xi}{dt} &= -\xi + f(\vartheta)(1-\xi) \\ \frac{d\vartheta}{dt} &= \vartheta_{0s} - k_p(\vartheta - \vartheta_s) - k_1\tau\varphi + N\vartheta_e - (1+N)\vartheta + f(\vartheta)(1-\xi) \\ \frac{d\varphi}{dt} &= \vartheta - \vartheta_s \end{aligned} \tag{2.5}$$

where

$$f(\vartheta) = k_0\tau exp\left(-\frac{E}{R(T_{ref} + \Delta T_a\vartheta)}\right)$$

It has been shown that the reactor controlled so does not exhibit input multiplicity, consequently the assumed set point is the only singular point of the closed-loop system. It was considered the local behaviour of the controlled CSTR in the neighbourhood of the extended set point $(\xi_s, \vartheta_s, \varphi_s)$.

By assuming $k_1\tau$ as the bifurcative parameter the Hopf locus by the 1-parameter family $(k_1\tau)^*$ can be obtained[4].

In fig.1 the stability/instability regions in the $\kappa_1\tau$ - κ_p plane are shown for a first order reaction evolving in the controlled CSTR under the following conditions[3]:

$k_0\tau = 4.807908E8$ $\quad \Delta T_a = 430K$ $\quad T_{0s} = 298.42K$ $\quad \frac{E}{R} = 8000K$ $\quad T_{ref} = 430K$

$\xi_s = 0.8$ $\quad N = 0.5$ $\quad T_e = 373.16K$ $\quad \vartheta_s = 0$

The stability region is the union of the two open sets S_1 and S_2 where

$$S_1 = \left\{(k_p, k_1\tau) \middle| k_p^* < k_p \le k_p^{**}; 0 \le k_1\tau < (k_1\tau)^*\right\}$$

$$S_2 = \left\{(k_p, k_1\tau) \middle| k_p > k_p^{**}; k_1\tau > 0\right\}$$

The instability region is given by:

$$S_0 = \left\{(k_p, k_1\tau) \middle| 0 < k_p < k_p^*; 0 < k_1\tau < \infty\right\}$$

$$S_3 = \left\{ (k_p, k_I\,\tau) \middle| k_p^{\,*} < k_p < k_p^{\,**}; (k_I\,\tau)^* < k_I\,\tau < \infty \right\}$$

In particular, in the region S_3 the Hopf-bifurcated periodic orbits are stable and the region is characterized only by periodic oscillations. In the region S_0 the system presents periodic oscillations mixed with windows of chaotic behaviour. The transition from periodic orbits to chaos is governed by period-doubling cascade and the chaos windows appear after the period-dubling.

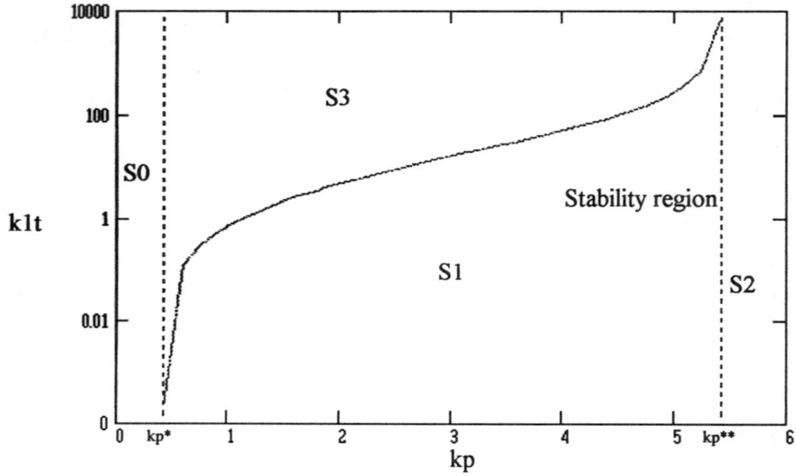

Figure 1: Hopf locus for the ideal model.

The ideal CSTR model is clearly the simplest choice of a chemical engineer during his/her process design. This model has been largely studied and design computations based on the ideal CSTRs cascade are still today applied to industrial processes, also from a dynamic point of view. Corrections to results are often made to take into account the non-ideal mixing. Nevertheless, since complex, and sometimes chaotic dynamic behaviour appears also using this simple model, a detailed study of the other models taking into account of imperfect mixing seems necessary.

3. Bifurcation analysis of a single non-ideal CSTR

Giona et al.[5] and Paladino[6] applied this analysis to two backmixed ideal CSTRs-in-series. That study implied some difficulties:

1. the impossibility to evaluate analitically the first Hopf bifurcation locus, so that the obtained results were purely numerical;
2. the fact that the convergence to the ideal model occurs for the backmixing ratio tending infinity, makes impossible to evaluate numerically the dynamic behaviour of this model for that limit.

On the other hand these works have pointed out some important facts:

1. chaotic regions found for the ideal model disappear for the non ideal one, also at very high values of the backmixing ratio;
2. at the same values of the backmixing ratio chaotic behaviour is detected in different regions unexpected with the ideal model.

This seems to suggest the impossibility to foresee, during design, where the chaotic regions find themselves in the parameters space. The two models, in fact, are equal from a physical point of view: stability and multiplicity analysis performed with the two different models brings to the same results and we have no elements to decide which model is more similar to the reality.

In order to clarify and verify these hypotheses we studied another non-ideal-model. Among the available ones, we analyzed that of Lo and Cholette[7] because of its simplicity. This property allowed us to calculate analitically the Hopf locus. Furthermore the convergence to the ideality occurs when the non-ideality parameters n and m tend to one: a very easy situation. Therefore a more clear and complete comparison with the ideal CSTR was possible.

3.1 Controlled Lo & Cholette CSTR model

The equations expressing mass and energy balance for the single non-isothermal non-ideal CSTR performing a first order reaction $A \to B$ can be written as (see figure 2):

$$mV\frac{dc'_A}{dt} = Q(c_{A0} - c'_A) - k_0\, exp\left(-\frac{E}{RT'}\right)c'_A mV$$

$$mV\rho c_p\frac{dT'}{dt} = nQ\rho c_p(T_0 - T') - \Delta H_r k_0\, exp\left(-\frac{E}{RT'}\right)c'_A mV - UA(T' - T_e) \qquad (3.1)$$

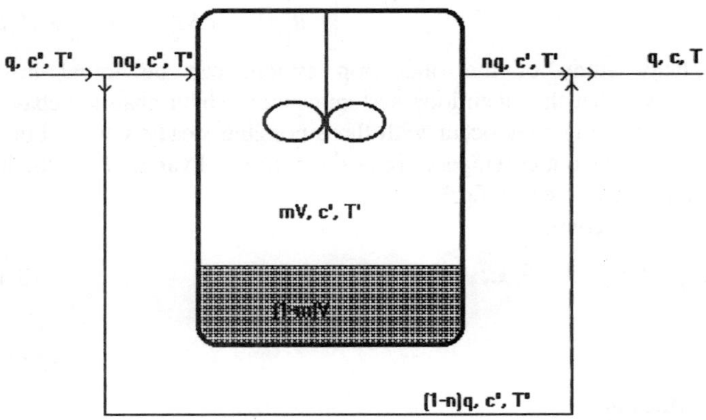

Figure 2: Open-loop Lo&Cholette model.

Using the mass and energy balances before and after the bypass-flow we obtain the following relations:

$$c'_A = \frac{1}{n}[c_A - (1-n)c_{A0}] \qquad T' = \frac{1}{n}[T - (1-n)T_0] \qquad dc'_A = \frac{1}{n}dc_A \qquad dT' = \frac{1}{n}dT$$

Equations (3.1) then become:

$$\frac{mV}{n}\frac{dc_A}{d\bar{t}} = nQ\left[c_{A0} - \frac{1}{n}[c_A - (1-n)c_{A0}]\right] - \frac{mV}{n}[c_A - (1-n)c_{A0}]k_0 e\,exp\left(-\frac{En}{R[T-(1-n)T_0]}\right)$$

$$\frac{mV}{n}\rho c_p\frac{dT}{d\bar{t}} = Q\rho c_p(T_0 - T) - \Delta H_r\frac{mV}{n}[c_A - (1-n)c_{A0}]k_0\,exp\left(-\frac{En}{R[T-(1-n)T_0]}\right) +$$

$$-UA\left[\frac{1}{n}(T-(1-n)T_0) - T_e\right] \tag{3.2}$$

This autonomous system of ordinary differential equations can be normalized by means of the previously stated manipulations and becomes:

$$\frac{m}{n}\frac{d\xi}{dt} = -\xi + \frac{m}{n}(n-\xi)k_0\tau\,exp\left[-\frac{En}{R\left[nT_{ref} + \Delta T_a(n-1)\vartheta_0 + \Delta T_a\vartheta\right]}\right] \tag{3.3}$$

$$\frac{m}{n}\frac{d\vartheta}{dt} = \left(1 + \frac{N}{n} - N\right)\vartheta_0 + N\vartheta_e - (1+\frac{N}{n})\vartheta + \frac{m}{n}(n-\xi)k_0\tau exp\left[-\frac{En}{R\left[nT_{ref} + \Delta T_a(n-1)\vartheta_0 + \Delta T_a\vartheta\right]}\right]$$

Stability and multiplicity study of the open-loop system can be found in the bibliography[7,8].(It is known that this open-loop system cannot exhibit chaotic behaviour but different operating conditions may occur with three possible steady states). For the closed-loop system we take the outlet temperature as the controlled variable and the inlet temperature as the manipulable one (see fig.3).

Considering the following PI-controller:

$$\vartheta_0 = \vartheta_{0s} - k_p(\vartheta - \vartheta_s) - k_1\tau\int_0^t(\vartheta - \vartheta_s)d\bar{t} \tag{3.4}$$

and by positing

$$\varphi = \int_0^t(\vartheta - \vartheta_s)d\bar{t} \tag{3.5}$$

system (3.2) is transformed into:

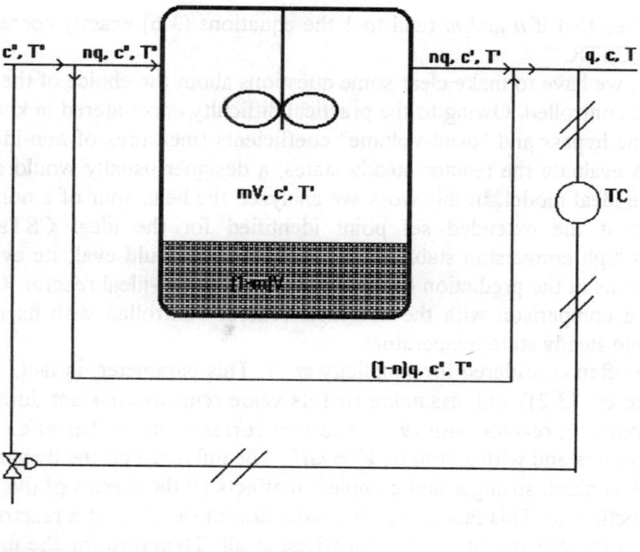

Figure 3: Controlled Lo&Cholette model.

$$\frac{d\xi}{dt} = -\frac{n}{m}\xi + (n-\xi) f(\vartheta,\varphi)$$

$$\frac{d\vartheta}{dt} = v - g\varphi - q\vartheta + (n-\xi) f(\vartheta,\varphi) \tag{3.6}$$

$$\frac{d\varphi}{dt} = \vartheta - \vartheta_s$$

where

$$f(\vartheta,\varphi) = k_0 \tau \cdot exp\left[-\frac{nE}{R(b + c\vartheta + d\varphi)}\right]$$

and

$$v = \frac{n}{m}\left(1 + \frac{N}{n} - N\right)\left(\vartheta_{0s} + k_p \vartheta_s\right) + \frac{n}{m} N \vartheta_e \qquad g = \frac{n}{m}\left(1 + \frac{N}{n} - N\right) k_1 \tau > 0$$

$$q = \frac{n}{m}\left[\left(1 + \frac{N}{n} - N\right)k_p + 1 + \frac{N}{n}\right] > 0 \qquad b = nT_r - \Delta T_a (1-n)(\vartheta_{os} + k_p \vartheta_s)$$

$$c = \Delta T_a (1 + (1-n)k_p) > 0 \qquad d = (1-n)\Delta T_a k_1 \tau > 0$$

It can be easily verified that if n and m tend to 1 the equations (3.6) exactly correspond to those of the ideal CSTR.

Before going on, we have to make clear some questions about the choice of the value of temperature to be controlled. Owing to the practical difficulty encountered in knowing the exact value of the bypass and "dead-volume" coefficients (measures of non-ideality) and consequently to evaluate the reactor steady states, a designer usually would design the reactor using the ideal model. In this work we analyzed the behaviour of a non-ideal reactor applying to it the extended set point identified for the ideal CSTR and corresponding to its high conversion stable steady state. So we could evaluate eventual limits of the ideal model in the prediction of the behaviour of a non-ideal reactor. On the other hand we did a comparison with the non-ideal reactor controlled with its proper high conversion stable steady state temperature.

Furthermore we often considered for simplicity $m=1$. This parameter, in fact, affects only the volume (see eq. (3.2)) and, assuming that its value remains constant during the process, the behaviour of a reactor with $m<1$ exactely corresponds to that of a smaller one without dead volume and with a volume $V'=mV$. The influence on the dynamics of n, on the other hand, is much stronger and complex: n affects all the therms of the model and above all the kinetic one. This fact makes the reduction of the case of a reactor with a bypass flow rate to another one with $n=1$ not trivial at all. Therefore for the moment we analyzed the behaviour of the model at various n and for $m=1$. Of course nothing prevents, in the future, from observing the influence of m too.

3.2 Input multiplicity study of the closed-loop system

By positing to zero equations (3.6) we obtain the following equations, which give the steady state values of the variables ϑ, ξ, φ.

$$
\begin{aligned}
&-\frac{n}{m}\xi_s+(n-\xi_s)f(\vartheta_s,\varphi_s)=0\\
&v-g\varphi_s-q\vartheta_s+(n-\xi_s)f(\vartheta_s,\varphi_s)=0\\
&\quad\vartheta=\vartheta_s
\end{aligned}
\tag{3.7}
$$

Combining the first and the second equation we obtain:

$\varphi_s=\dfrac{v-q\vartheta_s+\dfrac{n}{m}\xi_s}{g}$ and than, after a few passages, we obtain the equation (3.8)

$$-\frac{n}{m}\xi_s + k_o\tau(n-\xi_s)exp\left[-\frac{nE}{R(nT_{ref}+\Delta T_a\vartheta_s+\frac{(1-n)\Delta T_a}{\left(1+\frac{N}{n}-N\right)}\left(N\vartheta_e-\left(1+\frac{N}{n}\right)\vartheta_s+\xi_s\right)}\right]=0 \quad (3.8)$$

that gives the steady state conversion value corresponding to the chosen ϑ_s. It can be easily seen that this equation depends only on the non-ideality parameters *n* and *m*, but it does not depend on the control gains κ_p and $k_1\tau$. If this equation has only one solution, then input multiplicity cannot occur.

In figures 4 and 5 we have the plot of the function in equation (3.8) at different values of *n* and fixed *m*=1, assuming the same conditions of Pellegrini and Biardi and being the extended set point for the ideal $(\vartheta_s = 0)$ and non-ideal CSTR respectively. The curves have only one interception with the axis of abscissas, therefore no multiplicity can occur.

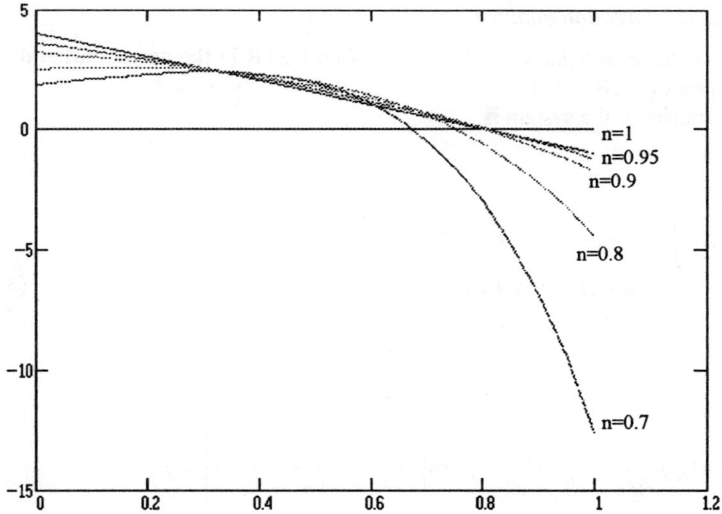

Figure 4: Graphics of equation 3.9 for the non-ideal model controlled with the ideal set point.

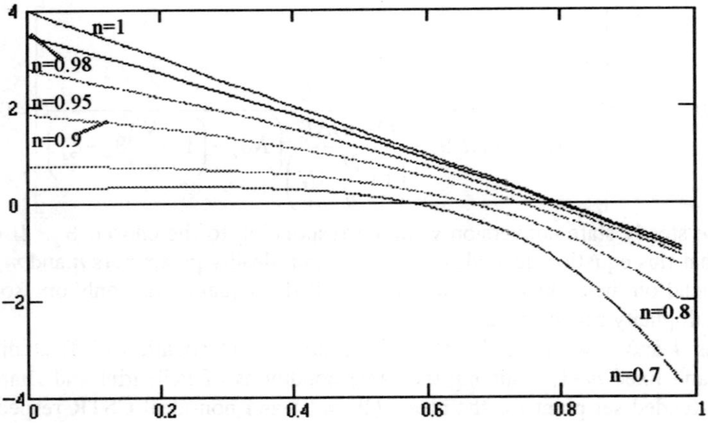

Figure 5: Graphics of equation 3.9 for the non-ideal model controlled with its proper solution.

3.3 Stability and bifurcation analysis

Let us consider the local behaviour of the controlled CSTR in the neighbourhood of the extended set point $(\xi_s, \vartheta_s, \varphi_s)$.
The Jacobian matrix of the system is:

$$J = \begin{pmatrix} -\left(\frac{n}{m} + A_{21}\right) & A_{12} & A_{13} \\ -A_{21} & -q + A_{12} & -g + A_{13} \\ 0 & 1 & 0 \end{pmatrix} \tag{3.9}$$

where

$$A_{12} = k_0\tau(n - \xi_s)\frac{nE \cdot c}{R(r + s\vartheta_s + z\xi_s)^2} exp\left[-\frac{nE}{R(r + s\vartheta_s + z\xi_s)}\right] > 0$$

$$A_{13} = k_0\tau(n - \xi_s)\frac{nE \cdot d}{R(r + s\vartheta_s + z\xi_s)^2} exp\left[-\frac{nE}{R(r + s\vartheta_s + z\xi_s)}\right] > 0$$

$$A_{21} = k_0\tau \cdot exp\left[-\frac{nE}{R(r + s\vartheta_s + z\xi_s)}\right] > 0$$

and

$$r = nT_{ref} + \frac{(1-n)}{\left(1+\frac{N}{n}-N\right)} N\Delta T_a \vartheta_e \qquad s = \Delta T_a - \frac{(1-n)\Delta T_a}{\left(1+\frac{N}{n}-N\right)}\left(1+\frac{N}{n}\right) \qquad z = \frac{(1-n)\Delta T_a}{\left(1+\frac{N}{n}-N\right)}$$

The invariant scalars of the jacobian matrix are:

$$det(J) = \left(\frac{n}{m} + A_{21}\right)(A_{13} - g) - A_{21}A_{13}$$

$$tr(J) = -\frac{n}{m} - A_{21} - q + A_{12}$$

$$m_2(J) = \frac{n}{m}(q - A_{12}) + qA_{21} - A_{13} + g$$

Necessary conditions for the stability of the extended fixed point are:

$$det(J) < 0$$
$$tr(J) < 0$$

In fact only in this case it is possible that three eigenvalues with negative real part exist.

We now analyze the instability of the extended fixed point: a Hopf bifurcation occurs when two complex conjugate eigenvalues cross the imaginary axis. This means that at certain values of $k_1\tau$ or κ_p, the two imaginary eigenvalues λ_1 and λ_2 have a vanishing real part, i.e. λ_1=+iβ, λ_1=–iβ and λ_1 is negative. In this case

$$tr(J) = \lambda_3; \quad m_2(J) = \beta^2; \quad \det(J) = \lambda_3\beta^2 \tag{3.10}$$

From equation (2.10) it follows that the Hopf locus is the set of $k_1\tau$ or κ_p satisfying the following conditions:

$$\frac{\det(J)}{tr(J)} = m_2(J) \tag{3.11}$$

$$P_J[tr(J)] = 0; \quad tr(J) < 0 \tag{3.12}$$

since λ_3=tr(J) and P_j is the characteristic polynomial of the Jacobian matrix.
From equation (3.11), by assuming $k_1\tau$ as the bifurcative parameter we obtain explicitly, after a few passages, the Hopf locus by the 1-parameter family $(k_1\tau)^*$ versus κ_p:

$$(k_1\tau)^* = \frac{(\frac{n}{m} + A_{21} + q - A_{12})(\frac{n}{m}(q - A_{12}) + qA_{21})}{\left[\frac{n}{m}\left(1+\frac{N}{n}-N\right)(A_{12} - q) + A_n(A_{21} - A_{12} + q)\right]} \tag{3.13}$$

satisfying conditions (3.12):

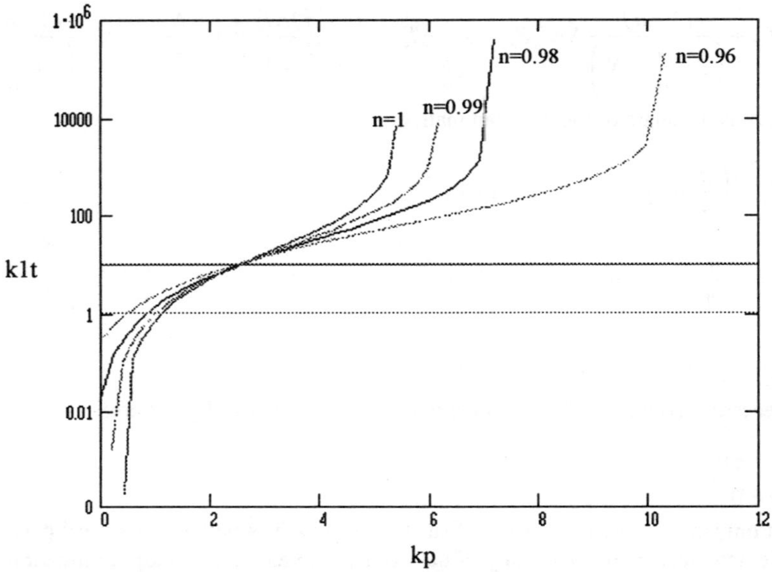

Figure 6a: Hopf locus for the non-ideal model controlled with the ideal set point.

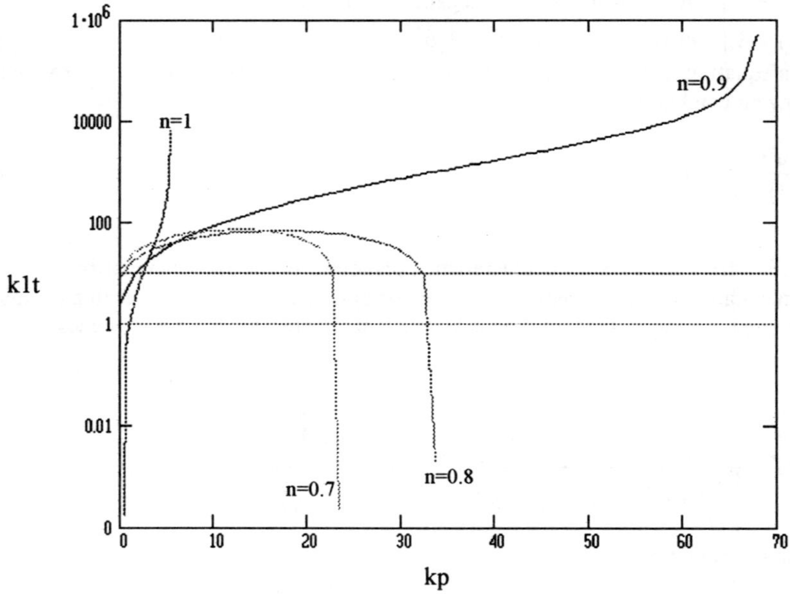

Figure 6b: Hopf locus for the non-ideal model controlled with the ideal set point.

$$P_j(A_{12} - A_{21} - q - \frac{n}{m}) = 0; \quad A_{12} - A_{21} - q - \frac{n}{m} < 0$$

and where

$$A_n = k_0\tau(n - \xi_s)exp\left[-\frac{nE}{R(r + s\vartheta_s + z\xi_s)}\right]\frac{nE \cdot \Delta T_a(1-n)}{R(r + s\vartheta_s + z\xi_s)^2} > 0$$

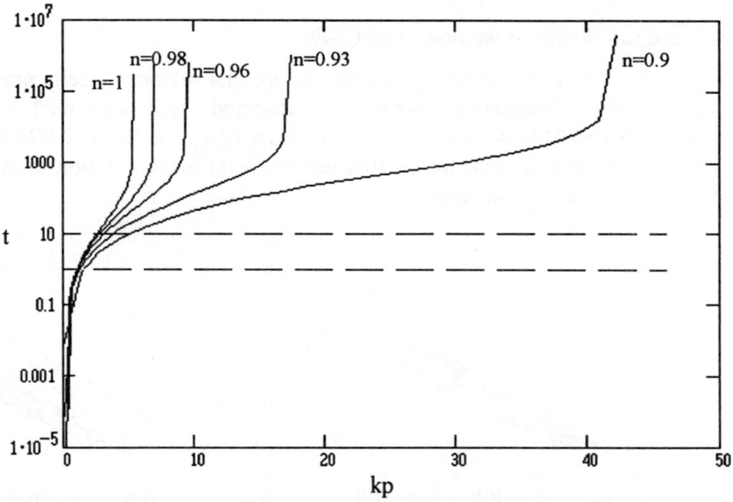

Figure 7a: Hopf locus for the non-ideal model controlled with its proper solution.

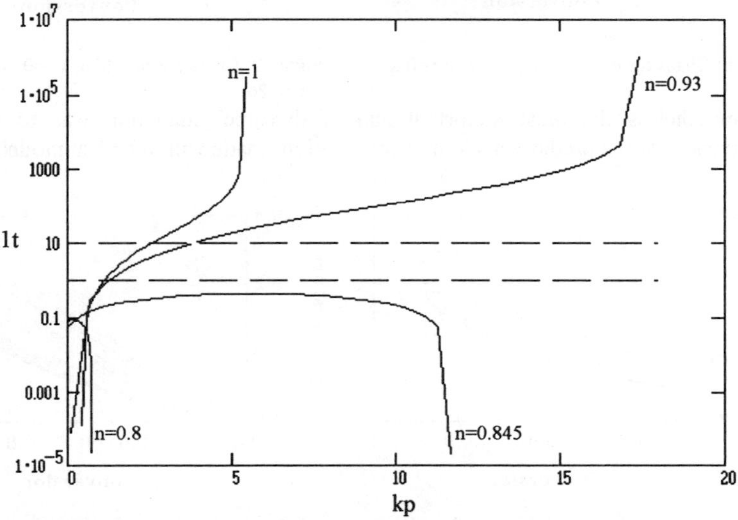

Figure 7b: Hopf locus for the non-ideal model controlled with its proper solution.

In fig.6a-b and 7a-b the stability region in the $\kappa_1\tau$ - κ_p plane is shown for a first order reaction evolving in the controlled CSTR under the Pellegrini and Biardi[3] conditions,at different values of *n* and for *m*=1. For figures 6a-b the set point is, for all *n*, the solution of the ideal CSTR ($\vartheta_s = 0$), while for figures 7a-b the set point exactly corresponds, at each *n*, to the value of the high conversion stable solution of the uncontrolled non-ideal model.

3.4 Dynamical simulation of the non-ideal CSTR

In this section we verify (and deny) the conjecture that region S_3 really presents only stable periodic orbits. Simulation results here presented were performed in the case $\vartheta_s = 0$. Figures 8 and 9 show that for very low values of κ_p and for n=0.976 the system can have chaotic behaviour. This means that region is no more a region of stability for limit cycles as for the non-ideal reactor.

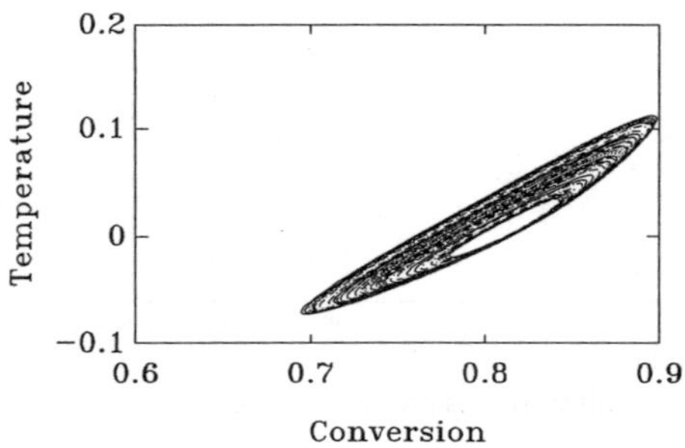

Figure 8: Chaos for n=0.976, κ_p=0.01, $k_1\tau$=0.2.

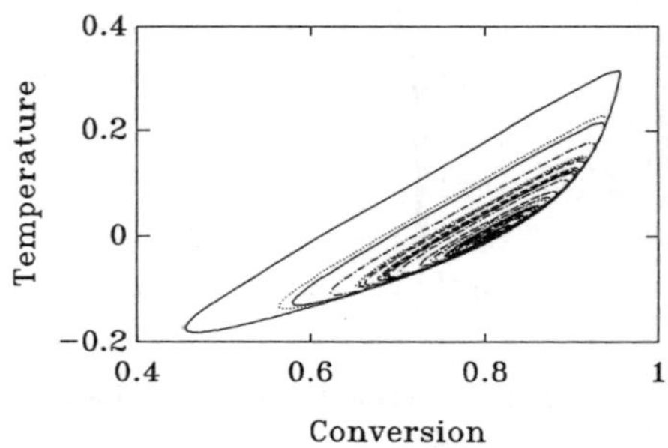

Figure 9: Chaos for n=0.976, κ_p=0.04, $k_1\tau$=0.26.

Nevertheless the most important aim of dynamic simulation was to verify the effectiveness of the prediction about chaotic regions made with the ideal model.

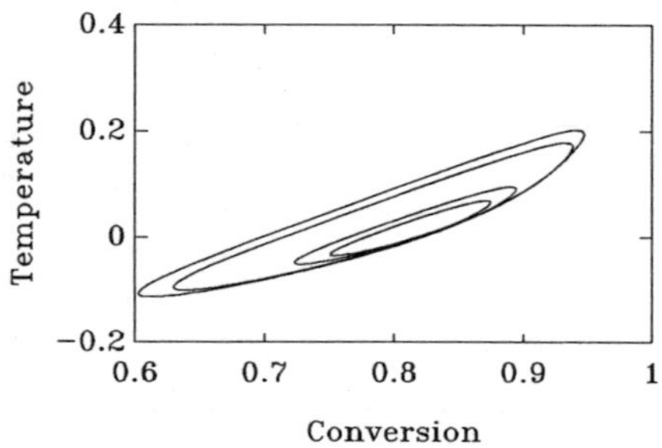

Figure 10a: Stable limit cycle for n=0.9955, κ_p=0.3, $k_1\tau$=0.36.

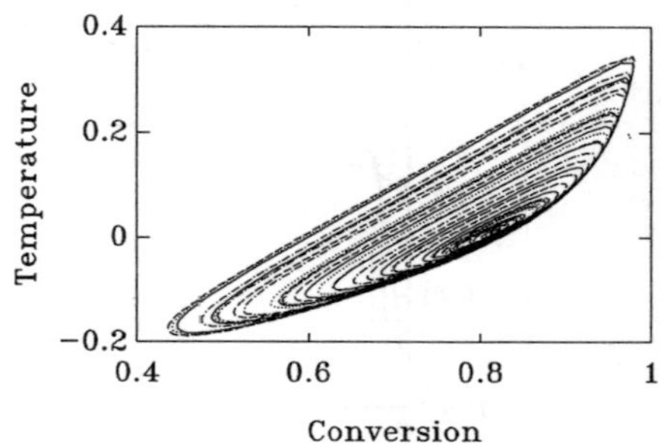

Figure 10b: Chaos for n=1, κ_p=0.3, $k_1\tau$=0.36.

In order to do this we did a comparison between the ideal model and a practically indistinguishable from ideality non-ideal one: in the first case we have (of course) n=1, in the second one we have n=0.9955 (m=1 as usual).

The conversion steady state corresponding to these two conditions are repectively: ξ_s=0.8 for n=1 and ξ_s=0.80117 for n=0.9955; the temperature is the controlled one ϑ_s=0 in both cases: the difference between the models is not appreciable at all.

Figures 10a and 10b show that even for n=0.9955 the chaotic window found by Pellegrini and Biardi for κ_p=0.3 disappears and another similar one appears for lower values of κ_p and where the ideal model presented only stable limit cycles (see fig.11a and 11b). Furthermore we observe that the proper solution of the open-loop non-ideal system for n=0.9955 is ϑ_s=-0.0023 and therefore the externally forced set point ϑ_s=0 is not so far from its natural one to justify these totally opposite results.

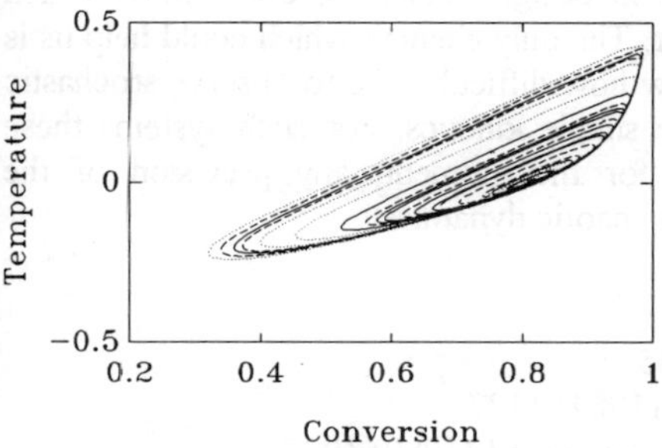

Figure 11a: Chaos for n=0.9955, κ_p=0.224, $k_1\tau$=0.36.

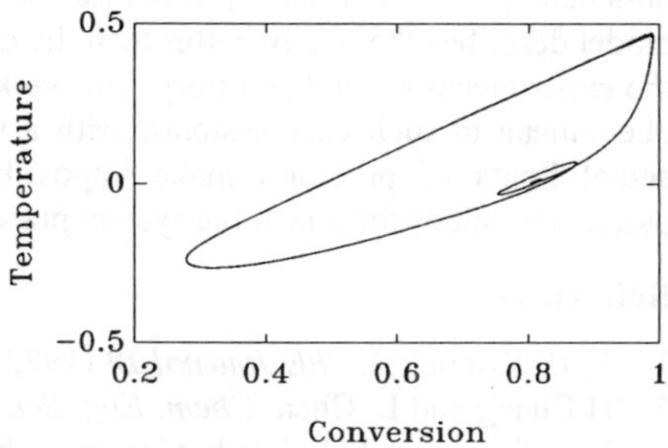

Figure 11b: Stable limit cycle for n=1, κ_p=0.224, $k_1\tau$=0.36

4. Conclusions

The present study has brought to our attention some important questions:

1. In both of the analysed cases, the stability region of the non-ideal model widens for low values of $k_1\tau$ and narrows for the high ones. This means that for physically more acceptable values of the control gains the system becomes more stable. In that sense the bypass coefficient can be thought as an order parameter. Nevertheless this property disappears as n is smaller than a certain value. This is probably due to the fact that as the bypass flow rate increases the value of the controlled temperature becomes more and more different from that of the reactor. Therefore the controlled parameter is not able to take into account the real behaviour of the reaction and the control becomes ineffective.
2. Chaotic behaviour has been found in the non-ideal model too: nevertheless sensitivity, typical of stochasctic phenomena, causes a continuous translation of the chaotic areas even for very low variations of the bypass coefficient and for values which are physically indistinguishable from the ideal case. An analogue behaviour was detected also in the case of two backmixed CSTRs, confirming our results.

Therefore the dynamic behaviour of the system is sensitive to the particular choice of the model.

Nevertheless the precision of the mathematical models describing processes in chemical engineering is seriously affected by the errors in the determination of the various parameters (above all the kinetics). Also in the luckiest cases that error is never less than 5%. In this paper we have seen that two models, whose difference is clearly less than that error and where all the thermodynamical and kinetic therms are assumed to be perfectly known, give opposite results in the chaotic behaviour analysis. This implies that our precision in the determination of all coefficients should be much larger than actually is.

So even if from the purely mathematical and analytical point af view the bifurcation analysis, performed with classical continuation methods or simply by successive simulations, gives no problem because of the extremely simple system under study, the obtained results have hardly practical validity in design: infact we don't know which model describes the reality better than the other. The only element which could help us is the experimentation in laboratory, but we know how difficult it is, to observe stochastic phenomena in such easy systems, with a very simple kinetics. For such systems these actual limits of precision make impossible, for the moment, any prevision of the parameter values for which the system presents chaotic dynamics.

References

1. L. B. Koppel, *AIChE Journal* **28** (1982) 935.
2. H.Chang and L. Chen, *Chem. Eng. Sci.* **39** (1984) 1127.
3. L. Pellegrini and G. Biardi, *Comput. Chem. Engng.* **14** (1990) 1237.
4. M. Giona and O. Paladino, *Comput. Chem. Engng.* **18** (1994) 877.
5. M. Giona, O. Patierno and O. Paladino, *AIChE Annual Meeting,* Miami Beach, Novembre 1992, Ext. AB, N.67-G.
6. O. Paladino, *Environmetrics Journal* **5** (1994) 57.
7. S. N. Lo and A.Cholette, *Chem. Eng. Sci.* **38** (1983) 367.
8. C. T. Liou and Y. S. Chien, *AIChE Journal* **36** (1990) 951.

CONTROL OF CHAOTIC REACTOR DYNAMICS

MASSIMILIANO GIONA
Dipartimento di Ingegneria Chimica, Universitá di Cagliari
Piazza d'Armi, 09123 Cagliari, ITALY

ORESTE PATIERNO
Dipartimento di Ingegneria Chimica, Universitá di Roma "La Sapienza"
Via Eudossiana 18, 00184 Roma, ITALY

and

STEFANO BOY
c/o Dipartimento di Ingegneria Chimica, Universitá di Cagliari
Piazza d'Armi, 09123 Cagliari, ITALY

ABSTRACT

We analyze the range of applicability and the limitations of parametric excitation control over chaotic dynamical systems. We propose an extension of the parametric excitation control of chaos in order to stabilize the evolution of the system towards steady states. Applications of this control method to chemical reactors are presented.

1. Introduction

One of the most interesting aspects in the theory of chaotic dynamical systems is the possibility of controlling chaotic evolution in order to stabilize unstable fixed points and periodic orbits. This topic was first developed by Ott, Grebogi and Yorke[1] in terms of parametric perturbations of chaotic evolution (the OGY method).

The OGY method is intrinsically based on the chaotic nature of the dynamic system. Parametric perturbation serves in the OGY method to annihilate the component of the dynamics in the direction associated with the unstable and central eigenvalues. The method was specifically developed for discrete dynamical systems and can therefore be applied in dealing with continuous systems (e.g. reaction kinetics or reactor dynamics) to Poincaré sections, with the consequence that only periodic orbits can be stabilized.

In recent years, research on the control of chaotic dynamical systems has undergone exponential growth, as shown by the review article of Chen and Dong[2].

In this article we analyze the application of the OGY method to chaotic reactor dynamics and undertake a critical examination of its range of validity in chemical reaction and reactor engineering problems in the presence of one-dimensional and multidimensional unstable eigenspaces. In particular, we analyze the application of these techniques to a PI-controlled CSTR, which, as discussed by Pellegrini and Biardi[3], exhibits chaotic behaviour for a wide region of control gains in the control parameter space. We propose a method of stabilizing chaotic reactor dynamics towards the unique set point induced by the action of the PI-controller (and not towards a periodic orbit, as in the classical application of the OGY technique). This method can easily be generalized to other controlled chaotic systems.

Finally, we discuss the structural differences underlying the application of parametric excitation methods (like the OGY method or the set-point stabilization method proposed here) with respect to classical feedback-control schemes.

2. Control of chaotic dynamics

2.1. The parametric excitation method

Following Devaney[4], a chaotic dynamical system is characterized by three properties: (a) sensitive dependence on the initial conditions, which implies that the maximal Ljapunov exponent is positive[5]; (b) topological transitivity on the limit attractor; (c) dense structure of the periodic orbits $\mathcal{C}_{per}$ on the limit attractor $\mathcal{A}$, i.e. $\mathcal{A} \subseteq \overline{\mathcal{C}}_{per}$, where $\overline{\mathcal{C}}_{per}$ is the closure of $\mathcal{C}_{per}$. The last two conditions imply that in a finite time interval the dynamics will visit any neighbourhood of any unstable periodic orbit with the sole exception of a few isolated orbits (e.g. unstable steady states in continuous dynamical systems).

On the basis of these properties, it is possible to develop a parametric control system which stabilizes the unstable periodic orbits of a given dynamical system by simply annihilating the components of the state vector lying in the unstable eigenspace of the orbit considered[1,6]. This simple idea is the only theoretical tool underlying OGY parametric control. The only basic limitation of the OGY method is the fact that it is applicable solely to discrete dynamical systems.

Let us therefore consider a generic discrete dynamical system

$$\mathbf{x}_{n+1} = \mathbf{f}(\mathbf{x}_n; \mathbf{p}) , \tag{1}$$

where $\mathbf{x}$ is the n-dimensional state vector and $\mathbf{p}$ an n_p-dimensional vector of parameters susceptible of external manipulation.

The dynamical system (1) can be associated with the *augmented dynamical system*

$$\mathbf{y}_{n+1} = \mathbf{F}(\mathbf{y}_n; \mathbf{p}) , \tag{2}$$

where

$$\mathbf{y}_n = \begin{pmatrix} \mathbf{x}_n \\ \mathbf{w}_n \end{pmatrix} , \quad \mathbf{F}(\mathbf{y}_n, \mathbf{p}) = \begin{pmatrix} \mathbf{f}(\mathbf{x}_n) \\ \mathbf{x}_n \end{pmatrix} . \tag{3}$$

The augmented dynamical system, Eq. (2), is represented by the union of dynamical system (1) and of its replica at the preceding instant of time.

Let us suppose that dinamical system (1) is chaotic for a given value of the parameter vector, say $\mathbf{p} = 0$, and let $\mathbf{x}^*$ be the desidered fixed point we wish to stabilize. The extension to n-cycles is straightforward since an n-cycle of $\mathbf{f}$ can always be regarded as a fixed point for its n-th iterative, $\mathbf{f}^{(n)}$. $\mathbf{x}^*$ has the corresponding fixed point $\mathbf{y}^* = (\mathbf{x}^*, \mathbf{x}^*)^T$ for the augmented dynamical system (2). Let us also set $\mathbf{z}_n = \mathbf{y}_n - \mathbf{y}^*$.

If we use $\mathbf{z}^*(\mathbf{p})$ to denote the fixed point as a function of the parameter vector $\mathbf{p}$, $\mathbf{z}^*(\mathbf{0}) = \mathbf{0}$, the evolution of the augmented dynamical system in the neighbourhood of $\mathbf{z}^*(\mathbf{p})$ can be expressed in the linearized form

$$\mathbf{z}_{n+1} - \mathbf{z}^*(\mathbf{p}) = \mathbf{J}\left[\mathbf{z}_n - \mathbf{z}^*(\mathbf{p})\right] , \tag{4}$$

where

$$\mathbf{J} = \left[\begin{array}{cc} \mathbf{A} & \mathbf{0} \\ \mathbf{I} & \mathbf{0} \end{array}\right] , \quad \mathbf{A} = \left.\frac{\partial \mathbf{f}}{\partial \mathbf{x}}\right|_{\mathbf{x}=\mathbf{x}^*,\mathbf{p}=\mathbf{0}} . \tag{5}$$

Our reason for considering the augumented dynamical system is that, independently of $\mathbf{f}$, system (2) admits the eigenvalue $\lambda = 0$ with algebraic multiplicity equal to n at least, and its geometric multiplicity is also at least equal to n (see Appendix A).

Locally, $\mathbf{z}^*(\mathbf{p})$ can be approximated in linear form as

$$\mathbf{z}^*(\mathbf{p}) \simeq \sum_{i=1}^{n_p} \left.\frac{\partial \mathbf{z}^*(\mathbf{p})}{\partial p_i}\right|_{\mathbf{p}=\mathbf{0}} p_i = \sum_{i=1}^{n_p} \mathbf{g}_i\, p_i \tag{6}$$

where $\mathbf{g}_i = \partial \mathbf{z}^*/\partial p_i|_{\mathbf{p}=\mathbf{0}}$, $(i = 1,..,n_p)$ are the *sensitivity vectors* associated with the dependence of the fixed point $\mathbf{z}^*(\mathbf{p})$ on $\mathbf{p}$.

By assuming that $\mathbf{A}$ can be put in diagonal form, and by taking into account Eq. (6), Eq. (4) can be rewritten as

$$\mathbf{z}_{n+1} - \sum_{i=1}^{n_p} p_i\, \mathbf{g}_i = \left(\sum_{i=1}^{n_u} \lambda_i^{(u)} \mathbf{e}_i^{(u)} \mathbf{f}_i^{(u)} + \sum_{j=1}^{n_s} \lambda_j^{(s)} \mathbf{e}_j^{(s)} \mathbf{f}_j^{(s)}\right) \left(\mathbf{z}_n - \sum_{i=i}^{n_p} p_i\, \mathbf{g}_i\right) , \tag{7}$$

where $n_u \leq n$ is the number of unstable and central eigenvalues $\{\lambda_i^{(u)}\}$ of $\mathbf{J}$ and $n_s \geq n$ (see Appendix A) the number of stable eigenvalues $\{\lambda_s^{(s)}\}$ of $\mathbf{J}$, $n_u + n_s = 2n$.

In Eq. (7) $\{\mathbf{e}_i^{(u)}\}, \{\mathbf{e}_j^{(s)}\}$ are respectively the eigenvectors (normalized to unity) related to the unstable and stable eigenvalues of $\mathbf{J}$, where $\{\mathbf{f}_i^{(u)}\}, \{\mathbf{f}_j^{(s)}\}$ is the corresponding normalized basis of eigenvectors associated with $\{\mathbf{f}_i^{(u)}\}, \{\mathbf{f}_j^{(s)}\}$ in the dual space (i.e. in the space of the linear functional acting on the considered vector space and equipped with a vector algebra). From the definition of dual basis it follows that

$$\begin{aligned} \mathbf{f}_i^{(u)}(\mathbf{e}_j^{(u)}) &= \delta_{ij} , \quad \mathbf{f}_j^{(s)}(\mathbf{e}_j^{(s)}) = \delta_{ij} \\ \mathbf{f}_i^{(u)}(\mathbf{e}_j^{(s)}) &= 0 , \quad \mathbf{f}_i^{(s)}(\mathbf{e}_j^{(u)}) = 0 , \end{aligned} \tag{8}$$

where δ_{ij} are the Kronecker symbols and the application of the linear functional $\mathbf{f} = \{f_\alpha\}$ to the vector $\mathbf{e} = \{e_\alpha\}$ is defined as

$$\mathbf{f}(\mathbf{e}) = \sum_{\alpha=1}^{2n} f_\alpha e_\alpha . \tag{9}$$

The principle of parametric excitation control of chaotic evolution is to perturb the system occasionally with the parameter vector $\mathbf{p}$ in such a way that the evolution vector $\mathbf{z}_{n+1}$ will have no components along the unstable and central eigenspaces (i.e. the eigenspace associated with eigenvalues $\{\lambda_i^{(u)}\}$ not lying inside the unit circle). This means that the application of $\mathbf{f}_i^{(u)}$ to $\mathbf{z}_{n+1}$ should be equal to zero for $i = 1, .., n_u$, i.e.

$$\mathbf{f}_i^{(u)}(\mathbf{z}_{n+1}) = 0 \qquad (i = 1, .., n_u) . \tag{10}$$

The last equation implies

$$\mathbf{G}\mathbf{p} = \mathbf{\Lambda}\mathbf{X}^{(n)} , \tag{11}$$

where $\mathbf{G} = \{G_{ij}\}$ is the $n_u \times n_p$ matrix

$$G_{ij} = \mathbf{f}_i^{(u)}(\mathbf{g}_j) , \tag{12}$$

$\mathbf{\Lambda} = \text{diag}(\Lambda_{11}, .., \Lambda_{n_u n_u})$ a $n_u \times n_u$ diagonal matrix,

$$\Lambda_{ii} = \frac{\lambda_i^{(u)}}{\lambda_i^{(u)} - 1} , \tag{13}$$

and $\mathbf{X}^{(n)} = \{X_i^{(n)}\}$ a n_u-vector of components

$$X_i^{(n)} = \mathbf{f}_i^{(u)}(\mathbf{z}_n) . \tag{14}$$

The matrix $\mathbf{G}$ can be called the *parametric controllability matrix* of the system. A fixed point $\mathbf{x}^*$ is said to be paramerically controllable with respect to the parameter vector $\mathbf{p}$ if, by varying $\mathbf{p}$, we can stabilize the system toward $\mathbf{x}^*$.

From Eq. (11) it follows that a necessary condition for the parametric controllability of $\mathbf{x}^*$ is that

$$\text{rank}(\mathbf{G}) = n_u . \tag{15}$$

In other words, to control a given unstable fixed point (or equivalently an unstable periodic orbit) with the OGY method we need at least n_u parameters such that the resulting sensitivity vectors $\{\mathbf{g}_i\}$ form a basis of the unstable eigenspace.

Let $\mathcal{P}(\mathbf{p})$ be the admissible region of the parameter space in which the parameter vector $\mathbf{p}$ may vary. The structure of the set $\mathcal{P}(\mathbf{p})$ depends on the physical constraints limiting the range of variability of the manipulable parameters. According to Eq. (11), the region of the state space within which the parametric excitation can act is given by the set $\mathcal{X}$ (which can be called *parametric controllability set*),

$$\mathcal{X} = \{\mathbf{z} \mid \mathbf{p} = \mathbf{G}^{-1}\mathbf{\Lambda}\mathbf{X}(\mathbf{z}) \in \mathcal{P}(\mathbf{p})\} , \tag{16}$$

where $\mathbf{X}(\mathbf{z}) = \{\mathbf{f}_i^{(u)}(\mathbf{z})\}$.

To sum up, the OGY method of controlling discrete chaotic dynamical systems can be expressed in the form

$$\mathbf{x}_{n+1} = \mathbf{f}(\mathbf{x}_n; \mathbf{p}_n) , \tag{17}$$

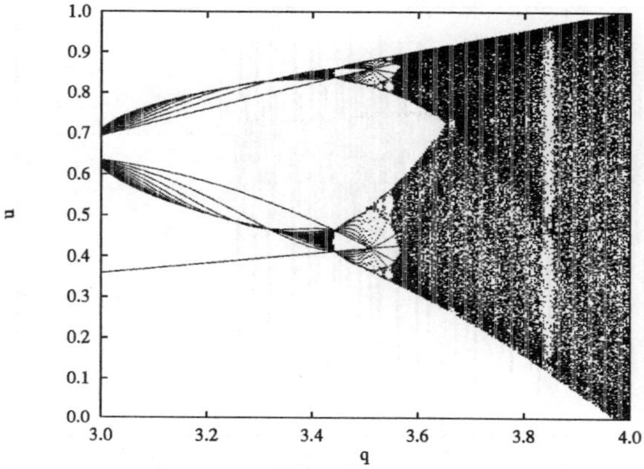

Figure 1: Bifurcation diagram u vs $q \in [3,4]$ of the coupled quadratic map of Eq. (20), with $D = 0.01$.

where the parametric perturbation $\mathbf{p}_n$ is given by

$$\mathbf{p}_n = \mathbf{G}^{-1}\mathbf{\Lambda X}^{(n)} \; \chi_{\mathcal{P}}(\mathbf{G}^{-1}\mathbf{\Lambda X}^{(n)}) \, , \tag{18}$$

$\chi_{\mathcal{P}}$ being the characteristic function of the set $\mathcal{P}$

$$\chi_{\mathcal{P}}(\mathbf{p}) = \begin{cases} 1 \text{ if } \mathbf{p} \in \mathcal{P} \\ 0 \text{ if } \mathbf{p} \in \mathcal{P} \, . \end{cases} \tag{19}$$

As an example of application, let us consider the evolution of a twodimensional system represented by the diffusional coupling of two onedimensional maps

$$\begin{aligned} u_{n+1} &= |g(u_n; q_1) - D(u_n - v_n)| \mod 1 \\ v_{n+1} &= |g(v_n; q_2) - D(v_n - u_n)| \mod 1 \, , \end{aligned} \tag{20}$$

where $g : [0,1] \to [0,1]$. As function g, we choose the quadratic family $g(u,q) = qu(1-u)$. Let us focus attention on the symmetric case $q_1 = q_2 = q$. This implies that there exist symmetric steady states $v^* = u^*$, $u^* = g(u^*, q) = (q-1)/q$, and equivalently for the periodic orbits. Figure 1 shows the bifurcation diagram of this map for a fixed value of the diffusional coupling D. Examples of application of coupled systems are given by Marek and Schreiber[7] and by all the literature on the synchronization of chaotic evolutions[8].

System (20) is a three parameter system (q_1, q_2, D). The Jacobian matrix in the neighbourhood of the symmetric steady state $u^* = (q-1)/q$ is given by

$$\mathbf{A} = \begin{bmatrix} -(\mu + D) & D \\ D & -(\mu + D) \end{bmatrix} \tag{21}$$

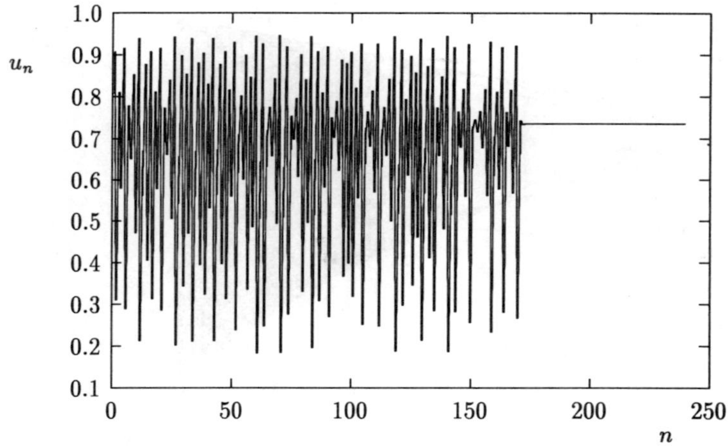

Figure 2: u_n vs n in the presence of parametric excitation ($q = 3.77$, $D = 0.01$).

where $\mu = |g'(u^*; q)| = |2 - q|$, and admits the eigenvalues $\lambda_1 = -\mu$, $\lambda_2 = -(\mu + D)$. Therefore, provided that $\mu > 1$ ($q > 3$), both the eigenvalues are unstable.

This implies that we need at least two parameters in order to apply OGY control. The diffusional coupling constant D cannot be used for this purpose since in the symmetric case $\mathbf{g}_D = \partial \mathbf{f}/\partial D = \mathbf{0}$. An appropriate choice of controlling parameters is represented by the pair (q_1, q_2). Figure 2 shows the stabilization of the sistem (20) toward the symmetric steady state with $D = 0.01$, $q = 3.77$, by applying the parametric excitation control, i.e. by setting $q_1 = q + p_1$, $q_2 = q + p_2$, and exciting (p_1, p_2) only when the dynamics passes close to the steady state (see Eq. (16)). The action of the controlling parameters is shown in figure 3. As can be seen, the parameter excitation develops an impulsive burst: once parametric control has annihilated the unstable components of the dynamics, the control action rapidly falls to zero.

The OGY control method has been widely applied in connection with continuous dynamical systems[9] and with chaotic evolutions in real physical systems[10,11]. In this case, the only trick is to consider a generic discretized map associated with the dynamics (e.g. deriving from a Poincaré local section or from the map of the local maxima/minima). This kind of approach was employed by Peng et al.[12,13] to stabilize chaotic chemical oscillator and by Bandyopadhyay et al.[14] to control the periodic orbits of nonisothermal CSTRs.

As an example of the application of these techniques, let us consider the Lorenz system $\dot{x} = \sigma(y - x)$, $\dot{y} = -xy + rz - y$, $\dot{z} = xy - bz$, with $\sigma = 10$, $r = 28$ and $b = 8/3$. The Lorenz system is characterized by onedimensional map for the maxima of the z-coordinate, which can be used to drive the parametric excitation. The parameter r was chosen as a control parameter. Figure 4 shows the periodic orbit of the Lorenz system obtained by applying the OGY method in order to stabilize the

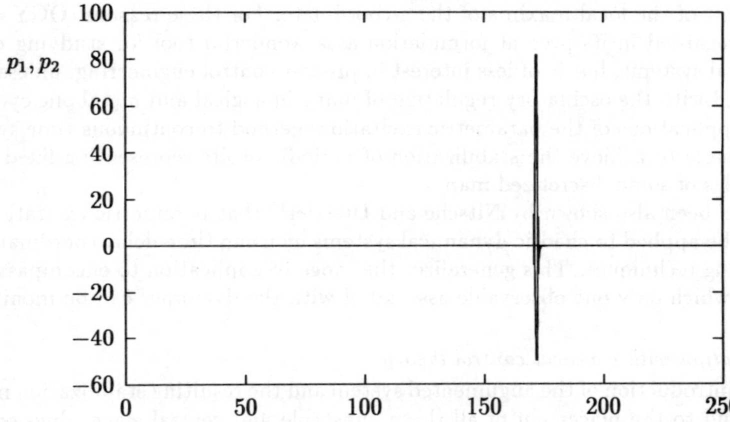

Figure 3: Behaviour of the parameters (p_1, p_2) for the same simulation as figure 2. p_1, p_2 are multiplied by 10^3.

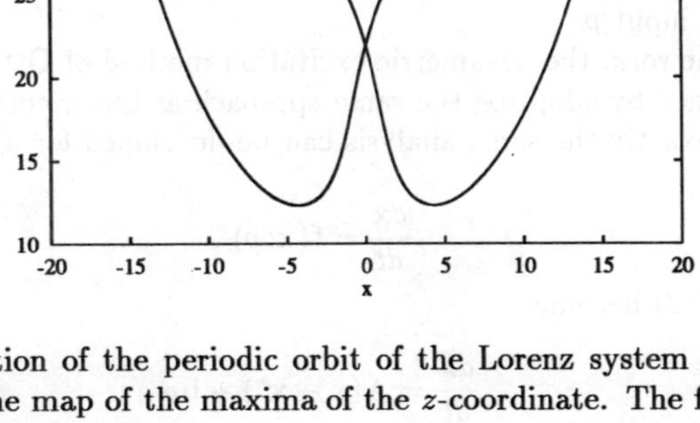

Figure 4: Stabilization of the periodic orbit of the Lorenz system associated with the fixed point of the map of the maxima of the z-coordinate. The figure shows the projection in the x-z-plane.

fixed point of the local maxima of the z-coordinate. For these reasons, OGY control can be regarded in its present formulation as a wonderful tool for studying chaotic dynamical systems, but is of less interest in process control engineering, unless one is concerned with the oscillatory regulation of many biological and metabolic cycles. In all the applications of the parametric excitation method to continuous-time systems, it is possible to achieve the stabilization of periodic orbits representing fixed points or n-cycles of some discretized map.

It has been also shown by Nitsche and Dressler[15] that parametric excitation control can be applied to chaotic dynamical systems by using time-delay coordinates and embedding techniques. This generalizes the range of application to encompass situations in which only one observable associated with the dynamics can be monitored.

2.2. Relation with classical control theory

The introduction of the augumented system and the resulting stabilization method correspond to the placement of all the n_u unstable and central eigenvalues equal to zero (see Appendix A). In this sense, OGY control is a peculiar case of the pole-placement method.

Indeed, parametric excitation control can be expressed in a form analogous to the classical pole-placement method adopted in linear control theory[16]. In order to simplify the notation, we assume that the parameter vector reduces a single component p.

Starting from Eq. (1), the linearized system in the neighbourhood of the point $(\mathbf{x}^*, p)$ reduces to

$$\mathbf{x}_{n+1} - \mathbf{x}^* = \mathbf{A}(\mathbf{x}_n - \mathbf{x}^*) + \mathbf{b}p \, , \tag{22}$$

where

$$\mathbf{b} = \left. \frac{\partial \mathbf{f}}{\partial p} \right|_{\mathbf{x}=\mathbf{x}^*, p=0} . \tag{23}$$

In control theory, Eq. (22) is the state-space representation of a system with a onedimensional input p.

In this framework, the parametric excitation method of Ott, Grebogi and Yorke can be generalized by adopting the same approach as linear controllability theory[17].

Of course, exactly the same analysis can be developed for a continuous-time dynamical system

$$\frac{d\mathbf{x}}{dt} = \mathbf{f}(\mathbf{x}; p) \, , \tag{24}$$

for which Eq. (22) becomes

$$\frac{d\mathbf{x}}{dt} = \mathbf{A}(\mathbf{x} - \mathbf{x}^*) + \mathbf{b}p \, . \tag{25}$$

The fundamental requirement for controlling a system towards $\mathbf{x}^*$ by means of the parameter p (considered as an input) is that the controllability matrix $\mathbf{C}$,

$$\mathbf{C} = (\mathbf{b}, \mathbf{A}\mathbf{b}, \mathbf{A}^2\mathbf{b}, .., \mathbf{A}^{n-1}\mathbf{b}) \tag{26}$$

should be nonsingular, i.e.

$$\mathrm{rank}(\mathbf{C}) = n \ . \tag{27}$$

If condition (27) is satisfied, it is possible to modulate the parameter p as

$$p = -\mathbf{k}^T(\mathbf{x} - \mathbf{x}^*) = -\sum_{i=1}^{n} k_i(x_i - x_i^*) \tag{28}$$

and to synthetize the gain vector $\mathbf{k}^T = (k_1, .., k_n)$ in such a way that the eigenvalues of the controlled system, $\mathbf{A} - \mathbf{b}\mathbf{k}^T$, admit a prescribed system of eigenvalues (*regulation poles*). This can be achieved by making use of the pole-placement techniques. The analysis developed for the case of single-parameter perturbation can be extended to a generic multidimensional parameter vector $\mathbf{p}$.

3. Stabilization of fixed points in chaotic controlled reactors

In the preceding section we have seen that the fundamental limitation of the OGY method lies in its ability to controll only the periodic orbits of continuous dynamical systems. From the theoretical standpoint of chaos theory this is not a problem since the structure of a strange attractor is organized around the unstable periodic orbits.

However, this limitation does represent the main drawback, especially in connection with engineering control of reactor dynamics, because oscillating behaviour is something that should be avoided and regime operation should be stabilized towards a steady-state regime of constant conversion.

In this section we analyze a practical example of a controlled CSTR in the presence of a first-order reaction for which it is possible, to achieve steady-state stabilization starting from chaotic behaviour.

3.1. The PI-controlled CSTR

It is known that feedback control may induce instabilities[18]. Pellegrini and Biardi[3] first analyzed the complex dynamic behaviour of a proportional-integral (PI) controlled CSTR in the presence of a first-order exothermal reaction. This system was subsequently studied by Giona and Paladino[19], who analyzed the onset of unstable behaviour of the set point. The analysis of chaos induced by control action has been further developed by Biardi et al.[20] with reference to tubular reactors.

Expressed in the dimensionless formalism of Giona and Paladino, the balance equation of a PI-controlled CSTR reads as

$$\begin{aligned}
\frac{d\xi}{d\tau} &= -\xi + f(\theta)(1-\xi) \\
\frac{d\theta}{d\tau} &= \theta_{os} - k_p(\theta - \theta_s) - k_1\varphi + N\theta_e - (1+N)\theta + f(\theta)(1-\xi) \\
\frac{d\varphi}{d\tau} &= \theta - \theta_s \ ,
\end{aligned} \tag{29}$$

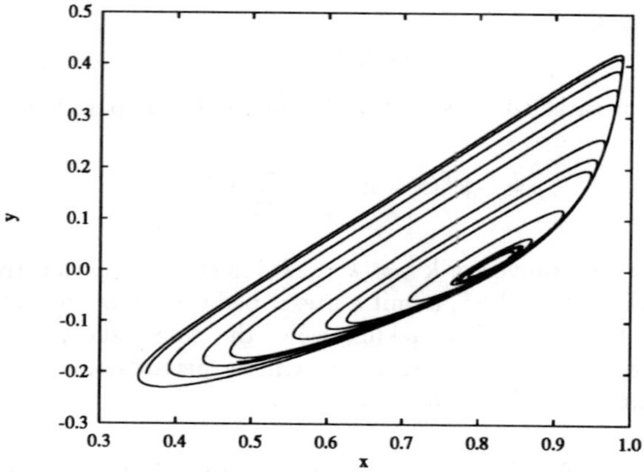

Figure 5: Projection in the ξ-θ-plane ($x = \xi$, $y = \theta$) of the evolution of the PI-controlled reactor in the absence of parametric-excitation control.

where $\xi = (c_{Ao} - c_A)/c_{Ao}$ is the conversion, $\theta = (T - T_{ref})/\Delta T_a$ the dimensionless temperature, $f(\theta)$ the Arrenhius term

$$f(\theta) = k_o t_r \exp\left(-\frac{E}{R(T_{ref} + \Delta T_a \theta)}\right) , \tag{30}$$

φ the internal variable associated with the integral action, and $\mathbf{x}_s = (\xi_s, \theta_s, \varphi_s)$ the set point controlled by the temperature value θ_s. The control rule acts on the temperature θ of the reactor. The parameters k_p and k_1 are the dimensionless control gains associated with the proportional and integral action.

The detailed analysis of the bifurcation properties of this system has been presented elsewhere[19]. In this framework it is important to point out the following properties: (a) independently of control gains k_p, $k_1 > 0$, the unique steady state is the set point $\mathbf{x}_s$, i.e. there is no input multiplicity; (b) there exists a wide region of the parameter space k_p-k_1 in which the set point is unstable; (c) the instability of the set point is governed by a Hopf bifurcation; (d) the Hopf-bifurcated periodic orbits lose stability towards a period-doubling cascade, which makes the system chaotic; (e) the bifurcation behaviour of the controlled reactor is equivalent to that of a unimodal quadratic map (period-doubling cascade, intermittency windows, unversality at the onset of chaos).

Let us consider the following set of parameters:

$$\begin{aligned} k_o t_r &= 4.807908 \cdot 10^8 \ , \ E/R = 8000K \ , \ N = 0.45 \ , \ \theta_s = 0.0 \\ \Delta T_a &= 200K \ , \ T_{ref} = 430K \ , \ T_e = 376.16K \ , \ T_{os} = 298.42K \ , \end{aligned} \tag{31}$$

and $k_p = 0.3$, $k_1 = 0.41$. In these operating conditions, the reactor exhibits chaotic oscillations as depicted in figure 5.

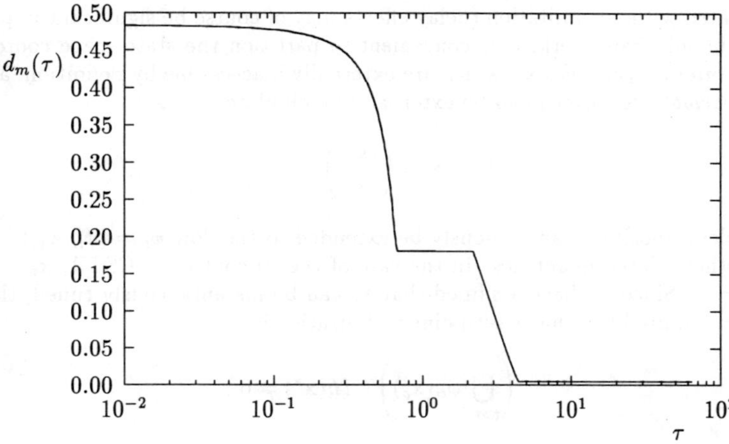

Figure 6: Minimal distance $d_m(\tau)$ as function of time τ of the internal projection of the trajectory of the PI-controlled reactor from the internal projection of the set point. The initial position is chosen at random on the limit attractor.

3.2. Set-point stabilization

The OGY method cannot be applied to stabilize continuous chaotic dynamical systems towards an unstable steady state. The simple reason for this limitation is the fact that unstable steady states are generically repellers: if $\mathbf{x}^*$ is an unstable steady state and $\phi_t(\mathbf{x}_o)$ is the phase flow associated with the dynamical system, Eq. (24), having $\mathbf{x}_o$ as the initial condition, there will always exist a neighbourhood $\mathcal{U}(\mathbf{x}^*)$ of $\mathbf{x}^*$ such that the phase flow will definitely not intersect $\mathcal{U}(\mathbf{x}^*)$.

In order to apply OGY parametric excitation in its local formulation to the steady-state stabilization of continuous systems, the two basic conditions in the OGY approach (topological transitivity and dense structure of the periodic orbits), ensured by the chaotic nature of the dynamics, should be replaced by the condition

$$\left(\bigcup_{t>t^*} \phi_t(\mathbf{x}_o) \right) \cap \mathcal{U}(\mathbf{x}^*) \neq 0 \, , \tag{32}$$

for sufficiently large t^* (i.e. asymptotically). Eq. (32) is not satisfied by a repelling steady state.

If we consider the PI-controlled CSTR, it is possible to derive a condition weaker than Eq. (32) in order to achieve set-point control. The structure of the state space of the PI-controlled reactor consists of the two quantities, the conversion ξ and the temperature θ, which are intrinsically related to the internal behaviour of the system and of the variable φ which is the memory coordinate associated with the integral action. In the limit of instantaneous control, the latter coordinate can be externally manipulated without limitation (delay effects may of course be significant in practice delay). In this framework, it is convenient to partition the state-space coordinates into the *internal variables* $\mathbf{x}_i$, which are externally inaccessible by definition, and the *control variables* $\mathbf{x}_c$, which can be externally manipulated

$$\mathbf{x} = \begin{pmatrix} \mathbf{x}_i \\ \mathbf{x}_c \end{pmatrix} . \tag{33}$$

Similar decomposition can obviously be extended to the flow $\phi_t = (\phi_{ti}, \phi_{tc})^T$ and to all the other related quantities. In the case of the PI-controlled CSTR, $\mathbf{x}_i = (\xi, \theta)^T$ and $\mathbf{x}_c = \varphi$. Since we have assumed that $\mathbf{x}_c$ can be instantaneously tuned, the only condition required to achieve set-point stabilization is

$$\left(\bigcup_{t>t^*} \phi_{ti}(\mathbf{x}_o) \right) \cap \mathcal{U}_i(\mathbf{x}^*) \neq 0 , \tag{34}$$

where $\mathcal{U}_i(\mathbf{x}^*)$ is the projection with respect to the internal coordinates $\mathbf{x}_i$ of the neighbourhood $\mathcal{U}(\mathbf{x}^*)$.

Figure 6 shows, in the case of the PI-controlled CSTR for the considered operating conditions, Eq. (31), that the minimal distance $d_m(\tau) = \min_{0 \leq \tau' \leq \tau} |\phi_{\tau' i}(\mathbf{x}_o) - \mathbf{x}_{si}|$ tends asymptotically to zero starting from an arbitrary initial position (this result is equivalent to Eq. (34)).

Under this condition, a control scheme can be designed to achieve the relaxation of the reactor towards the set point. Let us fix a given parametric controllability set of the form $\mathcal{K} = \{\mathbf{x} | \, |\mathbf{x} - \mathbf{x}_s| < \varepsilon'\}$ with the corresponding internal projection $\mathcal{K}_i = \{\mathbf{x}_i | \, |\mathbf{x}_i - \mathbf{x}_{si}| < \varepsilon\}$, where $\varepsilon' > \varepsilon > 0$.

Once the dynamics crosses the border of $\mathcal{K}_i$ moving inward, the control variable $\mathbf{x}_c$ can be "splashed" towards the set-point value $\mathbf{x}_{cs}$. It is thus possible to apply parametric-excitation control, following Eq. (28). As a control parameter, we can chose the gain of the integral controller, for which the controllability condition, Eq. (26), is satisfied, and apply the pole-placement method in order to derive the values of the gain vector $\mathbf{k}$. More precisely, the dimensionless heat-balance equation attains the form

$$\frac{d\theta}{d\tau} = \theta_{os} - k_p(\theta - \theta_s) - (k_1 + p)\varphi + N\theta_e - (1 + N)\theta + f(\theta)(1 - \xi) , \tag{35}$$

where

$$p = -k_\xi(\xi - \xi_s) - k_\theta(\theta - \theta_s) - k_\varphi(\varphi - \varphi_s) \tag{36}$$

As an example, figure 7 shows the application of this control rule to the PI-controlled CSTR and the convergence of the state variable towards the set-point value. The control gain vector $\mathbf{k}$ was determined by imposing the condition that all the regulation poles should be equal to $\lambda = -0.75$.

Detailed analysis of the parametric control and steady-state stabilization of PI-controlled CSTR has been developed by Giona and Patierno[21]. There are, however,

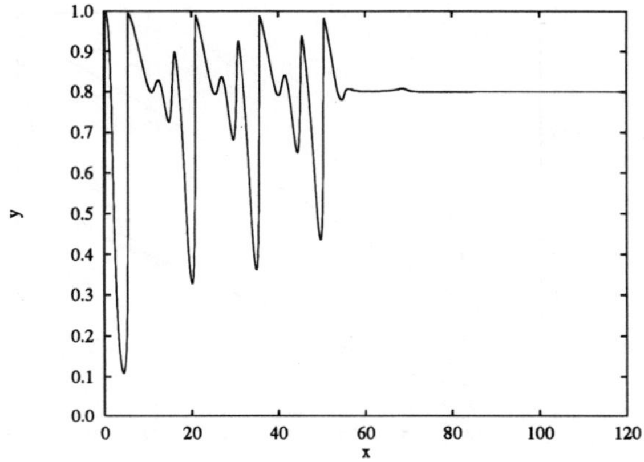

Figure 7: $\xi(\tau)$ vs τ ($x = \tau$, $y = \xi(\tau)$) for the controlled CSTR in the presence of parametric excitation control, derived by setting all the regulator poles equal to $\lambda = -0.75$. The radius of the control ball $\varepsilon = 1.0 \cdot 10^{-2}$. Parametric excitation control starts from $\tau = 50.0$.

a number of points to be addressed. The set-point stabilization presented here again makes implicit use of the chaotic nature of the reactor. In those regions of k_p-k_1 parameter space in which the PI-controlled CSTR oscillates, it is not possible to obtain steady-state stabilization with the methods discussed in this section, since Eq. (34) is no longer satisfied. The analysis of parametric excitation control developed in this section can be improved by further considering some practical aspects, and in particular the fact that the conversion ξ cannot easily be monitored on-line. This implies $k_\xi = 0$ in Eq. (36), and correspondingly the need to define stable regular poles with this constraint[21]. In the same way, a practical definition of the region of the phase space in which to activate parametric control should refer exclusively to the only internal variable which can be monitored on-line, i.e. the temperature θ. All these aspects represent a refining of steady-state stabilization and are therefore presented elsewhere[21].

4. On the structure of the control rule: parametric excitation vs feedback control

In their review article on the control of chaos[2], Chen and Dong stress the comparison between the parametric excitation method and classical engineering control schemes (e.g. global feedback control) as applied to chaotic dynamical systems. As the authors argue that the latter is in some ways a better approach to the solution of engineering problems involving chaotic evolutions, it is worth addressing this question with some care.

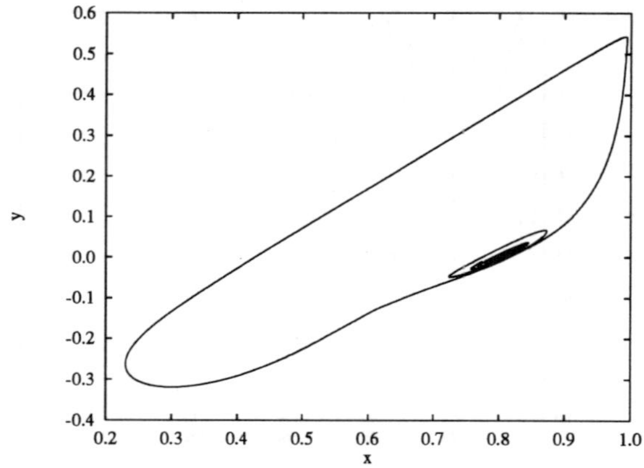

Figure 8: Projection in the ξ-θ ($x = \xi$, $y = \theta$) of the evolution of a PI-controlled CSTR in the presence of the parametric control (36) acting globally. The evolution stabilizes towards a spurious limit cycle. The gains k_ξ, k_θ, k_φ are the same as in the simulation of figure 7 and all the regulator poles are thus necessarily equal to $\lambda = -0.75$.

By the expression *classical engineering control*, Chen and Dong mean any kind of perturbation of a system (Eq. (24)) of the form

$$\frac{d\mathbf{x}}{dt} = \mathbf{f}(\mathbf{x};\mathbf{p}) + \mathbf{H}_c(\mathbf{x},t)\ , \tag{37}$$

where $\mathbf{H}_c(\mathbf{x},t)$ represents the action of the controller, which may depend on time t. The simplest version of this control scheme is $\mathbf{H}_c(\mathbf{x},t) = -\mathbf{K}(\mathbf{x}-\mathbf{x}^*)$, where $\mathbf{x}^*$ is the steady state to be stabilized. By applying this form of control, Cheng and Dong analyze the stabilization of the discrete Lozi map, of the Duffing oscillator, and of the Chua system[2].

Control rules of the form (37) represent an intrinsic modification of the dynamical system: the action of the controller is applied globally, i.e. for every value of the state vector $\mathbf{x}$. However, such global control rules are not necessarily suitable in the case of nonlinear systems. Indeed, the PI-controlled CSTR of the previous section is a typical example of a system in which proportional-integral control does not necessarily stabilize the set point; and for some values of the gains k_p, k_1 the system behaves chaotically. However, there is a stronger argument against the generalized use of global feedback in nonlinear systems.

Let us again consider the the PI-controlled CSTR, and carry out parametric control, Eq. (36), not only locally but globally. In the presence of global parametric control, the reactor evolution is trapped in a spurious limit cycle, figure 8. This example clearly shows that it is not possible to design a succesfully control rule based on local properties of a nonlinear system (see section 2.2) which should act globally.

On the other hand, the parametric excitation method is designed on the basis of local properties of the system, and consequently the control action is activated only in a proper neighbourhood of the orbit to be stabilized. In this framework, topological transitivity and the dense structure of the periodic orbits ensure that the system will visit all neighbourhoods of the chosen orbit, so that the chaotic nature of the dynamics plays the role of an dynamic carrier which places the system in a region of the state-space where control can be efficiently activated.

Problems arise in connection with the steady-state stabilization of continuous chaotic systems. An approach to this problem, involving the decomposition of the state vector into internal and control variables, has been presented for a typical reacting system. This method can be certainly improved upon and generalized to other engineering systems. Going back to the original question posed by Chen and Dong, we are convinced that the most suitable way to approach the control of generic chaotic systems would be by improving and generalizing parametric excitation methods in order to encompass practical engineering problems rather than by analysing specific *classical control rules* (in the sense of Chen and Dong) that work for a narrow class of chaotic systems.

5. Concluding remarks

We have presented a general analysis of parametric excitation methods to control chaotic evolution. The key problem in the engineering application of parametric excitation is the difficulty in achieving steady-state stabilization. This is due to the fact that unstable steady states are in general repellers, and therefore the dynamics will not pass close to them (as is required for switching parametric action). In the particular case of the PI-controlled CSTR in chaotic permanent oscillations, it is still possible to achieve steady-state stabilization. This is due to the fact that the proportional-integral action drives the dynamics arbitrarily close to the projection of the set-point with respect to the internal state variables. Indeed, it should be noted that such control action requires local application. A global application of parametric excitation will drive the system towards undesired limit cycles.

The approach developed for the PI-controlled CSTR can be generalized to other reacting systems. On the basis of the results outlined in this article, we are convinced that a promising direction for research on chaos control that could have a direct practical impact on engineering problems lies in the generalization of steady-state control for a generic continuous-time chaotic system. A possible way to achieve this goal should be sought in the class of *migration/excitation controls*, figure 9. This means designing a migration system[22,2] that will drive the dynamics arbitrarily close to the internal projection of the steady state and applying the parametric excitation method in the neighbourhood of the state state, as discussed in sections 2.2 and 3.2. In the case of the PI-controlled CSTR, the migration action is ensured by the linear proportional-integral action.

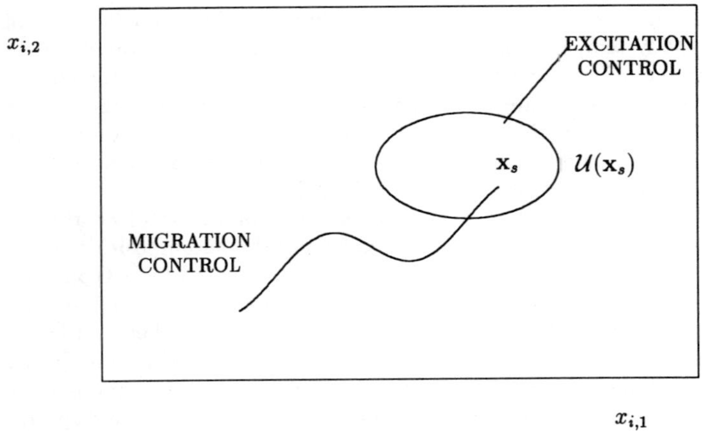

Figure 9: Schematic representation of a general migration/excitation control method for dynamical systems The migration control will drive the system in a neighbourhood $\mathcal{U}(\mathbf{x}_{si})$ of the projection on the internal state variables $\mathbf{x}_{si}$ of the set point $\mathbf{x}_s$ to be stabilized. Once the dynamics lies in $\mathcal{U}(\mathbf{x}_{si})$, parametric-excitation control can be applied as in the case of the PI-controlled CSTR (see Eq. (36)).

Appendix A

The characteristic equation of the dynamical augumented system in the neighbourhood of the fixed point $\mathbf{x} = \mathbf{x}^*$ and $\mathbf{p} = \mathbf{0}$ is given by

$$\Pi(\lambda) = \begin{vmatrix} \mathbf{A} - \lambda\mathbf{I} & \mathbf{0} \\ \mathbf{I} & -\lambda\mathbf{I} \end{vmatrix} = (-1)^n \lambda^n \det(\mathbf{A} - \lambda\mathbf{I}) = 0 \tag{A.1}$$

and therefore the eigenvalue $\lambda = 0$ has algebric multiplicity at least equal to n. It is easy to see that there exist at least n linearly independent eigenvectors $\{\varepsilon_1, .., \varepsilon_n\}$ associated with $\lambda = 0$ which are of the form $\varepsilon_i = (\mathbf{0}, \mathbf{e}_i)^T$, where $\{\mathbf{e}_i\}$ is the n-dimensional orthonormal Cartesian basis $\mathbf{e}_1 = (1, 0, .., 0), \mathbf{e}_2 = (0, 1, .., 0)$, etc.

References

1. E. Ott, C. Grebogi and J. A. Yorke, *Phys. Rev. Lett.* **64** (1990) 1196.
2. G. Chen and X. Dong, *Int. J. Bifurcation and Chaos* **3** (1993) 1363.
3. L. Pellegrini and G. Biardi, *Comp. Chem. Eng.* **14** (1990) 1237.
4. R. L. Devaney, An Introduction to Chaotic Dynamical Systems (Addison Wesley, Redwood City, 1989).

5. Hao Bai-lin, *Elementary Symbolic Dynamics and Chaos in Dissipative Systems* (World Sci., Singapore, 1989).
6. E. Ott, C. Grebogi and J. A. Yorke, in Chaos/XAOC Soviet-American Perspectives on Nonlinear Science, ed. D. K. Campbell (American Institute of Physics, New York, 1990).
7. M. Marek and I. Schreiber, *Chaotic Behaviour of Deterministic Dissipative Systems* (Cambridge University Press, Cambridge, 1991).
8. L. M. Pecora and T. L. Carroll, *Phys. Rev. Lett.* **64** (1990) 821.
9. J. Singer, Y-Z. Wang and H. H. Bau, *Phys. Rev. Lett.* **66** (1991) 1123.
10. W. L. Ditto, S. N. Rauseo and M. L. Spano, *Phys. Rev. Lett.* **65** (1990) 3212.
11. R. Roy, T. W. Murphy Jr,, T. D. Maier, Z. Gills and E. R. Hunt, *Phys. Rev. Lett.* **68** (1992) 1259.
12. B. Peng, V. Petrov and K. Showalter, *J. Phys. Chem.* **95** (1991) 4957.
13. V. Petrov, B. Peng. and K. Showalter, *J. Chem. Phys.* **96** (1992) 7506.
14. J. K. Bandyopadhyay, V. Ravi Kumar and B. D. Kulkarni, *AIChE J.* **39** (1993) 908.
15. G. Nitsche and U. Dressler, *Physica* **D 58** (1992) 153.
16. F. J. Romeiras, C. Grebogi, E. Ott and W. P. Dayawansa *Physica* **D 58** (1992) 165.
17. K. Ogata, *Modern Control Engineering* (Prentice Hall, Englewwod Cliffs, 1970).
18. H. C. Chang and L. H. Chen, *Chem. Engng. Sci.* **39** (1984) 1127.
19. M. Giona and O. Paladino, *Comp. Chem. Eng.* **18** (1994) 877.
20. L. Pellegrini, S. Albertoni and G. Biardi, *Chem. Engng. Sci.* **47** (1992) 2463.
21. M. Giona and O. Patierno, to be presented at the AIChE Annual Meeting, San Francisco 13-18 November 1994.
22. E. A. Jackson, *Phys. Rev.* **A 44** (1991) 4839.

ON THE NUMERICAL SIMULATION OF THE CHAOTIC BEHAVIOUR OF SOME DISTRIBUTED-PARAMETER SYSTEMS

G. CONTINILLO
Istituto di Ricerche sulla Combustione - C.N.R.
via Diocleziano 328 - 80124 Napoli

P. L. MAFFETTONE and S. CRESCITELLI
Dipartimento di Ingegneria Chimica, Università 'Federico II'
Piazza V. Tecchio - 80125 Napoli

ABSTRACT

The detailed numerical simulation of autoignition of coal stockpiles lead to observe non-steady regimes. To better investigate this phenomenon, two simplified distributed-parameter (one- and two-dimensional) models were developed which incorporate heat conduction, mass diffusion, and a one-step Arrhenius exothermic chemical reaction. Both models' equations were solved with straightforward finite-difference schemes. Clues of a chaotic behaviour were found for both models, resembling the one known as 'diffusion-induced chemical turbulence' earlier observed for some multi-step, isothermal reactions. The simpler (one-dimensional) model's PDEs were then approximated by a standard collocation technique, leading to a system of ODEs which was fed to a continuation package, and the non-dimensional activation energy parameter was explored among the four parameters of the model.

1. Introduction

Kuramoto[1] showed that diffusion-induced chemical 'turbulence' may arise in isothermal systems with at least two 'nontrivial' reacting species and mass diffusion along one space coordinate. Kim and Hlaváček[2] confirmed this finding for a distributed-parameter system including reaction, convection, and diffusion; they also showed that, in that system, chaotic behaviour is not found at sufficiently high values of the Peclet number. Aris[3] reported that complex oscillations may occur in a catalyst pellet when mass diffusivity is smaller than thermal diffusivity.

The analysis of bifurcation and the onset of chaos for distributed-parameter systems is a nontrivial issue. A possible approach consists of discretising the continuous problem, and conducting the computation on the resulting non-differential system until a steady state is reached. This approach, also referred to as 'brute force', was used for example by the group of Biardi[4] for the case of a tubular reactor with axial diffusion in the presence of a feedback controller[5,6]. A similar technique was also followed in the analysis of the diffusion-induced chaotic reignition of a fixed-bed non isothermal reactor[7]. The 'brute force' technique is rather straightforward to apply, but suffers from important limitations, in that non-stable limit sets (such as saddle points) cannot be detected[8]. On the other hand, there is currently no practical method for locating chaotic solutions other than brute force[9]. Once located, strange attractors can be characterised as such by means of further computations[10].

To obtain the complete bifurcational portrait of a system in a certain region of the parameter space, it is desirable to apply the parameter continuation techniques used for ODE sets pertaining to lumped-parameter systems. Actually, Juncu and Bildea[11] have investigated the computation of simple, double, and Hopf bifurcations with orthogonal collocation and parameter continuation for a diffusion-reaction distributed-parameter problem, favourably comparing their results with earlier data obtained with different techniques. They also claim the inadequacy of the pseudo-arclength continuation technique when applied to this system, that lead them to develop an alternative continuation method; this inadequacy was not confirmed in our experience.

In this work, the 'brute force' and the orthogonal collocation approach were employed to study the complex dynamic behaviour of a 1D distributed-parameter model of the spontaneous reignition of a coal stockpile; pseudo-arclenght continuation was applied to precisely locate the successive period-doubling bifurcations encountered in the simulations. Finite-difference simulations were then conducted for a 2D distributed-parameter model of the same phenomenon.

2. Model Equations

The evolutionary equations of a classic diffusion-reaction problem are considered. After algebraic manipulation, the governing equations are cast in the following nondimensional form:

Species

$$\frac{\partial Y}{\partial t} = \mathrm{Le}\frac{\partial^2 Y}{\partial x^2} - \Phi^2 Y \exp\left(-\frac{\gamma}{T}\right) \tag{1}$$

Energy

$$\frac{\partial T}{\partial t} = \frac{\partial^2 T}{\partial x^2} + \beta\Phi^2 Y \exp\left(-\frac{\gamma}{T}\right) \tag{2}$$

with initial and boundary conditions: $T(0,t) = T(1,t) = 0$; $Y(0,t) = 1$; $\partial Y/\partial x|_{x=1} = 0$ for $t > 0$, and $T(x,0) = Y(x,0) = 0$ at $t = 0$. Here Le is the ratio between mass and heat diffusivities, β is a nondimensional heat of reaction, Φ is the Thiele modulus, and γ is a nondimensional activation energy. The 2D equations are:

Species

$$\frac{\partial Y}{\partial t} = \mathrm{Le}\left(\frac{\partial^2 Y}{\partial x^2} + \frac{\partial^2 Y}{\partial y^2}\right) - \Phi^2 Y \exp\left(-\frac{\gamma}{T}\right) \tag{3}$$

Energy

$$\frac{\partial T}{\partial t} = \frac{\partial^2 T}{\partial x^2} + \frac{\partial^2 T}{\partial y^2} + \beta\Phi^2 Y \exp\left(-\frac{\gamma}{T}\right) \tag{4}$$

with initial and boundary conditions: $\partial Y/\partial x = 0, \partial T/\partial x = 0$ at $x = 0$; $T = 0, \partial Y/\partial y = 0$ at $y = 0$; $T = 0, Y = 1$ at $x = 1$ and at $y = \alpha$; $T = 0, Y = 0$ at $t = 0$, where α is the aspect ratio of the rectangular domain.

3. Numerical Methods

3.1. Finite-difference method

Although somewhat trivial, it is perhaps worth reporting the simple finite-difference schemes employed in the calculations. Explicit FTCS (Forward-Time-Centered-Space) schemes were adopted in the equations, except that for the source terms, where a semi-implicit scheme (implicit in the concentration) was preferred to ensure positivity in the species balance equation, and to avoid stability problems with the production term. Stability issues are thus confined to the explicit treatment of the second-derivative terms. The 1D model equations were discretised as follows, with a uniform spatial grid:

Species, node i

$$\frac{Y_i^{n+1} - Y_i^n}{\Delta t} = \mathrm{Le}\frac{Y_{i-1}^n - 2Y_i^n + Y_{i+1}^n}{\Delta x^2} - \Phi^2 Y_i^{n+1} \exp\left(-\frac{\gamma}{T_i^n}\right) \tag{5}$$

Energy, node i

$$\frac{T_i^{n+1} - T_i^n}{\Delta t} = \frac{T_{i-1}^n - 2T_i^n + T_{i+1}^n}{\Delta x^2} + \beta\Phi^2 Y_i^{n+1} \exp\left(-\frac{\gamma}{T_i^n}\right) \tag{6}$$

The difference equations for Y_i^{n+1} are decoupled from those for T_i^{n+1}, thus the former are solved first, and Y_i^{n+1} is fed to the latter, which are solved in turn. A simple control of the time step was incorporated by limiting the maximum expected relative change in the discrete unknowns by linear extrapolation from the two last time levels. The corresponding finite-difference equations for the 2D model are:

Species, node i, j

$$\frac{Y_{i,j}^{n+1} - Y_{i,j}^n}{\Delta t} = \mathrm{Le}\left(\frac{Y_{i-1,j}^n - 2Y_{i,j}^n + Y_{i+1,j}^n}{\Delta x^2} + \frac{Y_{i,j-1}^n - 2Y_{i,j}^n + Y_{i,j+1}^n}{\Delta y^2}\right) - \Phi^2 Y_{i,j}^{n+1} \exp\left(-\frac{\gamma}{T_{i,j}^n}\right) \tag{7}$$

Energy, node i, j

$$\frac{T_{i,j}^{n+1} - T_{i,j}^{n}}{\Delta t} = \frac{T_{i-1,j}^{n} - 2T_{i,j}^{n} + T_{i+1,j}^{n}}{\Delta x^2} + \frac{T_{i,j-1}^{n} - 2T_{i,j}^{n} + T_{i,j+1}^{n}}{\Delta y^2} + \beta\Phi^2 Y_{i,j}^{n+1} \exp\left(-\frac{\gamma}{T_{i,j}^{n}}\right). \tag{8}$$

3.2. Orthogonal collocation method

Equations (1,2) were also integrated with an orthogonal collocation scheme[12]. The collocation equations are:

$$\frac{dY_j}{dt} = \mathrm{Le} \sum_{i=1}^{n+2} B_{ji} Y_i - \Phi^2 Y_j exp\left(-\frac{\gamma}{T_j}\right) \tag{9}$$

$$\frac{dT_j}{dt} = \mathrm{Le} \sum_{i=1}^{n+2} B_{ji} T_i + \beta\Phi^2 Y_j exp\left(-\frac{\gamma}{T_j}\right), \tag{10}$$

$$T_1 = 0, T_{n+2} = 0, Y_1 = 1, \sum_{i=1}^{n+2} A_{n+2,i} Y_i = 0. \tag{11}$$

where subscript j refers to the jth of the n interior collocation points. The matrices **A** and **B** depend only on the choice of the polynomial approximation and are calculated once and for all at the beginning of the integration procedure. This numerical approach reduces the two PDEs to a set of $2n$ ODEs, where n is the number of collocation points. The set of ODEs is integrated with a fourth-order Runge-Kutta algorithm with time step control. So far, the polynomials were chosen with no optimization; the number of collocation points was changed from 3 to 15, and a good match with the finite-difference results was obtained with 10 to 12 collocation points. All the reported results were calculated with 12 collocation points. We do believe that a better choice of the polynomial and of the position of the collocation points can lead to a significant reduction of the number of collocation points required and thus of the order of the ODE system.

The use of the orthogonal collocation scheme allows a parameter-continuation analysis of the model. The set of 24 ODEs was implemented in the code AUTO[13], a continuation package based on pseudo-arclength parametrisation. Several kinds of bifurcation points were detected: limit points, Hopf bifurcations, folds of periodic branches, period-doubling bifurcations, bifurcations of limit cycles to tori. However, only one cascade of period doubling bifurcations, which leads to chaotic behavior is here reported in a following section.

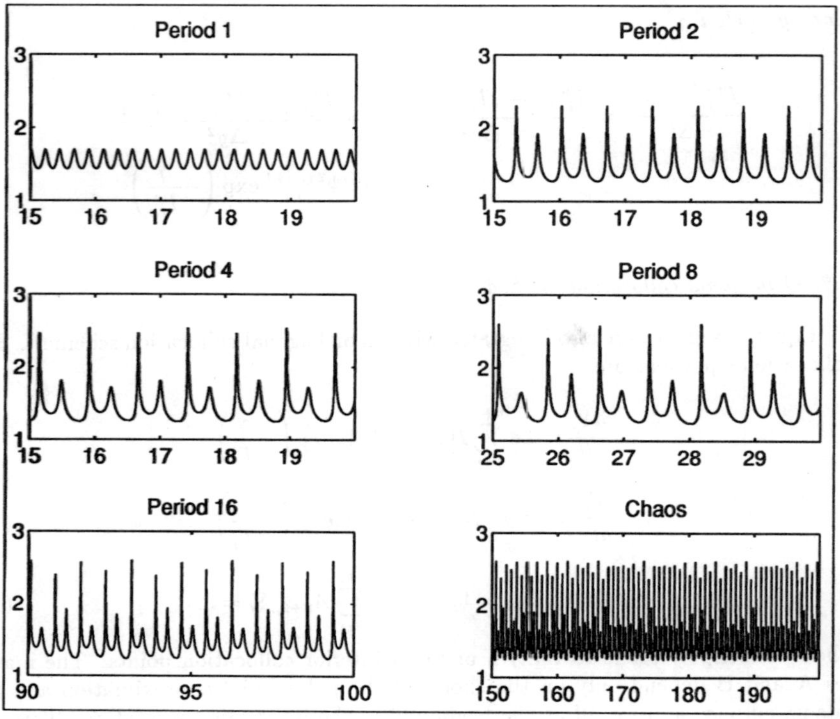

Figure 1: Period-doubling cascade as computed with finite-differences

4. Results and Discussion

As mentioned earlier, the application of the finite-difference approach to some selected sets of values of the parameters had shown oscillatory regimes along with clues of chaotic behaviour. In view of the highest sensitivity of the system to γ among the four (five for the 2D problem) parameters, this was selected as the bifurcation parameter in this study, while keeping the others constant to the following values: $\mathrm{Le} = 0.233, \Phi^2 = 70000, \beta = 4.287$.

The aperiodic oscillation was achieved via a period-doubling cascade as represented in Figure 1. Here the time series of the value taken by the nondimensional temperature at $x = 0.115$ are shown for six different values of γ.

The mussel-shaped limit set corresponding to $\gamma = 12.6$ is reported in Figure 2 as a phase portrait of the two variables (temperature and concentration) evaluated at the same spatial location. All of these results were obtained by means of finite-difference calculations, and are grid-independent for 25 or more spatial nodes.

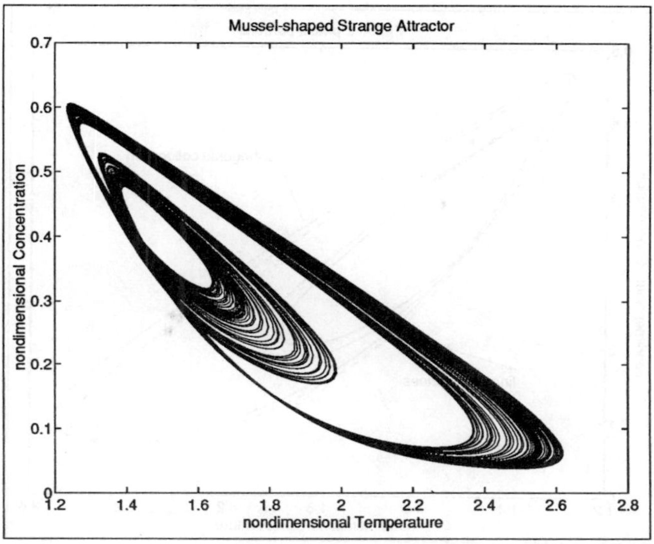

Figure 2: Phase-portrait at $\gamma = 12.60$

This approach could not locate the bifurcation points with reasonable accuracy. To this end, a continuation technique can be adopted; however, reliable continuation procedures are currently available for ODE sets only. Thus, the orthogonal collocation numerical fomulation was set up to reduce the PDE set to a higher number of ODEs.

Figure 3 compares the limit set found at $\gamma = 12.55$ with finite-differences (50 nodes) and orthogonal collocation (12 points). The agreement was considered satisfactory, and the parameter continuation was conducted with 12 points (24 ODEs).

Figure 4 reports the bifurcation diagram obtained with AUTO, constructed by plotting values of the nondimensional temperature at $x = 0.115$ versus γ. A certain degree of care must be used to interpret this diagram. Solid lines represent stable steady-state solutions. Dashed lines represent unstable steady-state solutions. Circles represent periodic solutions, filled if stable, hollow if unstable. When representing a periodic solution, the maximum value achieved by the variable is picked.

AUTO labels bifurcation points with cardinal numbers in the order they are detected, and characterises them in terms of the kind of bifurcation. For the sake of clarity, we removed all bifurcation points from the diagram but the two Hopf bifurcations (4 and 8) and four period-doubling bifurcation points (11, 13, 15 and 19).

For the selected set of values of the other three parameters, the system has stable steady-state solutions for $\gamma < 12.12944$ and $\gamma > 13.604$. At the Hopf bifurcation point 8 an unstable periodic branch emanates from the stable steady-state branch, which readily turns into stable with a fold, invisible in our representation: this behaviour

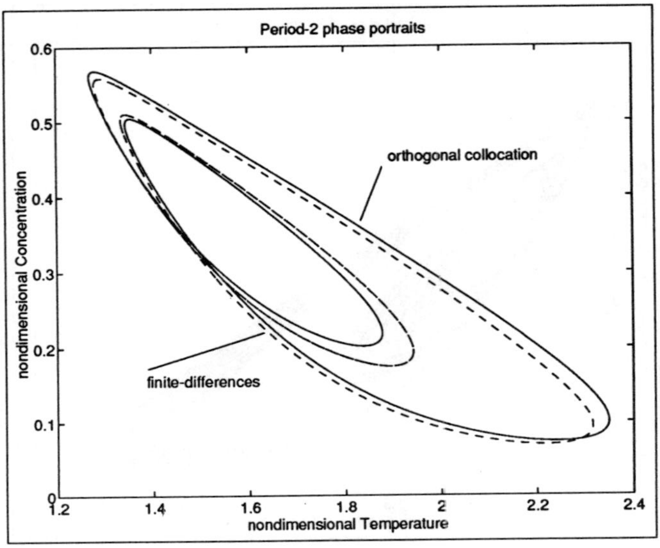

Figure 3: Comparison of phase-portraits obtained with finite-difference calculations (dotted line) and orthogonal collocation calculations (solid line) at $\gamma = 12.55$

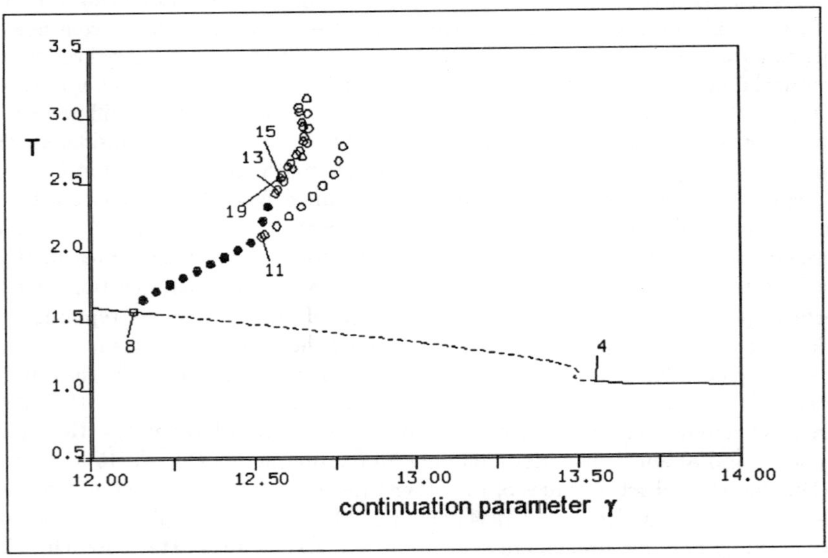

Figure 4: Bifurcation diagram obtained with pseudo-arclength continuation

is recognisable from the hollow circle at point 8. The period-2 branch emanating from point 11 is clearly visible whereas period-4, period-8 and period-16 branches are nearly superimposed in our representation. The bifurcation values are as follows: $\gamma_{11} = 12.52652$; $\gamma_{13} = 12.57657$; $\gamma_{15} = 12.58624$; $\gamma_{19} = 12.58832$. The evaluation of the Feigenbaum ratio

$$\delta = \frac{\gamma_n - \gamma_{n-1}}{\gamma_{n+1} - \gamma_n} \tag{12}$$

yields, for the last available triplet, the value of 4.649 which is already close to the universal limit value of 4.669 discovered by Feigenbaum for period-doubling routes to chaos[14].

The complex dynamic behaviour encountered for the 1D system is found, even more complex, in finite-difference simulations of the 2D system, as shown by preliminary calculations whose results will be reported and discussed in a forthcoming paper.

5. Conclusions and future work

Diffusion-reaction systems with a single-step exothermal reaction have been variously addressed in the literature, and complex oscillating behaviours have been reported. However, to our knowledge, aperiodic oscillations have never been found for such systems. This work points out that aperiodic (actually chaotic) behaviour is exhibited by such systems for at least one set of values of the parameters. The transition to chaos occurs through a period-doubling cascade as reconstructed with continuation analysis. The Feigenbaum constant is closely approached by the succession of period-doubling bifurcation values.

The non optimal choice of collocation points, i. e., the high dimensionality of the ODE system, determine a cumbersome continuation analysis: in fact, each time one wants to follow the branch that emanates from a period-doubling bifurcation, the memory requirements of the continuation procedure are nearly doubled, so that one may rapidly reach the hardware limits. A more efficient selection of the collocation points must be sought to overcome this limitation.

The continuation analysis should be extended to the 2D system. However, the extension of the numerical collocation procedure to 2D is not trivial[15].

The strange attractor still has to be thoroughly characterised, by means of standard nonlinear techniques (computation of the largest Lyapunov exponent; estimation of the fractal dimension of the attractor).

References

1. Y. Kuramoto, *Progr. Theor. Phys. Suppl.* **64** (1978) 3346.
2. S. H. Kim, and V. Hlaváček, *Physica D* **10d** (1984) 413.
3. R. Aris, *The Mathematical Theory of Diffusion and Reaction in Permeable Catalysts* (Clarendon Press, Oxford, 1975), Vol. 2.
4. G. Biardi, L. Pellegrini and C. Tablino Possio, in *Analysis, Simulation and Dynamics of Chemical Processes*, Edited by G. Continillo, P. G. Lignola and S. Crescitelli (Società Editrice di Chimica, Milan, 1994, in press)
5. L. Pellegrini, C. Tablino Possio, and G. Biardi, Proceedings of the *First Conference on Chemical and Process Engineering*, Firenze, 13-15 May 1993, p.378.
6. L. Pellegrini, S. Albertoni and G. Biardi, *Chem. Eng. Sci.* **47** (1992) 2463.
7. G. Continillo, P. L. Maffettone and S. Crescitelli, Proceedings of the *First Conference on Chemical and Process Engineering*, Firenze, 13-15 May 1993, p. 299
8. T. S. Parker, and L. O. Chua, *Practical Numerical Algorithms for Chaotic Systems*, (Springer-Verlag, New York, 1989), p.115.
9. T. S. Parker, and L. O. Chua, *ibidem*, p.136.
10. J. M. T. Thompson and H. B. Stewart, *Nonlinear Dynamics and Chaos* (Wiley, New York, 1986)
11. G. Juncu, and S. Bildea, *Computers Chem. Engng.* **16** (1992) 1059.
12. J. Villadsen and M. L. Michelsen, *Solution of Differential Equation Models by Polynomial Approximation* (Prentice-Hall, Englewood Cliff, 1978)
13. E. J. Doedel and J. P. Kernevez, *AUTO: A Software for Continuation and Bifurcation Problems in Ordinary Differential Equations*, California Institute of Technology, 1986.
14. M. J. Feigenbaum, *J. Stat. Phys.* **19** (1978) 25.
15. J. C. Eilbeck and J. E. Furter, in *Continuation and Bifurcations: Numerical Techniques and Applications*, Edited by D. Roose, B. De Dier, and A. Spence (Kluwer Academic Publishers, Dordrecht, 1990) p.30.

THE ROLE OF TIME DELAY ON THE STABILITY OF CHEMICAL SYSTEMS

LAURA PELLEGRINI, GIUSEPPE BIARDI
Dipartimento di Chimica Industriale e Ingegneria Chimica "G. Natta"
Politecnico di Milano - P.le Leonardo da Vinci, 32
Milano, 20133, Italy

and

CRISTINA TABLINO POSSIO, SERGIO ALBERTONI
Dipartimento di Matematica "F. Enriques" - Università di Milano - via Saldini, 50
Milano, 20133, Italy

ABSTRACT

Two different delayed chemical reacting systems are considered in order to analyse the role of time delay on their dynamic behaviour in comparison with the undelayed one.
Different techniques are considered for the numerical simulation and for the Lyapunov Characteristic Exponents computation of such infinite dimensional systems.
The existence of critical time delay values, for which destabilisation occurs, is proved for both the chemical reacting systems.

1. Introduction

In simulating the dynamic behaviour of nonlinear chemical systems one must often take into account the presence of a time delay, due to the system configuration itself (recycles, etc.) or inherent in the control system (delays due to measurement and actuator).
Since the few papers on the subject (Boe and Chang[1], Inamdar et al.[2]), agree on the destabilising effect of time delay, we have analysed some representative examples, with the aim of recognising the common features and verifing whether destabilisation occurs for physically reasonable values of time delay or if the matter is merely speculative. When we say that destabilisation occurs, we mean that the destabilisation of a regime occurs, i.e. that time delay can also convert a strange attractor to a periodic solution or vice versa.
The examples we have analysed concern chemical reacting systems and specifically CSTRs with time delay due to a recycle stream and to controller action.
A system with time delay is infinite dimensional and its simulation usually requires a remarkable numerical and computational effort.
There are two ways of approaching the problem; the first, that seems to be, at first sight, the simpler, determines the value of the dependent variables at the time $t - \tau_D$ using an interpolation algorithm (Inamdar et al.[2]). It is worth remarking that the problem to be solved may be intrinsically "stiff" because of the different magnitude

of the time constants involved (i.e. the intrinsic time constant of the system and the time delay). So it is impossible "a priori" to choose the integration step proportional to the time delay and a proper self adjusting step method has to be used to solve the Ordinary Differential Equations.

The second way consists of substituting the delay operator with a plug flow system that represents the physical device responsible for the delay. This approach requires the introduction of a Partial Differential Equation that can be reduced, by applying a finite difference method, to an ODEs system to be coupled to the original one. The spatial subdivisions are, in some way, the equivalent of the interpolation knots.

Even if this is apparently more complex it offers a few advantages, i.e. the same numerical tools developed for the dynamic analysis of large ODEs systems can be applied. In particular the computation of the Largest Lyapunov Characteristic Exponent, that is the basic tool for chaos detection, can be performed in the same way. The analysis of the examples will show the great influence of the time delay on the dynamic behaviour of both systems.

It is not possible to define a univocal trend.

Numerical computations have shown cases where high τ_D values can stabilise a periodic orbit on a steady state and make chaotic behaviour periodic. In other situations, on the contrary, a τ_D increase is able to destabilise a steady state to a periodic orbit. However, in both examples, for given values of the system parameters, chaos occurs in a precise range of τ_D values, showing how the stability properties are not monotone to such a system parameter.

2. Formulation of the problem

Let us consider a $n-$dimensional nonlinear system

$$\frac{d\boldsymbol{y}(t)}{dt} = \boldsymbol{F}(\boldsymbol{y}(t); \boldsymbol{u}(t-\tau_D)) \quad \boldsymbol{y} \in R^n, \boldsymbol{u} \in R^m \quad with \quad m \leq n \tag{1}$$

where, with this notation $\boldsymbol{u}$, we want to indicate those components of vector $\boldsymbol{y}$ whose value at the time $t - \tau_D$ is required, since the computational effort is proportional to the $\boldsymbol{u}$ dimension.

τ_D represents the time delay; we assume that in the process only one representative time delay is present.

To solve the system (1) for times greater than t, the function $\boldsymbol{u}(t)$ over the the whole interval $(t - \tau_D, t)$ must be given.

So, the system is infinite dimensional since an infinite set of independent numbers are required to specify an initial condition. If Eqs. (1) are linear in $\boldsymbol{u}(t-\tau_D)$, as in the example cases examined in the following, they can be rewritten as:

$$\frac{d\boldsymbol{y}(t)}{dt} = A\boldsymbol{y}(t) + B\boldsymbol{u}(t-\tau_D) + \boldsymbol{G}(\boldsymbol{y}(t)) \tag{2}$$

where $\boldsymbol{G}(\boldsymbol{y}(t))$ contains all the nonlinear terms.

The simulation of an infinite dimensional dynamic system requires the approximation

of the continuous evolution of the infinite dimensional system by a finite number of elements whose value is considered at discrete times. In this way a continuous infinite dimensional dynamic system is replaced by finite dimensional iterated mapping.
There is not just one method of solving it.
Since the nonlinear equations we use in the following sections to describe the dynamic behaviour of the example cases cannot be solved analytically, we have adopted two different numerical methods of approaching the problem:

1. The values of variables or, better, the $\boldsymbol{u}$ values must be stored for a time not less than τ_D. Since, just because of the time delay, Eqs. (1) are usually stiff, a variable step size algorithm is used to solve the system.
   So, to evaluate the exact values of $\boldsymbol{u}(t-\tau_D)$, a cubic spline interpolation/extrapolation method is applied to the stored values of the $\boldsymbol{u}$ variable.

2. The role of the time delay is assumed by a plug flow system that represents the physical device responsible for the delay. This is the technique first utilised by Soliman and Ray[3] and applied again by Boe and Chang[1].
   The infinite dimensional delayed system is reduced at first to an infinite dimensional undelayed system by introducing the following hyperbolic Partial Differential Equation
   $$\frac{\partial \boldsymbol{x}}{\partial t} = -\frac{1}{\tau_D}\frac{\partial \boldsymbol{x}}{\partial \xi} \qquad \boldsymbol{x} \in R^m \tag{3}$$
   with the boundary conditions:
   $$\boldsymbol{x}(t,0) = \boldsymbol{u}(t) \quad \text{for} \quad \xi = 0 \tag{4}$$
   The coupled solution of Eqs. (3) plus Eqs. (2) with the position
   $$\boldsymbol{u}(t-\tau_D) = \boldsymbol{x}(t,\xi)_{|\xi=1} \tag{5}$$
   allows the cancellation of the delay operator of Eqs. (2). A finite difference method is applied to the system (3) along the spatial (delay) coordinate in order to reduce the hyperbolic equations to an ODEs system, normally of large dimension; with s spatial subdivisions we pass to a $m \cdot s$ ODEs system:
   $$\frac{d\boldsymbol{x}_i}{dt} = -\frac{1}{\tau_D \Delta\xi}(\boldsymbol{x}_i - \boldsymbol{x}_{i-1}) \qquad i = 1, s \tag{6}$$

Both of these techniques have been successfully used to simulate the delayed nonlinear system we have studied, in order to make sure that the numerical simulation is faithful to the original equations of the model.
As reported by Farmer[4], we can consider some indicators of a "proper simulation": a) the behaviour of the simulated system must agree with the analytical solution, if one; b) the trajectories must asymptotically converge when the resolution (number of interpolation knots or of spatial subdivisions) increases; c) the simulation by different

methods must give similar results.
In our work both point b) and c) have been verified; so we assume that the model equations are well simulated.

3. The Lyapunov Characteristic Exponents

For what concerns the existence, uniqueness and general properties of the spectrum of Lyapunov Characteristic Exponents in infinite dimensions, you can refer to Rouelle[5] and Farmer[4].
As for the simulation problem, the computation of the Lyapunov Characteristic Exponents (LCE) is necessarily finite dimensional so that the methods used for finite dimensional dynamic systems can be applied.
We briefly review the definition and the basic concepts related to the Lyapunov Characteristic Exponents before describing the specific procedures followed to compute LCE for delayed systems.
The LCE can be computed as:

$$\lambda = \lim_{n\to\infty} \frac{1}{n\tau} \sum_{i=1}^{n} \left(\ln \left| \frac{d_i}{d_0} \right| \right) \tag{7}$$

where d_0 and d_i represent, respectively, the distance between nearby trajectories in phase space at the initial time and at a generic time $i\tau$.
The sensitivity analysis to the initial conditions can then be performed by simultaneously integrating the generic autonomous nonlinear system

$$\frac{d\boldsymbol{y}}{dt} = \boldsymbol{F}(\boldsymbol{y}) \tag{8}$$

and the linearised system

$$\frac{dH}{dt} = \frac{\partial \boldsymbol{F}}{\partial \boldsymbol{y}} H \quad with \quad H(0) = I \tag{9}$$

where H is the $n \cdot n$ matrix representing a set of infinitesimal separation vectors that define an infinitesimal volume element that evolves according to (9).
For further details in the computation you can refer to Wolf et al.[6]
When taking into account a time delay, it is clear that when the simulation of the delayed system is accomplished following the procedure given at point 2. of the previous section, the procedure is analogous provided that the computational effort related to large systems is taken into account.
For the case under study the computation of the LCE requires the integration of $(n + m \cdot s)(n + m \cdot s + 1)$ equations.
Fortunately, the largest Lyapunov Characteristic Exponent, that is sufficient to characterise the system dynamic behaviour properly, can be computed by itself without computing all the other $n + m \cdot s - 1$ exponents; so only $2(n + m \cdot s)$ equations must be numerically solved.

Alternatively, if the procedure suggested in point 1. is followed, i.e. the spline interpolation method is applied for simulation, the method proposed by Farmer[4] can be successfully used.
In this case the evolution of a set of infinitesimal separation vectors can be evaluated according to:

$$\frac{dH(t)}{dt} = \frac{\partial F(t)}{\partial \boldsymbol{y}} H(t) + \frac{\partial F(t)}{\partial \boldsymbol{y}_{\tau_D}} H(t - \tau_D) \tag{10}$$

where for the nomenclature, you can refer to (1).
Also the largest Lyapunov Characteristic Exponent has been evaluated following the two procedures to check the exactness of the computations.

4. Examples cases and discussion

The first example concerns a CSTR with a recycle stream that was already studied, even if only performing a phase space analysis, by Inamdar et al.[2] The dimensionless

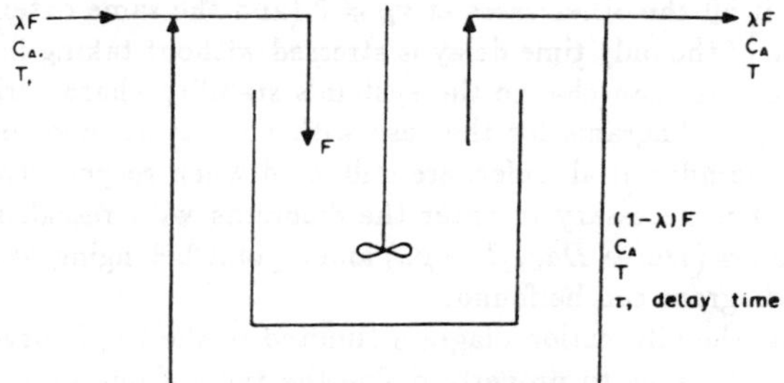

Figure 1: CSTR with recycle and time delay

equation representing the mass and the heat balances for the system shown in Fig. 1 can be written as

$$\begin{cases} \dfrac{d\zeta(t)}{dt} &= -\zeta(t) + (1-\lambda)\zeta(t-\tau_D) + Da\exp(\vartheta(t)/(1+\vartheta(t)/\gamma))(1-\zeta(t)) \\ \dfrac{d\vartheta(t)}{dt} &= -\vartheta(t) + (1-\lambda)\vartheta(t-\tau_D) - \beta(\vartheta(t)-\vartheta_c) \\ & \qquad +BDa\exp(\vartheta(t)/(1+\vartheta(t)/\gamma))(1-\zeta(t)) \end{cases} \tag{11}$$

ξ and ϑ stand respectively for the dimensionless extent of conversion and temperature in the reactor. Notice that the system (11) is of the same type as (2) where $G(y(t))$ contains the nonlinear terms of generation and u represents the variables $\xi_{|t-\tau_D}$ and $\vartheta_{|t-\tau_D}$.
In absence of time delay the equations for the system with a recycle stream can be

reduced to the same form with no recycle, provided that $Da_u = Da/\lambda$, $\beta_u = \beta/\lambda$ and $t_u = \lambda t$ are introduced:

$$\begin{cases} \dfrac{d\zeta}{dt_u} = -\zeta + Da_u \exp(\vartheta/(1+\vartheta/\gamma))(1-\zeta) \\ \\ \dfrac{d\vartheta}{dt_u} = -\vartheta - \beta_u(\vartheta - \vartheta_c) + BDa_u \exp(\vartheta/(1+\vartheta/\gamma))(1-\zeta) \end{cases} \tag{12}$$

All the cases reported by Inamdar et al.[2] have been analysed by adopting both the spline method and the procedure of adding a PDE (plug flow system). Following the two different techniques, we have obtained quite similar results that, unfortunately, do not agree with the ones by Inamdar et al.

In their work a comparison is made between simulation results for $\lambda = 1$ and $\tau_D = 0$ and for $\lambda \neq 1$ and $\tau_D \neq 0$.

In our opinion, on the contrary, it is more meaningful to assume as a reference situation the case with no time delay but with a fixed recycle value ($\tau_D = 0$ and $\lambda \neq 1$) and compare with it all the other cases at $\tau_D \neq 0$ (and the same extent of recycle).

In this way the role of the only time delay is stressed without taking into account the influence of recycle, that can change the system's stability characteristics. In fact, though the bifurcation diagrams for the case with no recycle reported by Uppal et al.[7], and to which Inamdar et al. refer, are still valid when recycle stream exists, we must rember that it is necessary to enter the diagrams with rescaled values of the bifurcation parameters ($Da \to Da_u$, $\beta \to \beta_u$) and a point belonging to another region of the bifurcation diagram can be found.

Figure 2 represents the bifurcation diagram (limited to the Hopf curve) in the plane Da_u vs β_u; its use allows us to properly define the type of operation of the system (on an equilibrium point or a periodic orbit) for a given set of values for Da, β and λ ($B = 22$).

We are convinced of the consistency of our results when compared with those by Inamdar et al. because, besides being obtained with two completely different methods, they comply, when $\tau_D \to 0$, with the system stability characteristics shown in the bifurcation diagram.

We have analysed the effect of time delay on the system's dynamic behaviour, starting from different regions of Figure 2.

As the first step we have verified the τ_D influence in the region of stable operation outside the Hopf curve: even for very high τ_D values we have always experimented the same stable solution also for the delayed operation.

Different patterns of behaviour can be found inside the Hopf curve. In most of the situations (see for instance $Da_u = 0.6744$ and $\beta_u = 6.$) a τ_D increase can stabilise the system on a steady state; it must be pointed out, however, that the τ_D values for which this happens have no meaning from a physical viewpoint ($\tau_D > 1$).

However much we have tried, the only case inside the Hopf curve for which destabilisation of the periodic regime occurs is for $Da_u = 0.22$ and $\beta_u = 6.1$, a point near to the transition from subcritical to supercritical Hopf bifurcation (point A on Fig. 2).

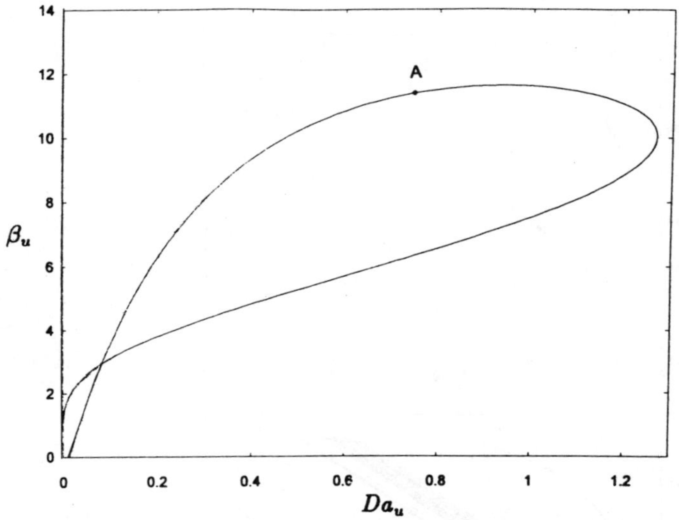

Figure 2: Bifurcation diagram β_u vs Da_u: point A marks the transition between supercritical (right) and subcritical (left) Hopf bifurcation

In Figure 3 the transition from a limit cycle to a strange attractor, and to a periodic solution again, is shown.
We know for certain that Fig. 3.b represents chaotic behaviour from the LCE analysis, as reported in Fig. 4. Also in this case the value for which the destabilising action occurs is too high to be interesting according to operative considerations.
The remarkable point is that the type of stability is no longer a monotone function of time delay but a critical value of time delay seems to exist for which destabilisation occurs. The same type of behaviour was suggested by Boe and Chang[1].

Let's now examine the second example where the controller action is responsible for time delay.

$$\begin{cases} \dfrac{d\zeta(t)}{dt} = -\zeta(t) + Da\exp(\vartheta(t)/(1+\vartheta(t)/\gamma))(1-\zeta(t)) \\[2ex] \dfrac{d\vartheta(t)}{dt} = -k_I\tau\phi(t-\tau_D) - k_P(\vartheta(t-\tau_D)-\vartheta_{set}) - \vartheta(t) - \beta(\vartheta(t)-\vartheta_c) \\ \qquad\qquad +BDa\exp(\vartheta(t)/(1+\vartheta(t)/\gamma))(1-\zeta(t)) \\[2ex] \dfrac{d\phi(t)}{dt} = \vartheta(t) - \vartheta_{set} \end{cases} \tag{13}$$

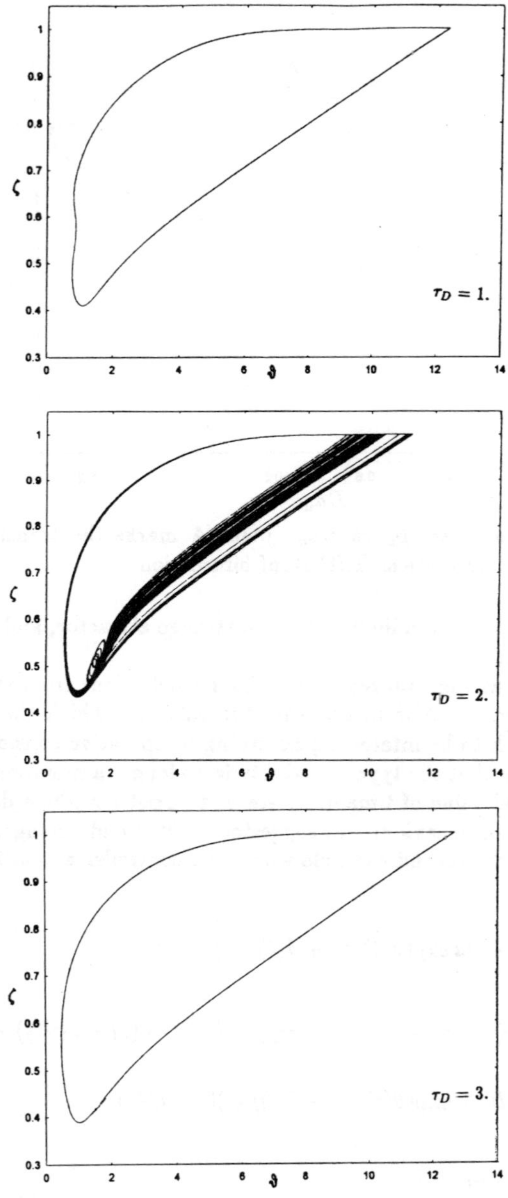

Figure 3: Transition from a limit cycle to chaos and vice versa ($Da_u = 0.22$, $\beta_u = 6.1$, $\lambda = 0.75$, $B = 22.$)

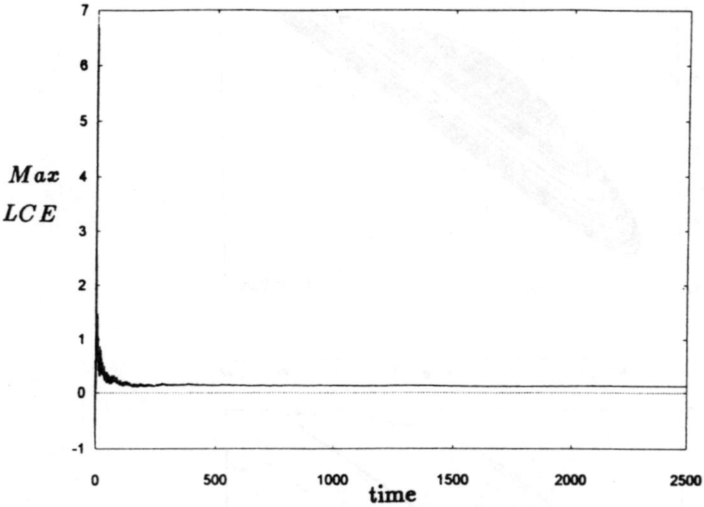

Figure 4: LCE vs time for the case in Fig. 3.b

For the values of parameters you can refer to Pellegrini and Biardi[8].
In must be pointed out that, unlike the preceeding example the undelayed system can exhibit chaotic behaviour by itself because it has three degrees of freedom.
Also in this example a stabilising effect of τ_D may be found since a proper τ_D value can convert a chaotic pattern of behaviour to a periodic one, as shown in Fig. 5, but the related values are too high to be realistic.
On the contrary, even a low value for the time delay can destabilise a steady state to a periodic orbit: the greater the $k_I\tau$ value, the lower the correspondig τ_D value necessary to destabilise the set point; the proportional gain seems to have little influence. What is more interesting, since is analogous to the previous example, is the possibility to destroy the stability of some limit cycles exhibited by the undelayed model. As in the reacting system previously analysed, critical τ_D values can be found: a chaotic pattern of behaviour appears between such values (Fig. 6), that are acceptable from an operative point of view.

References

1. E. Boe and C. H. Chang, *Chem. Engng. Sci.* **44** (1989) 1281.
2. S. R. Inamdar, V. Ravi Kumar and B. D. Kulkarni, *Chem. Engng. Sci.* **46** (1991) 901.
3. M. A. Soliman and W. H. Ray, *Int. J. Contr.* **15** (1972) 609.

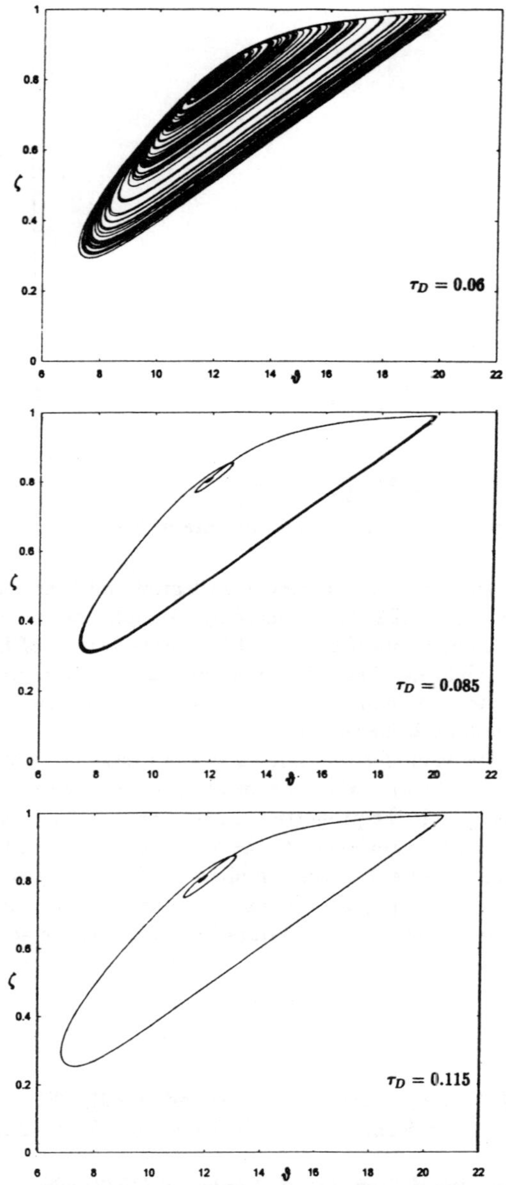

Figure 5: Transition from chaos to a limit cycle ($k_P = 0.3$, $k_I\tau = 0.3$)

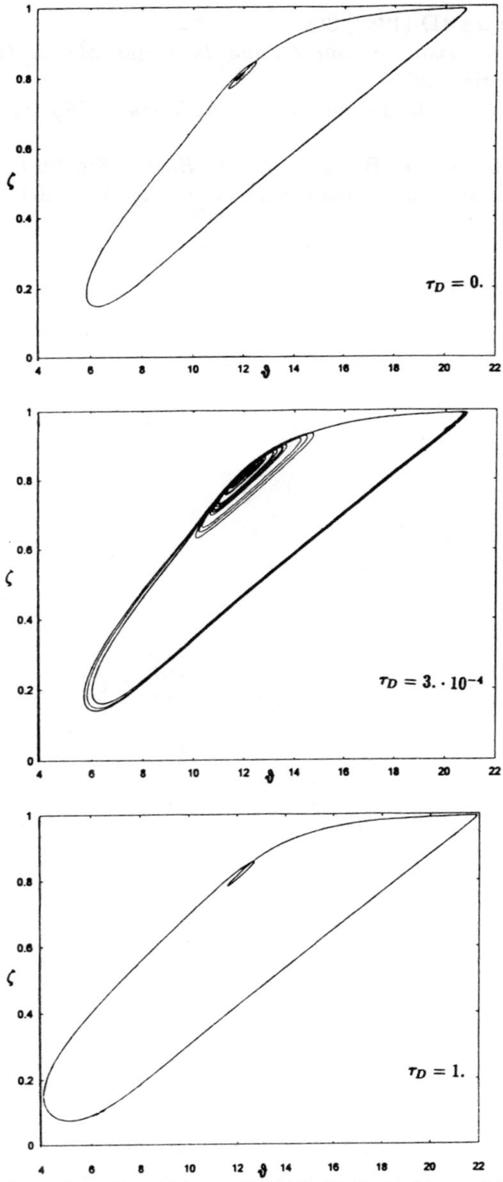

Figure 6: Periodic - chaotic - periodic behaviour ($k_P = 0.24$, $k_I\tau = 0.14$)

4. J. D. Farmer, *Physica* **4D** (1982) 366.
5. D. Rouelle, *Characteristic Exponents and Invariant Manifolds in Hilbert Spaces*, (IHES preprint, 1980).
6. A. Wolf, J. B. Swift, H. L. Swinney and J. A. Vastano *Physica* **16D** (1985) 285.
7. A. Uppal, W. H. Ray and A. B. Poore, *Chem. Engng. Sci.* **29** (1974) 967.
8. L. Pellegrini and G. Biardi, *Computers Chem. Engng.* **14** (1990) 1237.

BIFURCATION ANALYSIS OF CHEMICAL REACTING SYSTEMS

ALFONSO CUNDARI, FRANCESCO PAOLO DI MAIO
Dept. of Chemical Engineering, University of Calabria,
Rende (CS), 87030, Italy

and

PIER GIORGIO LIGNOLA
Dept. of Engineering, Second University of Naples,
Aversa (CE), 81031, Italy

ABSTRACT

A new method for the analysis of complex reacting systems, based on the application of bifurcation and continuation theory, is presented. The approach developed gives more information than the traditional methods based on time dependent simulations. The technique capabilities are shown by direct application to the analysis of a kinetic scheme. Results obtained indicate that the procedure is able to locate and to classify the mathematical nature of the boundaries of regions with distinct qualitative behaviours. Results given by applying the method to a large complex kinetic scheme are also reported.

1. Introduction

Chemical reacting systems can exhibit a great variety of different dynamic behaviours, owing to their high non linearity. Simple and complex wave form periodic oscillations, monotonic and damped approach to a steady state are the most frequently observed[1]. Classification of the various behaviours shown by such systems by varying control parameters appears expedient for their thorough understanding. Experimentally, regions in parameter space, with distinct qualitative dynamics, are determined by fixing all parameters but one, which is changed by discrete quantity[2]. The varying controlled variable is named bifurcation parameter and such a role is assumed by all the parameters in turn. Observations of the system response permit to position the boundaries of the various regions (bifurcation sets) and to trace the bifurcation map. The accuracy with which these contours are drawn depends on the step size used for varying bifurcation parameter. The described method is also used to reproduce the bifurcation map[3] by means of numerical simulations. This approach is time consuming and gives limited information. Hence, it appears convenient to develop a new method, able to give more information in a less tedious way.

The present work points out some of the potential advantages of applying recent mathematical studies, such as bifurcation analysis and continuation theory, to the study of chemical reacting systems. The main purpose is that of developing a general method for the bifurcation analysis of complex reacting schemes, involving hundreds of reactions and chemical species. However, in order to tune the procedure, the attention was focused on reduced kinetic schemes. In particular, the advantages of this method were established by its application to the simple kinetic scheme developed by Wang and Mou[4], for acetaldehyde oxidation.

2. Wang and Mou model

The Wang and Mou model[4] involves only three variables, yet it exhibits most, if not all, of the dynamic varieties found in more complex systems and is able to reproduce qualitatively the ignition diagram experimentally evaluated for the acetaldehyde-oxygen combustion process in continuous stirred tank reactor[2] (CSTR). The model includes the following five reactions:

$$Y \rightarrow X \tag{I}$$
$$X + Y \rightarrow fX \tag{II}$$
$$X \rightarrow 2X \tag{III}$$
$$X \rightarrow S_1 \tag{IV}$$
$$X \rightarrow S_2 \tag{V}$$

where Y represents the reactant concentration and X the branching radical concentration. S_1 and S_2 are stable products. In reaction (II), f represents the branching factor assumed to be equal to two in the model. The mass balances written for X and Y, together with the energy balance, give rise to a set of three ordinary differential equations, describing mathematically the chemical transformation in CSTR. The natural parameters of the system are the initial pressure P, the oven temperature T and the residence time τ. Wang and Mou studied the model for fixed residence time $\tau = 4$ s, by integrating numerically the differential equations for different values of the parameters P ($200 < P < 700$ torr) and T ($500 < T < 650$ K) and observing the dynamic behaviour of the system. In such a way they were able to divide the parameter plane (T-P) in five distinct regions with different dynamic behaviour. However, the accurate determination of the boundaries forced Wang and Mou to perform a great number of numerical simulations with no definition of the mathematical form of the bifurcation sets. Variation of the solution structure in phase space was determined by taking very small increments of the bifurcation parameter in the proximity of the observed transitions.

3. The proposed approach

The method developed in the present work, based mainly on the application of continuation theory[5,6], enables one to evaluate numerically the contours (bifurcation sets) of the regions with no need of carrying out a full study in the time domain.

Continuation is characterized by a particular interest in the parameter dependence of the branches and in the multiplicity of solutions. It enables one to trace the solution curve of the system

$$\mathbf{F}(\mathbf{x}, \lambda_1, \lambda_2, \cdots, \lambda_K) = 0 \tag{1}$$

as one or more parameters λ_i are varied. Assuming that a first solution is available, continuation allows one to evaluate further solutions until a fixed parameter value is reached. Recently the attention was focused on its relationship with bifurcation theory. As a matter of fact, modern continuation algorithms are capable of detecting, during the tracing of a branch, possible bifurcation points from which other secondary branches may

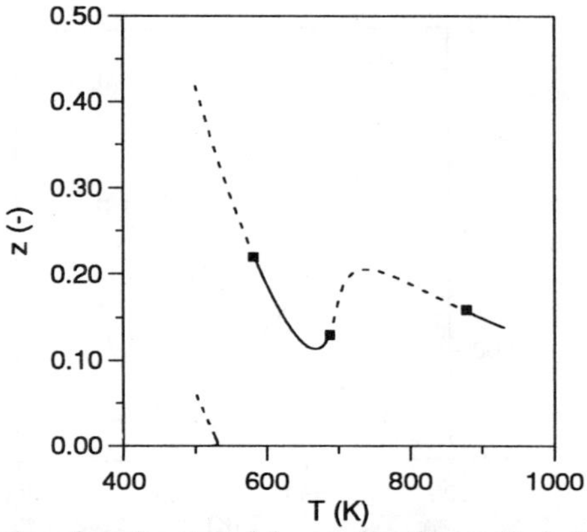

Figure 1: Stationary solutions branches traced by continuation method. The diagram, obtained at pressure P = 912 torr, presents one limit point on the lower curve and three Hopf points on the upper one. Note that from each Hopf point bifurcates a branch of periodic solution, not reported in the figure. Dashed line indicate unstable solutions.

bifurcate. Hence, continuation makes it possible to discover all interconnected branches thus providing a global view of the existing structure of solutions. Among the many numerical codes implementing continuation algorithms[7,8,9], that appeared in the literature, the AUTO package[7], written by Doedel and Kernevez, was chosen to perform the bifurcation analysis presented in this work.

For the proposed approach it is essential to find the bifurcation diagram presenting the most complex parameter dependence and the greatest number of singular points that allow a two parameter continuation. Among the free parameters available, the initial reactor pressure P was firstly fixed and the oven temperature has been taken as the bifurcation parameter. Figure 1 reports the bifurcation diagram (z,T) obtained at the pressure of 912 torr, z being defined as the dimensionless temperature $(T_{reactor}\text{-}T)/T$. The locus of stationary points is represented by two separated curves. A limit point is found on the lower branch whereas the upper branch presents three Hopf bifurcation points indicated by filled squares. Each Hopf bifurcation point gives rise to a branch of oscillatory solutions, not reported in the diagram. The proposed method allowed to detect unstable solutions that are reported as dashed lines in Fig. 1 and which are undetectable by the traditional investigation process based on time dependent simulations.

Taking the pressure as the bifurcation parameter one obtains a single branch with a greater number of singular points. For example, the bifurcation diagram at the constant temperature of 610 K presents four limit points and four Hopf points. It is interesting to note that two parameter continuation of singular points, detected using pressure or temperature as the bifurcation parameter, gives the same bifurcation sets. According to this remark a two parameter continuation, using the initial pressure as the second free parameter, was carried out in order to trace the curve described by the limit point and by the Hopf bifurcation point located at lowest temperature. The two bifurcation varieties

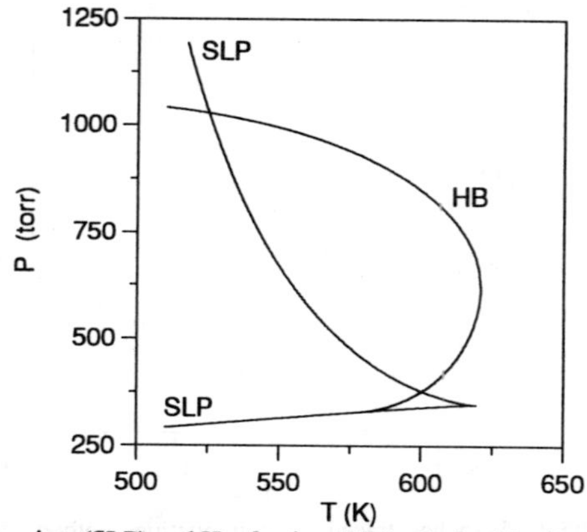

Figure 2: Loci of limit points (SLP) and Hopf points (HB), obtained by 2 parameters continuation.

obtained are reported in Fig. 2. Comparing Fig. 2 with the bifurcation map reported by Wang and Mou, one notices that the curve SLP (locus of limit points) is the bifurcation set separating the slow combustion region (I) from the region of periodic ignitions (II), whereas the curve HB (locus of Hopf bifurcation points) separates the periodic cool flames region (IV) from the *jumps* (V). Moreover, the two parameter continuation enabled to locate a multistability region at a pressure below 400 torr ($320 < P < 380$ torr, $575 < T < 620$ K), enclosed by SLP and HB curves. The presence of this multistability region was also pointed out by Wang and Mou, but their work was not able to define contours of this region. However, even though the information obtained is of high quality, it was not possible to locate, by means of continuation of singular points on stationary solution branch, the bifurcation sets that in the Wang and Mou map separate internal regions. Hence, continuation of periodic solutions branch, bifurcating from lowest temperature Hopf point, was operated in order to detect other singular points, capable of characterizing internal regions boundaries.

The bifurcation diagram at constant pressure $P = 684$ torr is reported in Fig. 3. In this diagram, filled circles indicate stable periodic solutions, whereas empty circles represent unstable periodic solutions. The position of each point is the maximum z value reached by the oscillatory solution in phase space. Two limit points and one period doubling point were found on this secondary branch. A period doubling cascade was also detected for decreasing temperature starting from the PD bifurcation set. Unfortunately, the AUTO program was able to perform the two parameter continuation only for the periodic limit point located at lowest temperature (indicated by OLP_1 in the insert of Fig. 3). The inability to trace the bifurcation sets relative to the other two singular points was due: (i) to AUTO package limitation to handle period doubling points and (ii) to the particular form of periodic branch in proximity of OLP_2. The locus of points (PD and OLP_2) for which AUTO does not admit a two parameter continuation was traced by performing several runs at different values of the initial reactor pressure P and by interpolating the singular points evaluated in such a way. Note that recent numerical codes[8] admitting period doubling points as regular solutions of enlarged systems, overcome such

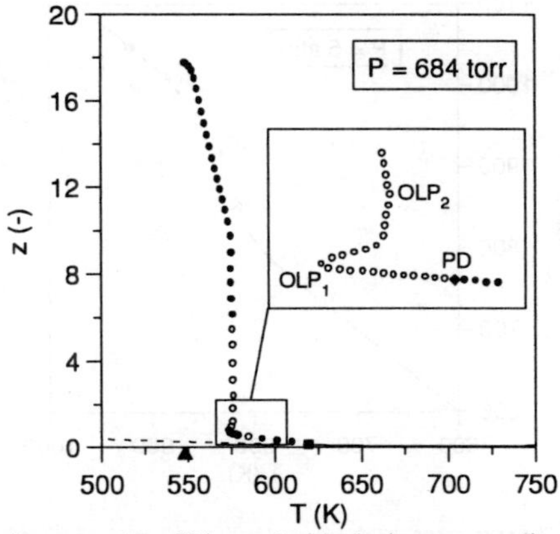

Figure 3: Bifurcation diagram at P = 684 torr. Inside the insert, a qualitative zoom is reported.

limitations.

Figure 4 reports the complete bifurcation map for the Wang and Mou model, as obtained by applying the new method. The reported diagram covers a wider range of pressure and shows new features both at high pressure ($P > 1150$ torr) and at low pressure ($P < 380$ torr). In Fig. 4 the nature of each bifurcation set is also reported. This bifurcation map can also be seen as a projection on parameter plane of the singular points belonging to the 2-dimensional surface generated by the defining equations. Hence, the order in which one finds the bifurcation varieties moving along the surface at constant pressure or temperature is not necessarily the same that appears if moving in the *P*-*T*

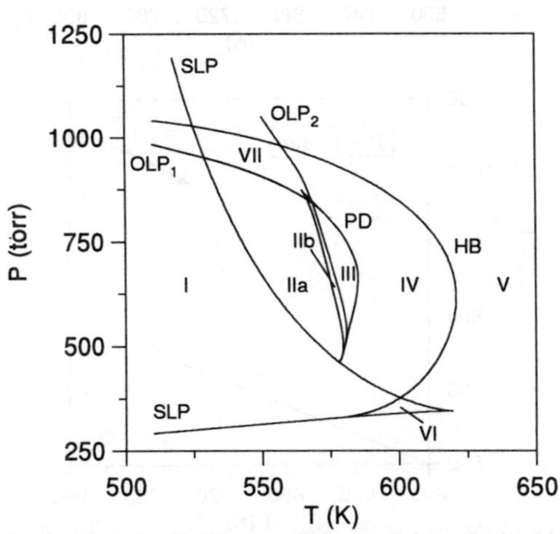

Figure 4: Complete bifurcation map for the Wang and Mou model, as obtained by the proposed method. Diagram reports region with distinct dynamic behaviour (I, II, III, IV, V) and multistability regions (VI, VII). Bifurcation varieties are reported as: SLP stationary limit points; OLP oscillatory limit points; PD period doubling; HB Hopf bifurcations.

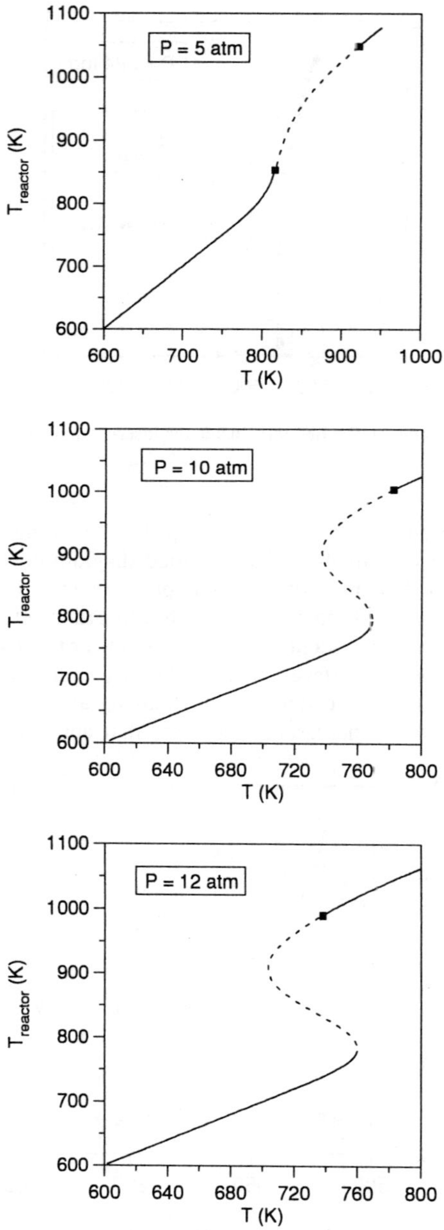

Figure 5: Bifurcation diagrams for the complex reacting system methane/oxygen at several pressures.

plane. For example, traversing the parameter plane (Fig. 4) at the constant pressure $P = 684$ torr from the right, one might erroneously deduce that the OLP_2 variety is crossed before OLP_1. Instead, from Fig. 3 one can see that OLP_1 is reached before OLP_2.

The results obtained with the application of bifurcation theory to the analysis of this simple kinetic scheme was very interesting and the method appeared so promising that an extension to more complicated models of reacting systems, such as acetaldehyde/oxygen and methane/oxygen, was operated. Even if this last investigation is only at beginning, the authors were able to trace the most typical bifurcation diagrams. Figure 5 reports the curves of stationary solutions for a mathematical model of methane oxidation in CSTR[10], at three different pressures. The kinetic scheme includes 22 chemical species involved in 74 reactions. From the figures reported one can see that at higher pressures the diagram presents the characteristic sigmoid curve which indicates the presence of an intermediate pressure at which the bifurcation diagram has a vertical tangent. Note also that for P = 12 atm the Hopf bifurcation point moves inside the sigmoid creating an interval of parameter values for which the system presents bistability. It would have been practically impossible to obtain the same information by resorting to the traditional methods that are based on the study of the system in the time domain.

4. Conclusions

In the present work a new method for the study of complex reacting systems was develope based on the application of bifurcation theory. The discussion and analysis of results indicate that, with respect to simulation processes, the proposed method enables the chemical engineer to trace, with great accuracy, the contour of the various regions of different behaviour, and to gain a general idea of the global behaviour of the reacting system. In particular, the major advantages featured by the new approach as compared to the classical methods, based on time dependent simulations, are:

1. the ability to locate and to classify the mathematical nature of bifurcation varieties;
2. the ability to determine unstable solutions
3. the ability to locate multistability regions otherwise not identifiable;
4. the ability to find singular degenerate points,
5. the rapidity with which global information are gained.

The method was tuned on the simple kinetic scheme reported by Wang and Mou, and appeared able to reproduce the bifurcation map as obtained by traditional methods. The first applications of the method to methane oxidation in CSTR show that the advantages pointed out for the Wang and Mou model are retained even if the complexity of the reacting systems is increased.

Acknowledgements

The work was performed with support of MURST of Italy and of CNR - Institute of research on membrane and on modelling of chemical reactors (Rende).

References

1. S.K. Scott, *Chemical chaos* (Oxford Science Publications, London, 1991).
2. P. Gray, J.F. Griffiths, S.M. Hasko and P.G. Lignola, *Combust. Flame* **43** (1981) 175.
3. F.P. Di Maio, P.G. Lignola and P. Talarico, *Comb. Sci & Tech.* **91** (1993) 119.
4. X.-J. Wang and C.Y. Mou, *J. Chem. Phys.* **83** (1985) 4554.
5. H.B. Keller, *Lectures on numerical methods in bifurcation theory* (Springer-Verlag, New Delhy, 1987).
6. W.I. Zangwill and C.B. Garcia, *Pathways to solutions, fixed points and equilibrium* (Prentice-Hall, Englewood Cliffs, 1981).
7. E. Doedel and J.P. Kernevez, *AUTO: software for continuation and bifurcation problems in ordinary differential equations* (Institute of Technology, Pasadena (CA), 1986).
8. U. Feudel and W. Jansen, *Int. J. Bifurcation and Chaos* **2** (1992) 773.
9. R. Seydel, *From Equilibrium to Chaos* (Elsevier, New York, 1988).
10. F.P. Di Maio and P.G. Lignola, *Proceedings of the Joint meeting of the Italian and Spanish sections of The Combustion Institute* (Stresa, 1993) II.2.

CHAOS SYNCHRONIZATION APPLICATION TO SECURE COMMUNICATIONS

A. DE ANGELI, R. GENESIO and A. TESI
Dipartimento di Sistemi e Informatica, Università di Firenze
Via di S.Marta,3 - 50139 Firenze, Italy

ABSTRACT

The problem of the synchronization of two or more chaotic systems is presented. For a special class of chaotic systems, a solution to the problem of the structure decomposition for transmission is given and an example is analyzed. For discrete-time chaotic systems the property of dead-beat synchronization is introduced and discussed. A possible application of synchronized chaotic circuits in the secure communication context is investigated.

1. Introduction

The study of chaotic phenomena has known in the last decade a strong increase in a variety of research branches: complex behaviours have been discovered and analyzed in chemical reactors, electrical circuits, lasers, biological systems, fluido dynamics, etc... .

Despite this widespread interest, with some notable exceptions [1,2], the domain of chaos was considered just an academic one, without deeper implication for real life operations. Chaos was generally considered rather an unwanted phenomenon, often hazardous for operation of real physical systems, causing their malfunctioning, thus being a regime of operation to be avoided. In fact, several methods developed by scientists studying chaos, if used in application at all, were used to design *chaos-free* systems.

There exist some problems, though, where the complexity and the richness of a chaotic signal is actually desirable if not a necessity. An example is the secure communication problem, where the information to be transmitted is 'masked' by means of a 'coding' operation between the information itself and a 'secret key', a pseudo-random signal that must be complex enough to generate an effective coding, and simple enough to be communicated without errors to the receiver[3].

The chaotic system here is a 'secret key generator' that presents two interesting features: the chaotic output is *complex* and in general *not reconstructable*, while it is *completely* defined by a few parameters and initial conditions.

Thus, a receiver that can *synchronize* its chaotic circuit on the transmitter one will generate the same secret key used for the coding of the information and provide a successful decoding.

The problem of chaos synchronization is however more general. Its natural applications are those situations where a chaotic motion is desirable but we need to fix or to influence the characteristics of the trajectories[4].

In the following, the problem of the synchronization is investigated and some solutions are proposed for a special class of chaotic circuits, both continuous-time and discrete-time. The problems and some possible schemes for the application of chaos synchronization in the secure communication context are presented.

2. Synchronization of Chaotic Circuits

The problem of chaos synchronization has been recently posed in a work by Pecora and Carroll[5], where this problem is theoretically investigated and a sufficient condition for the synchronization to occur is given.

A *drive* system transmits some of its state variables to a *response* system, that reproduces partially or totally the drive system. The two systems *synchronize* when the trajectories of the state variables of the response system tend asymptotically to the trajectories of the drive system.

A sufficient condition for the synchronization to occur[5], is that the conditional Lyapunov exponents of the response system are strictly negative. Unfortunately, this condition does not furnish any hint on the choice, given a chaotic system, of its suitable structure decomposition for the synchronization.

Here, we concentrate on a special class of chaotic circuits, i.e. the systems that can be represented in Lur'e form[6]

$$\begin{aligned} y(s) &= L(s)u(s) \\ u(s) &= h(y(s)) \end{aligned} \tag{1}$$

where $L(s)$ is a proper rational transfer function, $h(\cdot)$ is a nonlinear block and $u(s)$ and $y(s)$ are *scalar* signals. It has been shown[6] that a large number of chaotic systems can be represented in this form or in a slightly more general one: the Chua's circuit, the Lorentz system, the Rossler system among continuous-time system and the Hénon and the Lozi maps among discrete-time systems.

Given the system S, defined by (1), we consider a system $\hat{S}$ composed by a feedback interconnection of a linear subsystem $\hat{S}_L$ and a nonlinear one $\hat{S}_N$ described by

$$\hat{S}_L: \quad r(s) = F(s)x(s)$$

$$\hat{S}_N: \begin{cases} y(s) = G_{11}(s)u(s) + G_{12}(s)r(s) \\ x(s) = G_{21}(s)u(s) + G_{22}(s)r(s) \\ u(s) = h(y(s)) \end{cases} \tag{2}$$

where $F(s)$ is a proper rational transfer function, $x(s)$ and $r(s)$ are scalar signals and $G(s) = [G_{11}(s)\ G_{12}(s)\ ;\ G_{21}(s)\ G_{22}(s)]$ is a 2×2 matrix of proper rational transfer functions.

The system $\hat{S}$ is said to be *equivalent* to S if the two systems have the same dynamical behaviour, i.e. motions from each identical initial condition are equal.

Systems equivalent to S are described by the following

Proposition 1 Let n,m and l denote the order*of $L(s)$,$F(s)$ and $G(s)$ respectively. Then, system $\hat{S}$ is equivalent to S if

$$i) \quad n = m + l \tag{3}$$

$$ii) \quad L(s) = G_{11}(s) + \frac{F(s)G_{12}(s)G_{21}(s)}{1 - F(s)G_{22}(s)} \tag{4}$$

■

Equations (2) define a general partition of the original system S. To study its synchronization properties, consider a receiver system $\tilde{S}$ of analogous structure driven by $x(s)$ as

$$\tilde{S}_L: \quad \tilde{r}(s) = F(s)x(s)$$

$$\tilde{S}_N: \begin{cases} \tilde{y}(s) = G_{11}(s)\tilde{u}(s) + G_{12}(s)\tilde{r}(s) \\ \tilde{x}(s) = G_{21}(s)\tilde{u}(s) + G_{22}(s)\tilde{r}(s) \\ \tilde{u}(s) = h(\tilde{y}(s)). \end{cases} \tag{5}$$

Synchronization occurs when $\tilde{x}(s)$ tends to $\hat{x}(s)$ as $t \to \infty$. Following this approach, a number of sufficient conditions for the synchronization to occur have been found, depending on the structure of the linear blocks $F(s)$ and $G(s)$ [8,9].

The main feature of this decomposition method is that we obtain *one-parameter families* of synchronizing systems. This resulting degree of freedom can be used for other design requirements, i.e. synchronization speed or special frequency shapes for the obtained systems.

Example 1 Consider the Chua's circuit[10] defined by the following three equations

$$\frac{dv_1}{dt} = \frac{1}{C_1}[G(v_2 - v_1) - f(v_1)] \tag{6}$$

$$\frac{dv_2}{dt} = \frac{1}{C_2}[G(v_1 - v_2) + i_3] \tag{7}$$

$$\frac{di_3}{dt} = -\frac{1}{L}v_2. \tag{8}$$

(9)

where

$$G = \frac{1}{R}$$

and

$$f(v_1) = G_b v_1 + \frac{1}{2}(G_a - G_b)\{|v_1 + E| - |v_1 - E|\} \quad . \tag{10}$$

This circuit can be represented in the basic structure (1) by choosing $y = v_1$ and defining the new time variable $(G/C_2)t$.

*As concerns to the matrix of transfer functions G, we will refer to the standard McMillan degree[7].

$R = 1/G$ (Ω)	L (mH)	C_1 (nF)	C_2 (nF)	G_a (μS)	G_b (μS)	E
1680	18	10	100	−753	−396	1

Table 1: Values of the Chua's circuit parameters.

Assuming the parameters values shown in Table 1, we get

$$L(s) = \frac{10\,(s^2 + s + 15.68)}{s^3 + 11s^2 + 15.68s + 156.8} \tag{11}$$

and

$$h(y) = 0.6653y + 0.3204\,\{|y+1| - |y-1|\} \quad . \tag{12}$$

Consider this special drive-response decomposition, called *singular* (SI) in Tesi et al.[9]

$$G = \begin{bmatrix} (\alpha s + \beta)/(s+\gamma) & (\alpha s + \beta)/(s+\gamma) \\ (\alpha s + \beta)/(s+\gamma) & (\alpha s + \beta)/(s+\gamma) \end{bmatrix} \tag{13}$$

where α,β and γ are real numbers and s is the complex variable.

In this example, the one-parameter family of totally self-synchronizing systems is

$$F(s) = \frac{(\gamma - 10)s^2 + \gamma s + 15.86(\gamma - 10)}{10(s^2 + s + 15.68)}; \tag{14}$$

$$G_{11}(s) = G_{12}(s) = G_{21}(s) = G_{22}(s) = \frac{10}{s+\gamma} \quad . \tag{15}$$

According to the results presented in De Angeli[8], the sufficient conditions for the synchronization to occur are matched for $\gamma > 12.701$. Numerical simulations show that the synchronization is achieved in the larger interval $\gamma > 7$.

A special case is represented by the well-known equivalent system[11,12,13] which is directly obtained from the state space description by putting $x = v_1$ and $r = v_2$ and identifying the subsystems $\hat{S}_N$ and $\hat{S}_L$ with equations (6) and (7)-(8) respectively. This system corresponds to the choice $\gamma = 10$ in (14) and has been shown to synchronize by Dedieu et al.[13].

3. Discrete Time Synchronization

This approach to drive-response decomposition has been successfully applied in the discrete-time environment. The extension to this case of the structure decomposition approach and of the conditions on synchronization are immediate, just considering the difference on the stability regions for the discrete-time environment.

Moreover, consider the following special structure. Let $G_{11}(z) = 0$, $F(z) =$

$G_{12}(z) = 1$ in equations (2). Being $1/z$ the unitary delay element, we obtain

$$
\begin{array}{ll}
\hat{S}_L: & r(z) = x(z) \\
\hat{S}_N: & \left\{ \begin{array}{lcl} y(z) & = & r(z) \\ x(z) & = & G_{21}(z)u(z) + G_{22}(z)r(z) \\ u(z) & = & h(y(z)) \end{array} \right.
\end{array}
\tag{16}
$$

or

$$
\left\{ \begin{array}{lcl} x(z) & = & G_{21}(z)u(z) + G_{22}(z)x(z) \\ u(z) & = & h(x(z)) \end{array} \right.
\tag{17}
$$

The system (17) is equivalent to the original system (1), according to Proposition 1, if

$$
L(z) = \frac{G_{21}(z)}{1 - G_{22}(z)}
\tag{18}
$$

Let

$$
L(z) = \frac{N(z)}{D(z)}; \qquad G_{21}(z) = \frac{N_{21}(z)}{D_{21}(z)}; \qquad G_{22}(z) = \frac{N_{22}(z)}{D_{22}(z)}
\tag{19}
$$

and fix

$$
D_{22}(z) = D_{21}(z) = z^h.
\tag{20}
$$

From (18) we obtain

$$
N_{21}(z) = N(z); \qquad N_{22}(z) = z^n - D(z); \qquad h = n.
\tag{21}
$$

Notice that, due to the structure of the matrix of transfer functions G, the McMillan degree of the resulting decomposition is n.

Let us analyze the synchronization properties of the obtained system. Define, as in (5), a receiver system

$$
\left\{ \begin{array}{lcl} \tilde{x}(z) & = & G_{21}(z)\tilde{u}(z) + G_{22}(z)x(z) \\ u(z) & = & h(x(z)) \end{array} \right.
\tag{22}
$$

and consider the error system, obtained subtracting equations (22) from (17)

$$
\left\{ \begin{array}{lcl} \Delta x(z) & = & G_{21}(z)\Delta u(z) + G_{22}(z)0 \\ \Delta u(z) & = & 0 \end{array} \right.
\tag{23}
$$

where $\Delta x = x - \tilde{x}$ and $\Delta u = u - \tilde{u}$. By looking at equations (23), where $G_{22}(z)0$ simbolically indicates the unforced response of $G_{22}(z)$, we notice that the error Δx will reach exactly zero in n steps, i.e. an exact synchronization will be achieved in a finite time.

Definition 1 Systems that achieve exact synchronization in a finite number of steps are called *dead-beat synchronizing.*

Proposition 2 Given a Lur'e discrete-time system (1) there exists a dead-beat synchronizing decomposition defined by equations (17) and (21). ■

The importance of the property of dead-beat synchronization is immediate. Due to the deterministic nature of chaotic motions, in fact, once that dead-beat synchronization has been achieved, the two systems will remain synchronized *regardless of the presence of the synchronization signal.* In other words, to dead-beat synchronize two systems corresponds to move the initial conditions of the receiver *exactly* on the trajectory of the transmitter. From that moment on, the two systems will follow the same trajectory.

Example 2 Consider the Hénon map, defined by the following equations

$$\left\{ \begin{array}{lcl} \xi_{n+1} & = & \zeta_n + 1 - \alpha\xi_n^2 \\ \zeta_{n+1} & = & \beta\zeta_n \end{array} \right. . \tag{24}$$

This system exhibits chaotic behaviour in a large neighborhood of the parametric configuration $\alpha = 1.4$ and $\beta = 0.3$[14]. By rearranging equations (24) and defining $y = \xi$ and $h(y) = 1 - \alpha y^2$, we obtain the Lur'e form

$$\begin{array}{lcl} y & = & \dfrac{z}{z^2-\beta}\, u \\ \\ u & = & 1 - \alpha y^2 \quad . \end{array} \tag{25}$$

The corresponding dead-beat partition is then defined by

$$\begin{array}{lcl} x & = & \dfrac{1}{z}\, u + \dfrac{\beta}{z^2}\, x \\ \\ u & = & h(x) \quad . \end{array} \tag{26}$$

In this example, dead-beat synchronization is achieved in two steps[8].

4. Application to Secure Communication

The interest in the chaos synchronization problem stems from the possible applications in the secure communication context. Here, the idea is to cypher the information to be transmitted with a chaotic signal and to use the synchronization property of some chaotic systems by building a receiver which is able to reconstruct the chaotic signal used for coding, thus recovering the information[11,12,13,15].

4.1. Chaotic masking.

In this approach the information is simply *added* to a masking chaotic signal. The receiver must then be able to synchronize driven by this 'information disturbed' signal. To this purpose, some strict hypotheses are required both on the shape of the frequency response of the receiver response system and on the amplitude ratio of the masking and information signals (in Kocarev et al.[11], the response system is a high-pass filter and the amplitude ratio of the two signals is $-50dB$.) Once that the synchronization occurs, the decoding is made simply by subtracting the rebuilt masking signal from the received one. A certain amount of distortion of the recovered information is however unavoidable.

4.2. Chaotic switching.

Now we consider the information as a binary sequence. The transmitter is composed by a chaotic system, with a parameter μ variable as a function of the information value. Thus, for a bit 1 of information, the variable parameter will assume the value μ_1, corresponding to the chaotic attractor $\mathcal{A}_1$, while, for a 0 bit, the parameter will be μ_0 and the transmitted attractor $\mathcal{A}_0$.

The receiver is composed of *two* chaotic circuits, one in the μ_0 configuration (thus synchronizing during the transmission of the attractor $\mathcal{A}_0$) and the other in the μ_1 configuration (synchronizing to $\mathcal{A}_1$).

The decoding occurs by simply identifying the synchronized receiver circuit.

This approach does not require any particular hypothesis on the signals amplitude or on the filters shape. It will however result in a very slow transmission speed, since for any single bit of information the decoding process must wait for one of the two receiver circuits to synchronize.

4.3. Chaotic Modulation.

It is the most complex approach presented up to now[12]. The information signal is 'modulated' by means of an invertible function $c(\cdot,\cdot)$ of a chaotic signal and of the information itself. The decoding occurs by applying the inverse of the c transformation to the received signal and to a rebuilt modulation signal. The receiver must be able to synchronize by using as driving signal the received one.

As in the chaotic masking approach, some strong hypotheses are required on the signals and on the receiver frequency shape. The main feature of this method is that the masking is very efficient, in the sense that the information signal is 'spread' over the frequency content of the chaotic carrier. Moreover, depending on the modulation function c, the 'selectivity' of the coding can be stressed, thus increasing the security of the transmission, but also the difficulties of a practical implementation of the circuits.

5. Numerical Synchronization

The main practical problem for the realization of synchronizing chaotic systems is due to the necessity of a perfect copy of the transmitter circuit in the receiver. This requirement is due to the fact that chaotic trajectories are strongly sensitive both to small parametric variations on the circuit and to initial conditions. In analog electrical circuits, small technological variations on components values are unavoidable as well as noise and distortion due to the transmission channel. Simulations show that such disturbances may result in an unsuccessful synchronization process[8].

In a digital environment these problems can easily be solved, since it is possible to build numerical circuits *with zero parametric error* and to overcome channel noise and distortion.

A large class of difference equations exhibiting chaotic behaviour are available, thus allowing digital schemes to be investigated. The transmission schemes presented above can be easily implemented for numerical circuits and has been tested by means of numerical simulations[8].

5.1. Dead-beat synchronization.

The property of dead-beat synchronization can be applied to design a transmission scheme exhibiting a very high selectivity of decoding. We considered a 'two steps' transmission. During the first step, only a 'synchronization' signal is transmitted, allowing the dead-beat synchronizing receiver to achieve a zero-error synchronization. Then, the information is transmitted, coded as described in the chaotic modulation method. During this second step, the receiver chaotic circuit will continue to be synchronized to the transmitter, due to the deterministic nature of chaotic oscillations, thus producing the decoding signal.

The features of this method are obvious. Consider a decoding attempt made with a receiver not perfectly equal to the transmitter. During the synchronization step, the two circuits will synchronize, in a sense, but with a nonzero final error. Due to this small difference, during the second step, the trajectories will quickly become uncorrelated, resulting in an incomprehensible decoded signal.

This scheme has been tested, in numerical simulations, by using the Hénon map as a chaos generator. The results obtained are surprisingly good, both for the obtained selectivity and for the quality of the decoded signal, that shows practically no distortion at all[8].

References

1. A.V. Holden (editor). *Chaos.* Manchester Un. Press, Manchester, 1986.
2. J.H. Kim and J. Stringer (editors). *Applied Chaos.* Wiley, New York, 1992.
3. Proceedings of the IEEE. *Special Issue on Cryptography,* **32**, 1988.

4. F. Mossayebi, H.K. Quammar and T.T. Hartley, "Adaptive Estimation and Synchronization of Chaotic Circuits", Phys. Letters A, **161**, pp. 255-262, 1991.
5. L.M. Pecora and T.L. Carroll, "Synchronization in chaotic systems", *Physical Review Letters*, **64**, pp. 821-824, 1990.
6. R. Genesio and A. Tesi. Harmonic balance methods for the analysis of chaotic dynamics in nonlinear systems. *Automatica*, **28**, pp. 531-549, 1992.
7. H.H. Rosenbrock. *State Space and Multivariable Theory.* Nelson, London, 1970.
8. A. De Angeli, "Sincronizzazione di Sistemi Caotici. Metodologie e Applicazione alle Comunicazioni Sicure.", *Tesi di Dottorato*, Università di Bologna, 1994.
9. A. Tesi, A. De Angeli and R. Genesio, "On the system decomposition for synchronizing chaos", *to appear in International Journal of Bifurcation and Chaos*, **4**, 1994.
10. T.S. Parker and L.O. Chua. *Practical and Numerical Algorithms for Chaotic Systems.* Springler-Verlag, New York, 1989.
11. Lj. Kocarev, K.S. Halle, K. Eckert, L.O. Chua and U.Parlitz, "Experimental demonstration of secure communications via chaotic synchronization", *Int.J. of Bifurcation and Chaos*, **2**, pp. 709-713, 1992.
12. K.S. Halle, C. W. Wu, M. Itoh and L.O. Chua, "Spread spectrum communications through modulation of chaos", *Int.J. of Bifurcation and Chaos*, **3**, pp. 469-477, 1993.
13. H. Dedieu, M.P. Kennedy and M. Hasler, "Chaos shift keying: modulation and demodulation of a chaotic carrier using self-synchronizing Chua's circuits", *IEEE Trans. on Circuits and Systems (Part II)*, **40**, pp. 634-642, 1993.
14. S.N. Rasband. *Chaotic Dynamics of Nonlinear Systems.* Wiley Inter-Science, New York, 1990.
15. K.M. Cuomo and A.V. Oppenheim, "Circuit implementation of synchronized chaos with applications to communications", *Physical Review Letters*, **71**, pp. 65-68, 1993.

Time-Series Analysis and Neural Networks

Time-series analysis plays a fundamental role in studying and identifying complex systems from the measurement of a limited number of physical observables.

The articles presented in this section cover different topics in this field: nonlinear forecasting methods, neural networks and multifractal analysis.

The problem of predicting the future evolution of a chaotic system from the knowledge of its past history is one of the milestones of the applied theory of chaos (see H.D.I Abarbanel, R. Brown, J.J. Sidorowich and L.S. Tsimring, "The Analysis of Observed Chaotic Data in Physical Systems", *Rev. Mod. Phys.* **65**, 1993, 1331). The approach followed by Colace and A.R. Giona refers to fuctional reconstruction methods and to the application of orthogonal polynomials with respect to the invariant measure.

Neural networks represent a new a rapidly expanding field of investigation which may furnish promising results in process identification and control, as discussed by Baratti et al. and Bravi et al. in connection with distillation columns and crystallizer dynamics.

Multifractal theory and the identification of singular measure structures is a somewhat less known topic in applied sciences and engineering although it may have important applications in the description of the statistical properties of dynamical systems (Halsey et al., *Phys. Rev.* **A 33**, 1986, 1141), in the study of the fine-structure of turbulent flows (C. Meneveau, K.R. Sreenivasan, *J. Fluid Mech.* **224**, 1991, 429) and in the characterization of the distribution of reacting centres on a catalyst (R. Gutfraind, M. Sheintuch and D. Avnir, "Fractal and Multifractal Analysis of the Sensitivity of Catalytic Reactions to Catalyst Structure", *J. Chem. Phys.* **98**, 1991, 6100) The article by M. Giona focuses on a methodological aspect of multifractal analysis consisting in deriving information on multifractal scaling properties directly from time-averaged quantities thus bypassing box-counting methods which, as discussed by Greenside, Wolf, Swift and Pignataro ("Impracticality of a box-counting algorithm for calculating the dimensionality of a strange attractor, *Phys. Rev.* **A 25**, 1982, 3453), may be critical and subject to significant error with the increasing of the embedding dimension.

FUNCTIONAL RECONSTRUCTION OF CHAOTIC CHEMICAL KINETICS: AN ALGORITHM FOR FAST MEASURE BASED RECONSTRUCTION

LORENZO COLACE
Dipartimento di Ingegneria Elettronica, Universitá di Roma "La Sapienza"
V.Eudossiana 18, 00184 Roma, ITALY

and

ALESSANDRO R. GIONA
Dipartimento di Ingegneria Chimica, Universitá di Roma "La Sapienza"
V.Eudossiana 18, 00184 Roma, ITALY

ABSTRACT

A generalised method of functional reconstruction of chaotic chemical kinetics is proposed. Good local prediction of chaotic time series may be achieved if the detailed structure of the invariant measure associated to a dynamic system is taken into account. The method is based on the ergodic properties of the dynamic systems; the vector field associated to the kinetics is expressed in terms of the polynomial system orthonormal to the invariant measure. Functional reconstruction is a generalised Fourier expansion whose coefficients are expressed in terms of the hierarchy of moments and of the hierarchy of functional moments. This method of functional reconstruction has some practical properties because the expansion coefficients are directly obtained in terms of ergodic summations and they are independent, so least square fitting is unnecessary. The measure based reconstruction is applied to the single observable time-series analysis. Important parameters to be considered are the reconstructed phase-space dimension and the time delay. While the estimation of the former can be easily effected by means of known algorithms (e.g. the Grassberger-Procaccia algorithm for correlation dimension), a criterion based on the properties of correlation function is proposed for the evaluation of the optimal time delay. Prediction and reconstruction results seem to confirm the validity of this criterion. To perform the parameters evaluation, the reconstruction and the prediction of time-series, a new algorithm is proposed. Our efforts have the aim of performing fast processing so that reconstruction and prediction can run in real time for a wide field of application. Performances expressed in terms of prediction time, prediction error and processing time requirements are evaluated.

1. Introduction

Since the first works on turbulence and complex system [1,2], it has been known that simple non-linear dynamic systems can give rise to very complex temporal behaviour usually referred to as deterministic chaos. Chaotic behaviour has been observed in many non-linear dynamic systems. Mechanical and electromechanical systems can often be described by equations exhibiting chaotic behaviour [3,4]. Chaos has been reported in many chemical reactions systems [5,6,7] and in the dynamic evolution of reactors [8,9] often as a consequence of spatial concentration gradients [10] or of diffusional coupling [11]. Chaotic behaviour characterises the manifestation of turbulence and plays a fundamental role in the theoretical understanding of turbulent transport phenomena [12,13,14]. The complex behaviour in purely deterministic dynamic systems set strong limits on prediction-making capabilities; on the other hand, it opened up new perspectives in time series analysis, where one of the most interesting topic is the reconstruction of system dynamics from the knowledge of time series. It is clear that the assumption of a deterministic explanation for the time series $\{x(t_i)\}$ represents a strong dynamical constraint to be used in order to construct an efficient predictor. In general, we may say that a time series $\{x(t_i)\} : \{x(t_i)\} \in R^d$ has a deterministic explanation if there exists a dynamical system defined in $R^n (n \geq d)$

$$\frac{dy(t)}{dt} = f(y(t)), \quad y \in R^n \tag{1}$$

and a mapping $\phi : R^n \rightarrow R^d$ such that

$$x(t_i) = \phi(y(t_i)), \quad t_i \in R^+, \quad t_{i+1} > t_i \tag{2}$$

Criteria for establishing the deterministic nature of a time series have been proposed by Takens [15] and by Suighara and May [16]. Time series analysis has strong practical and theoretical implications in reaction and reactor modelling and identification. In this case, given the physical origin of the problem, the deterministic nature of the temporal evolution can be assumed in almost all cases. In dealing with chaotic kinetics the problem posed in time series analysis are:

a) the determination of the dynamical invariants [17](e.g. fractal dimension, entropies, Ljapunov exponents)

b) the reconstruction of the attractor in terms of an equivalent mathematical model [18]

c) the prediction of future evolution from the knowledge of the past history [19]

All these problems can be unified if we can produce, from the knowledge of a time series $\{x(t_i)\}$, a reconstructed functional approximation having the same invariant structure as $\{x(t_i)\}$, from which the future evolution can be predicted at least in some limited temporal range. There are several methods for the reconstruction of chaotic time series because there are different ways to represent non-linear functions. Polynomials are often a good representation because their coefficients can be easily fit to minimise least square and because they arise naturally in Fourier expansion. Polynomials have been recently used for reconstruction [13,20,21]. Rational approximation (ratio of two polynomials) has sometimes been used [22]. Recently radial basis functions have been used as interpolation scheme [23]. A further approach consists in adopting neural networks [24,25,26,27]. In this work, the problem of functional reconstruction of oscillatory and chaotic chemical reactions is analysed by means of the Measure-Based approach [28]. Let us consider before the problem of predicting the evolution of a time series $x(t_i)$ associated to a single observable coming from a deterministic dynamic system, only from the knowledge of its past history. The theoretical contribution of Takens [29] and Packard [30] et al. demonstrates that the local (temporally local) forecasting problem can be solved expanding the time series in an n-dimensional reconstructed space: $x(t_i), x(t_i-\tau), ..., x(t_i-(d_r-1)\tau)$, where d_r is the dimension of the reconstructed phase space and τ is a time delay. In this paper we formulate a consistent reconstruction and prediction method using a generalisation of Measure-Based (M-B) reconstruction and provide an algorithm which consists of: a pre-processing procedure to perform the evaluation of d_r and τ, a set-up calculation to acquire the necessary information about the dynamics, and an iterative procedure to perform on-line reconstruction.

2. M-B Reconstruction

A vector field $f(x)$ can be written as a sum of orthonormal polynomials in [a,b] with respect to a weight function $p(x)$, according to the Fourier expansion:

$$f(x) = \lim_{N\to\infty} \sum_{k=0}^{N} c_k P_k(x) \tag{3}$$

where c_k are given by:

$$c_k = \int_a^b p(x) f(x) P_k(x) dx \tag{4}$$

The partial sum:

$$f_N(x) = \sum_{k=0}^{N} c_k P_k \tag{5}$$

is an approximation of $f(x)$. If we consider for estimating the error the L_2 metrics:

$$\int_a^b p(x)[f(x) - f_N(x)]^2 dx \tag{6}$$

the partial sum is the best approximation. M-B reconstruction is an approximation similar to equation (5), i.e. is a sum of polynomials orthonormal to weight function $p(x)$. Particular care is required in the choice of $p(x)$ because when $p(x)$ is chosen, the polynomial system is automatically fixed. M-B reconstruction is based on the property that for ergodic dynamic systems there exists a unique probabilistic invariant measure $\mu_I(x)$ such that, for a summable function G defined in the phase space, we have:

$$\lim_{N\to\infty} \frac{1}{N} \sum_{i=1}^{N} G(x(t_i)) = \int_C G(x) d\mu_I(x) \tag{7}$$

where C is the limit set of the time series. Roughly speaking, the invariant measure $\mu_I(x)$ of a set represents the frequency of the occurrences of the dynamic in this set. In other words we can write:

$$d\mu_I(x) = d_I(x) d^d(x) \tag{8}$$

and $d_I(x)$ is a probability density. In particular $d_I(x)d^dx$ is the probability that a point of the dynamics exists in the volume element d^dx in a d-dimensional Cartesian space. (Note that $x = (x_1, ..., x_d)$ is a d-dimensional vector). Therefore, it is possible to construct a system of polynomials $\Pi = \{\Pi_k(x)\}$ orthonormal on the limit set with respect to the weight function represented by the invariant measure $\mu_I(x)$. From this position, one can reconstruct the vector field in a generalised Fourier expansion with respect to the elements of Π. The theoretical advantages gained with this method are that:

a) the expansion coefficients can be directly obtained in terms of ergodic summations in closed form, without requiring the solution of a potentially numerically unstable linear system [28];

b) the expansion coefficients are independent [28].

The formalism of M-B reconstruction is fully developed by Giona et al [28] for one, two and three-dimensional phase space.

3. Generalisation of M-B Reconstruction

In the present analysis, the formalism developed by Giona et al.[28] is generalised in such a way it can be simply implemented automatically and for any dimension of the reconstructed phase space. Generalised M-B reconstruction is a generalised Fourier expansion with respect to the polynomial system $\Pi = \{\pi_i(x)\}$ satisfying the condition:

$$\int_C \pi_i(x)\pi_j(x)d\mu_I(x) = \delta_{ij} \qquad x \in R^d \tag{9}$$

where δ_{ij} is the Kroenecker notation. In the development of the formalism of M-B it is useful to introduce the following positions. Let $d = n$ be the dimension of the time series. To each polynomial monomer $m_k(x)$ of degree ν we can associate its vector code $\underline{c}_k$ of d integers:

$$m_k(x) = \prod_{i=1}^{d} x_i^{l_i} \qquad \sum_{i=1}^{d} l_i = \nu \qquad \underline{c}_k = (c_k(1), ..., c_k(d)) \tag{10}$$

where:$c_k(i) = l_i$ and $x = (x_1, ..., x_d)$.

Among the codes $\underline{c}$ a hierarchical ordering can be defined: c_k precedes c_h ($c_k < c_h$) if the first non equal component $c_k(j)$ is less than $c_h(j)$. Moreover, the space of the codes can be equipped with the usual vector algebra. The ordering among the vector codes induces an ordering into Π such that if $\underline{c}_k < \underline{c}_h$, then $\pi_k < \pi_h$. Therefore, every polynomial $\pi(x)$ can be written as:

$$\pi_i(x) = \frac{1}{N_i}\sum_{j=1}^{i-1} \alpha_{ij}\pi_j(x) + m_i(x) = \sum_{j=1}^{i} \beta_{ij} m_j(x) \tag{11}$$

where:

$$N_i = \sqrt{<\pi_i(x), \pi_i(x)>} = \sqrt{\int_C \pi_i^2(x) d\mu_I(x)} \quad (12)$$

The coefficients α_{ij} can be evaluated by means of the Gram-Schmidt orthogonalisation procedure:

$$\alpha_{ij} = -\sum_{h=1}^{j} \beta_{jh} M(\underline{c_i} + \underline{c_h}) \quad (13)$$

where the moments $M(\underline{c_k})$ read:

$$M(\underline{c_k}) = \int_C m_k(x) d\mu_I(x) = \lim_{N\to\infty} \frac{1}{N} \sum_{i=1}^{N} m_k(x_i) \quad (14)$$

Once the coefficients α_{ij} have been determined, the normalisation coefficient N_i can be evaluated from Eqs. (11) and (12):

$$N_i^2 = M(\underline{c_i} + \underline{c_i}) - \sum_{j=1}^{i-1} \alpha_{ij}^2 \quad (15)$$

and thus the coefficients β_{ij}, defined by Eq. (11) can be obtained. Since $\beta_{11} = 1$, Eqs. (13-14) constitute an iterative process for determining the orthonormal system Π. Eqs. (14-15) imply that the polynomial system Π can be evaluated directly from the hierarchy of moments $M = \{M(\underline{c_k})\}$, i.e. from ergodic summations. The second step entails the evaluation of the expansion coefficients of the vector field to be reconstructed. If the time series x_i has a deterministic explanation, then there exists a vector field f such that:

$$\dot{x}(t) = f^{(p)}[x(t)] \quad (16)$$

where $f^{(p)} = (f_1^{(p)}, ..., f_d^{(p)})$ and $x \in \Re^d$. In this case we can approximate f with a polynomial reconstruction up to order p expressed as a series expansion in Π:

$$f_h^{(p)}(x) = \sum_{i=1}^{N(p,d)} \gamma_{hi} \pi_i(x) \quad (17)$$

The expansion coefficients γ_{hi} are the Fourier coefficients with respect to Π:

$$\gamma_{hi} = \int_C f_h(x) \pi_i(x) d\mu_I(x) = \sum_{j=1}^{i} \beta_{ij} \int_C f_h(x) m_j(x) d\mu_I(x) \quad (18)$$

The integral appearing in Eq.(18) can be evaluated from the time series $\{x_i\}$. If the time series is sampled with sampling time h_s then:

$$f_h(x_i) = D_h(x_i) = \frac{x_{i+1,h} - x_{i,h}}{h_s} \quad (19)$$

Of course, for $D_h(x)$ we can choose a more refined version of numerical differentiation (e.g. a five point expression). In any case, assuming a given expression for $D_h(x)$, the expansion coefficients γ_{hi} can be expressed in terms of the functional moments $F = \{\theta_h(\underline{c}_k)\}$:

$$\theta_h(\underline{c_k}) = \int_C D_h(x)\pi_k(x)d\mu_I(x) = \lim_{N\to\infty}\frac{1}{N}\sum_{i=1}^{N} D_h(x_i)\pi_k(x_i) \tag{20}$$

as:

$$\gamma_{hi} = \sum_{j=1}^{i} \beta_{ij}\theta_h(\underline{c_j}) \tag{21}$$

The expansion coefficients γ_{hi} are uniquely determined by the hierarchies of moments M and of functional moments F, and are independent. In this way, any kind of least square fitting becomes unnecessary. (This property has another interesting consequence, namely the independence of the coefficients γ_{hi}, which means that, if we want to reconstruct a kinetics up to a polynomial order p, having already obtained a lower approximation of order $p' < p$, we need only compute the new expansion coefficients of order $p'+1, ..., p$, because the preceding ones are identical).

The reconstruction method as outlined above may be applied in two different practical situations: when the dimension d of the observable time series is equal to the dimension of the phase space (complete time series analysis) and when $d = 1$ (single observable time series analysis). Our work deals with single observable time series analysis.

4. Parameters Estimation

Takens' theorem generically establishes that the prediction of the future evolution of a dynamic system is theoretically possible from the knowledge of a single observable, but its practical application opens several problems which are still matter of scientific debate. The theoretical contribution of Takens demonstrates that the local forecasting problem can be solved expanding the time series in an n-dimensional reconstructed space: $x(t_i), x(t_i-\tau), ..., x(t_i-(d_r-1)\tau)$, where d_r is the dimension of the reconstructed phase space and τ is a time delay. The open questions are principally connected with the choice of the parameters of the M-B predictor, i.e. the optimal value of the reconstructed phase space dimension d_r and that of the optimal delay τ, necessary in the definition of the reconstructed time series. The problem of optimal phase space dimension finds answer in Takens' theorem, according to which d_r should be given by:

$$d_r = 2D + 1 \tag{22}$$

where D is the dimension (in the sense of Hausdorff dimension) of the limit attractor of the dynamic system. A simple and rapid method for evaluating the dimension D (or at least one approximation for it) is the Grassberger-Procaccia algorithm [31].

Indeed Grassberger-Procaccia algorithm furnishes an estimate of the correlation dimension D_2 which is stictly less or equal than the fractal dimension D. The choice of the optimal time delay has been discussed by several authors [32,33] in relation to phase portrait reconstruction and some preliminary criteria have been proposed. Incidentally, with regard to the optimal time delay, there is no detailed analysis focused on prediction performances. We show that the optimal time delay can be chosen, with good results as regards forecasting performances, from the analysis of the autocorrelation function associated to the time series, as a function of the pseudo period of the oscillation and of the dimension of the reconstructed phase space.

4.1. Reconstructed Phase Space Dimension

As outlined above, the choice of the reconstruction dimension d_r is substantially solved by Takens' theorem, according to Eq.(22). Therefore, in an adaptive realisation of a predictor, estimation is required of the dimension D of the dynamic system from which the time series has been extracted. A satisfactory estimate of D can be calculated from the scaling of the Grassberger-Procaccia (G-P) correlation function:

$$C(r,q;\tau) = \lim_{N\to\infty} \frac{1}{N^2} \sum_{i=1}^{N} \sum_{j=1}^{N} \eta[r - d_e(w_i^{(q)}, w_j^{(q)})] \sim r^D \tag{23}$$

where η is the Heaviside step function, d_e is the Euclidean metrics and $w_i^{(q)}$ are given by:

$$w_i^{(d_r)} = \begin{pmatrix} x_i \\ x_{i-k} \\ \dots\dots \\ \dots\dots \\ x_{i-(d_r-1)k} \end{pmatrix} \tag{24}$$

In practice, one evaluate D for increasing values of q until it saturates to a constant value. The question arises in the simplest and most rapid implementation of the G-P algorithm since its algorithmic complexity is $O(N^2)$. In the sense of a M-B predictor, the evaluation of D is only a preliminary step so that in the perspective of a real time application of a functional reconstructor, the G-P correlation function must be evaluated from a relative small number of data points N. Even in this case, the choice of τ has to be adequate. As it will outline below, we found it convenient to choose $\tau = T_p/4$ where T_p is the pseudo period of the auto-correlation function of the time series. We assumed N=1000 in order to process about 250 pseudo periods of oscillations, which is a fairly reasonable time for setting up the parameters needed in M-B reconstruction. (It should be noted that the usual evaluation of D deals with N=30000-100000).

4.2. Optimal Time Delay

The question of the optimal time delay is of central importance in reconstruction and prediction from a single observable. The choice of τ influences the structure of the reconstructed attractor $\mathcal{C}_r$ and the dynamic properties of the reconstructed vector field. If τ is too small, the components of the reconstructed vector $w_{i,dr}$ are closely correlated and practically not independent of each other; if τ is too large, the non- correlation of the components of $w_{i,dr}$ implies a complex functional form of the reconstructed vector field. In order to bring these consideration into the form of quantitative criteria, an estimation of the proper time scale involved in the dynamics is needed. It appears clear from these intuitive arguments that the temporal behaviour of the correlation function associated to the time series $\{x_i\}$ conveys information on the temporal scale of the time series to be reconstructed. This idea was pointed out by Nicolis and Nicolis [19] and has its roots in classical signal theory. Broomhead and King [17] and later Hao Bai Lin [33] suggest that the optimal time delay must be a function of the reconstructed phase space. Fraser and Swinney [32] criticise the approach based on the analysis of the correlation functions and suggest that τ should be chosen from the behaviour of the two point mutual information. This approach has recently been challenged by other authors (see Hao Bai Lin, [33]). It may be stated that the choice of the optimal time delay in the reconstruction of chaotic evolution is a controversial subject. This is partially due to the fact that the notion of optimal time delay with no further specification is a fairly confusing concept. To be more explicit, the definition of an optimal time delay in reconstruction depends on what one is interested in, e.g. the evaluation of dynamic invariants (dimension, Lyapunov exponents), the prediction of its future evolution, or the global reconstruction of the attractor. In each of these cases, it is possible to define an optimal value of τ, all these values being in principle different from one another. In the case of time series prediction, the estimation of optimal time delay may be performed from the analysis of the auto-correlation function. The optimal time delay was found to coincide with a quarter of the period of the auto-correlation function $C(t)$:

$$\tau_{opt} = \frac{T_p}{4} \tag{25}$$

It must be borne in mind that for chaotic signals the auto correlation function is not periodic and decays with the increasing of the time shift. In the case of chaotic evolution, it is convenient to define the pseudo period T_p as the averaged distance between the minima of $C(t)$. Moreover, we found that the choice of τ must take into account the value of the dimension d_r. It is intuitive that, with the increasing of the reconstructed phase space, the optimal value of τ must decrease because the temporal distance between the first and the last coordinate must be of the same order of magnitude as $T_p/4$. We therefore expect that the natural generalisation of eq.(25) would be:

$$\tau_{opt}(d_r - 1) = T_p/4 \tag{26}$$

Equation (26) is a semi-empirical criterion for choosing τ and numerical simulations are needed in order to test its validity.

5. Numerical Simulations

In this section we shall report and discuss the results obtained with M-B reconstruction. In particular the optimal choice of d_r and τ and prediction performances will be evaluated. At the end of the reconstruction, we have the reconstructed dynamics represented by a set of non-linear differential equations; so we can then integrate this system to obtain the reconstructed time series and compare it to the original one.

Generally, the quality of a reconstruction method may be expressed in terms of both the distance between the original vector field and its reconstruction, and the prediction time. Moreover, for chaotic time series, a good dynamic reconstruction, expressed by low values of the distance $d^{(p)}$ between the original vector field f and its p-order reconstruction $f(p)$, does not necessarily imply good forecasting capacity due to exponential divergence. For this reason, we found useful to analyse the forecasting power of a reconstruction method by introducing the averaged prediction time $T(\epsilon)$, defined as the averaged time with respect to the invariant measure μ_I in which the reconstructed dynamics (starting from the same initial position) differs from the corresponding values of the original time series $\{x_i\}$ by a value less than ϵ:

$$T(\epsilon) = < t(x,\epsilon) >_I = \int_C t(x,\epsilon) d\mu_I(x) \tag{27}$$

$$t(x,\epsilon) \in R^+, \qquad d_e(\phi_x^{(p)}(t_i), x_i) < \epsilon \quad for \quad t_i < t(x,\epsilon), \qquad t_i = ih_s \tag{28}$$

where $\phi_x^{(p)}$ is the phase flow associated to the reconstructed vector field $f(p)$ and $d_e(f.g)$ is the Euclidean metrics

$$d_e(f,g) = \sqrt{\frac{1}{d}\sum_{h=1}^{d}[f_h - g_h]^2} \tag{29}$$

In order to analyse the properties of M-B reconstruction, let us consider the dynamic model of Table I, which represents well known chaotic differential schemes.

The Lorenz model is one of the simplest ordinary differential equation models for the onset of turbulence [1]; Rossler [7] and Varma [8] are qualitative models of reaction kinetics; the Brussellator model has been introduced [6] as a prototype of the influence of diffusional coupling in the transition to chaotic behaviour.

As discussed in the previous section, it is convenient to estimate d_r according to Eq.(22) where D is evaluated for increasing values of q (see Eq.(23)) until it saturates to a constant value. The values thus obtained for the dynamic systems considered are shown in figure 1, and are in close agreement with the known values [31].

Table 1: Dynamical models

Lorenz	Varma
$\begin{cases} \dot{x} = a(y-x) \\ \dot{y} = bx - y - yz \\ \dot{z} = xy - cz \end{cases}$ a=10 b=28 c=8/3	$\begin{cases} \dot{x} = -(y+z) \\ \dot{y} = x \\ \dot{z} = a(y-y^2) - bz \end{cases}$ a=0.375 b=0.230
Rossler	**Brussellator**
$\begin{cases} \dot{x} = -(y+z) \\ \dot{y} = x + ay \\ \dot{z} = b + xz - cz \end{cases}$ a=0.2 b=0.2 c=5.7	$\begin{cases} \dot{x} = a - (1+b)x + x^2y \\ \dot{y} = bx - x^2y \end{cases}$ a=2.0 b=5.5

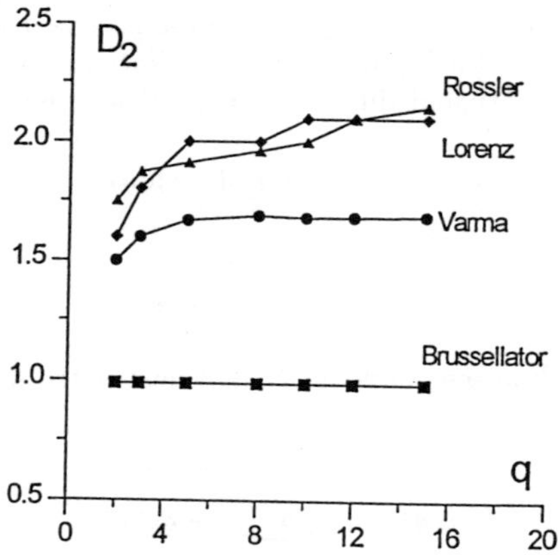

Figure 1: $D_2(q)$ vs q according to the scaling law of Eq.(23).

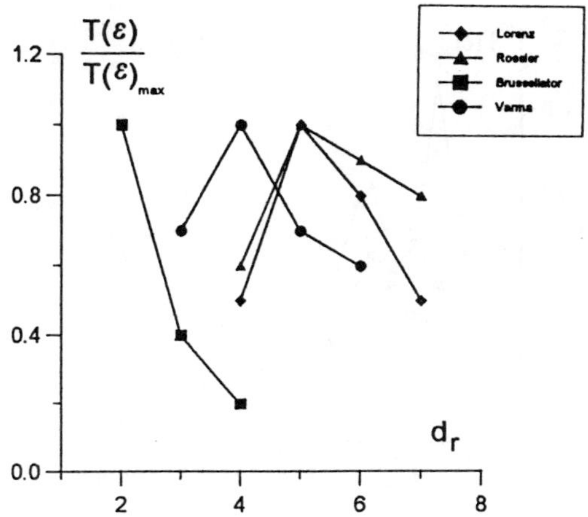

Figure 2: Normalised prediction time vs the reconstructed phase space dimension d_r.

The influence of d_r on prediction performances is shown in figure 2. From these data, and bearing in mind the values of D of figure 1, we may conclude that Takens' criterion is substantially satisfactory for functional reconstruction and prediction.

As regard the choice of the optimal time delay, the semiempirical criterion of Eq.(26) is considered. Numerical simulations are provided to verify its validity. Figure 3 shows the dependence of $T(\epsilon)$ for a fixed value of the error e as a function of the adimensional time delay θ:

$$\theta = \frac{4\tau(d_r - 1)}{T_p} \tag{30}$$

The figure shows that, for all the system considered, the optimal time delay may be well predicted by Eq.(26).

Now let us consider the prediction performances of the measure based reconstruction to test its applicability. Figure 4 show the prediction times of the four systems considered as functions of the error.

It is clear that the system of Varma (see Fig.4) provides the best performances; prediction time with an error less than 1%, is about 10s. Because of the sampling period of 10-3s, this means that we can predict about 10000 samples of the time series with an error less than 1%. For the Brussellator model a prediction time is obtained of about 4s corresponding to the forecast of about 4000 samples with error less than 1%. Worse prediction performances are obtained for more complex dynamic systems. The Rossler model gives a prediction time of about one second, corresponding to one thousand samples. For the Lorenz model, the prediction time falls to 0.3s with an error of 10% and as low as to 0.1s with an error of 0.1%,corresponding respectively

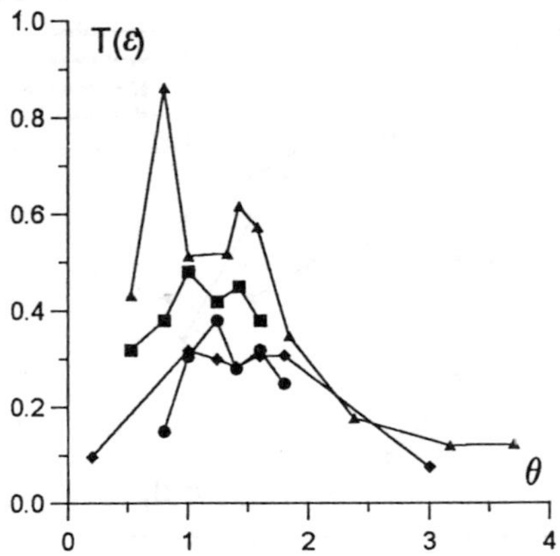

Figure 3: Prediction time $T(\epsilon)$ for a fixed value of ϵ vs the adimensional time delay θ.

to 600 and 400 samples of time series. The numerical results are in good agreement with the characteristic feature of the dynamic system: sensitive dependence on initial conditions. Given two distinct initial conditions arbitrarily close to one another, the trajectories emanating from these initial conditions diverge, at a rate characteristic of the system, until for all practical purposes they become uncorrelated. Consequently, long term predictions cannot be made for chaotic systems but only short term. As shown in figure 4, the prediction error increases with the increasing of the interval of prediction. The comparison of the prediction results of the different systems points out the difference between fully developed (e.g. Lorenz) and oscillatory chaos (e.g. Varma). Oscillatory chaos is characterised by small local instabilities (small value of the maximum positive Ljapunov exponent) and therefore the prediction time is longer.

6. On-line Adaptive Algorithm

The numerical simulations of the previous section has been obtained by means of a modular program which runs with a fixed number N of data points, supposed to be available. In this case the algorithm consists of four operations as shown in figure 5. At first, the evaluation of the reconstruction parameters τ and d_r is performed; then it is possible to evaluate the hierarchies of moments M and F from which the expansion coefficients can be calculated. Finally the integration of the reconstructed vector field is performed to yield the samples of the reconstructed time series. It is important to notice that the four procedures operate over a number N of data points.

This algorithm has been used essentially to test the criteria of the choice of τ and

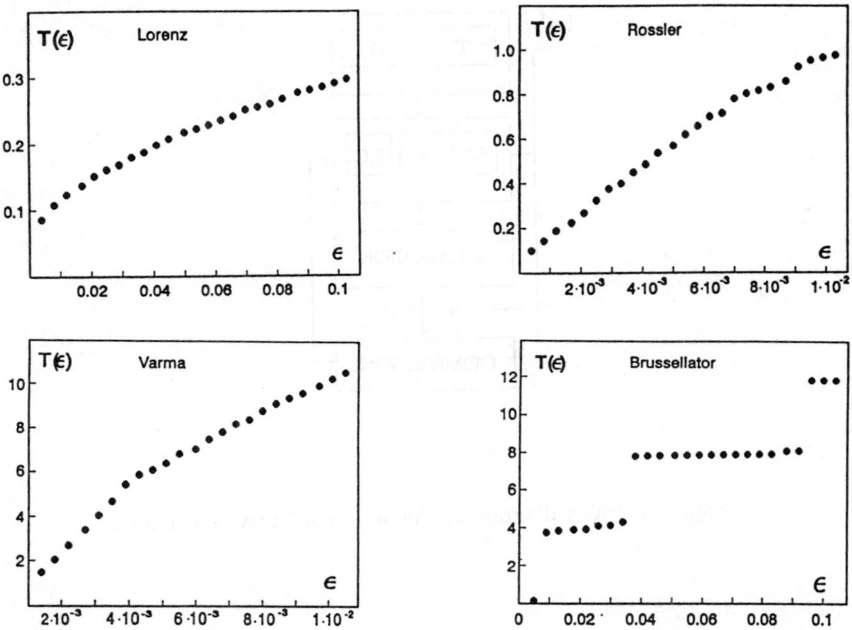

Figure 4: Prediction time $T(\epsilon)$ (in s) for different dynamical models.

d_r and the prediction performances of the M-B reconstruction in a preliminary stage of the work; for practical applications a different solution may be used: the on-line adaptive algorithm. The new algorithm grounds on the idea that, according to section 3, the expansion coefficients and the polynomial system can be directly evaluated from the hierarchies of the moments M and F which can be calculated in real-time updating the ergodic summations during the data flow from a data acquisition system. The block diagram of the real-time algorithm is shown in figure 6.

The set-up procedure yields the optimal values of τ and d_r and constitutes a preliminary stage of the algorithm. The iterative procedure performs the M-B recon-

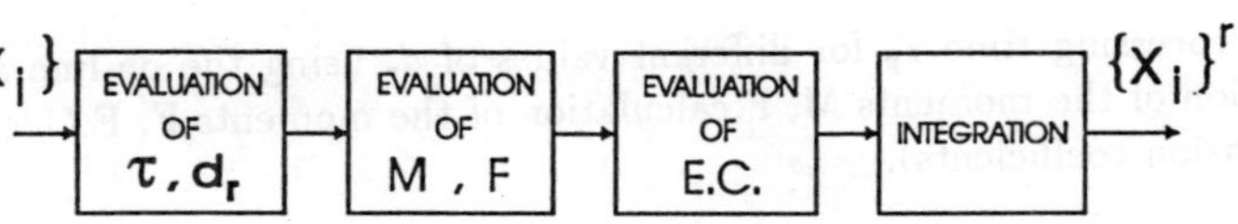

Figure 5: Block diagram of M-B reconstruction.

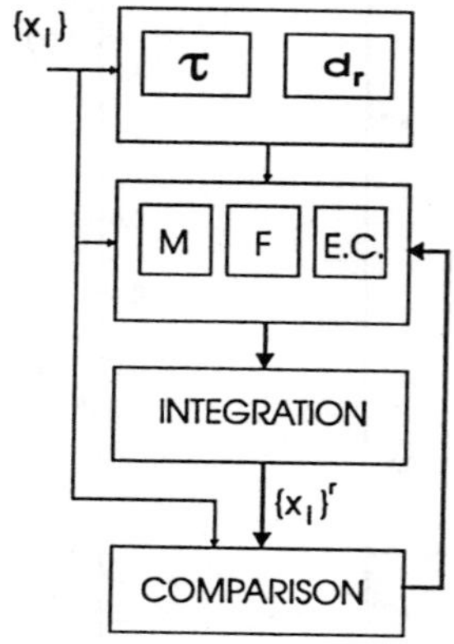

Figure 6: Block diagram of the on-line adaptive algorithm.

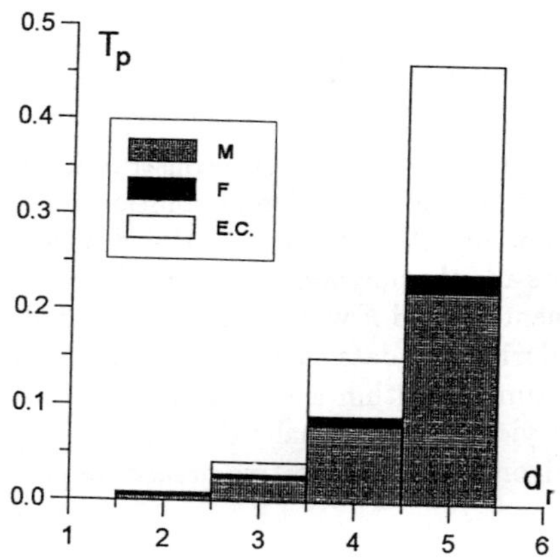

Figure 7: Processing time T_p for different values of d_r using the on-line algorithm. (M:calculation of the moments M, F:calculation of the moments F, E.C.:calculation of the expansion coefficients).

struction updating the functional moments M and F and then the expansion coefficients EC, while the integration procedure produces the reconstructed time series $\{x_i^r\}$. The update of the reconstructed vector field is performed whenever the comparison between $\{x_i\}$ and its reconstruction $\{x_i^r\}$ warns that prediction performances are going to deteriorate. Figure 7 shows, for different phase space dimension, the processing time required for a single step of the iterative procedure when the algorithm runs on a PC based on a 66MHz 486DX processor.

The algorithm may be used for practical application when the processing time is at least one order of magnitude less then the prediction time $T(\epsilon)$. Bearing in mind the prediction times showed in figures 4-7, we may conclude that the algorithm may be successfully used in oscillatory chaos (e.g.Varma and Brussellator models). Dealing with developed chaos such as Lorenz and Rossler models, a faster processing is needed. So the same on-line adaptive algorithm has been implemented on an IBM RISC 6000 which provide a processing time over fifty times smaller, thus allowing real-time reconstruction of a wide range of both oscillatory and developed chaos.

7. Conclusions

This work has presented a general treatment of functional reconstruction of chaotic dynamics based on series expansion in the polynomial system orthonormal to the invariant measure, also referred as to Measure Based Reconstruction. We have presented a detailed analysis of the parameters necessary to build up a predictor of chaotic evolutions from a single observable time series. The optimal time delay can be estimated in terms of the pseudo period of the auto correlation function of the time series and of the phase space dimension; the reconstructed phase space dimension may be estimated with the Takens' theorem through the G-P algorithm. Finally a real-time adaptive algorithm is presented. The results presented here constitute a fairly complete analysis on the applicability of functional reconstruction methods for prediction of chaotic kinetics. This results lead to practical criteria easily implementable in the design of an efficient predictor which can operate adaptively with respect to the time series to be processed. As a consequence, it is possible to consider the application of M-B reconstruction in control unit whose input is a complex but deterministic time series.

Acknowledgements

The authors are grateful to M.Giona for reading through the manuscript and making useful suggestions.

References

1. E.N. Lorenz, *J.Atmos.Sci.* **20** (1963) 130.
2. D. Ruelle and F. Takens, *Comm.Math.Phys* **20** (1971) 167.

3. P.J. Holmes and F.C. Moon,*J.of Appl.Mech.* **50** (1983) 1021.
4. P.J. Holmes, *Phyl.Trans.Roy.Soc.* **A292** (1979) 419.
5. J.C. Roux, A.Rossi, S. Bachelart and C. Vidal, *Phys.Lett.* **77A** (1980) 391.
6. J.L. Hudson and J.C. Mankin, *J.Chem.Phys.* **74** (1981) 6171.
7. I.R. Epstein, *Physica 7D* **47** (1983).
8. C. Kalhert, O.E. Rossler. and A. Varma, in *Modeling of chemical reaction system*, (Springer V., Berlin, 1981).
9. D.V. Jorgensen and R. Aris, *Chem.Eng.Sci.* **38** (1983) 45.
10. P.J. Nandapurkar, V. Halvacek and Van Pompay, *Chem. Eng.Sci.* **41** (1986) 2747.
11. I. Shreiber and M. Marek, *Phys. Lett.* **91** (1982) 263.
12. J.M. Ottino, *'The kinematics of mixing: stretching, chaos and transport* (Cambridge Univ.Press, Cambridge, 1989).
13. C.E. Lacy, M. Sheintuch, and A.E. Dukler, *A.I.Ch.E.J.* **37** (1991) 481.
14. I. Procaccia, *J.Stat.Phys.* **36** (1984) 649.
15. F. Takens, in *Dynamical System and Bifurcations*, (Springer Verlag, Berlin, 1985).
16. G. Suighara and R.M. May, *Nature* **344** (1990) 734.
17. D.S. Broomhead and G.P. King, *Physica* **20D** (1986) 217.
18. J. Cremers and A. Hubler, *Z.Naturforsch.* **42a** (1987) 797.
19. C. Nicolis and G. Nicolis, *Proc. Natl. Acad. Sci.* **83** (1986) 536.
20. J.P. Crutchfield and B.S. McNamara, *Complex Systems* **1** (1987) 417.
21. J.D. Farmer and J.J. Sidorowich, *Phys.Rev.Letters* **59** (1987) 845.
22. B.J. Bayly, I. Goldhirsh and S.A. Orszag, *Applied and Computational Mathematics* (Princeton U, 1987).
23. M. Casdagli, *Phisica* **D35** (1989) 335.
24. B.E. Ydstie, *Computers chem. Engng.* **4** (1990) 583.
25. N. Bhat and T. McAvoy, *Computers chem. Engng.* **4** (1990) 573.
26. J.L. Hudson, M. Kube, R.A. Adomaitis, I.G. Kevrekidis, A.S. Lapedes A.S, *J.Eng.Sci.* **8** (1990) 2075.
27. I. Grabec, (Dynamics Days, Dusseldorf, 1990).
28. M. Giona, F. Lentini and V. Cimagalli, *Phys. Rev.* **A44** (1991) 6.
29. F. Takens, in *Detecting strange attractors in turbulence*, ed D.A.Rand and L.S.Young (Springer Verlag, Berlin, 1981).
30. N.H. Packard, J.P. Crutchfield, J.D. Farmer, R.S. Shaw. *Phis.Rev.Letters* **45** (1980) 712. ge attractors' Physica D 9,189, 1983.
31. P. Grassberger and I. Procaccia, *Physica* **D9** (1983) 189.
32. A.M. Fraser and H.L. Swinney, *Phys. Rev.* **A33** (1986) 1134.
33. Hao Bai Lin, *Elementary symbolic dynamics and chaos in dissipative system'World Scientific* (World Scientific, Singapore, 1989).
34. O. Gurel and D. Gurel, *Oscillations in chemical Reactions* (Springer V., Berlin,1983).

NEURAL NETWORKS: DEVELOPMENT OF AN OBSERVER FOR DISTILLATION COLUMNS

ROBERTO BARATTI
Dip. di Ingegneria Chimica e Materiali, Università degli Studi di Cagliari, Piazza D'Armi, Cagliari, 09123, Italy

ALBERTO SERVIDA
Istit. di Chimica Industriale, Università degli Studi di Genova Corso Europa 30, Genova, 16130, Italy

and

GIORGIO VACCA
SARAS Raffinerie Sarde Sarroch, (CA) 09108, Italy

ABSTRACT

Preliminary results on the application of Artificial Neural Networks (ANN) as estimators of the outlet compositions from a distillation column are presented. The ANN process modeling is applied to a case of relevance for the oil refineries, that is, the butane splitter tower. The performance of the ANN observer with respect to the dynamic prediction capabilities is also assessed. For the case under examination, where there exhists multiple inputs of various nature (i.e., compositions, temperatures, volumetric flowrates and so on) the rescaling of the inputs (normalization) makes the training process more robust avoiding numerical difficulties related to the ill-conditioned nature of the parameter estimation task.

1. Introduction

In the industrial practice, it is not uncommon that the process engineer has to deal with controlling "difficult-to-measure" state variables. For instance, the real-time measurement of outlet concentrations in chemical plants represents the typical situation of "difficult-to-measure" controlled variable because of the large time-delayed response exhibited by the conventional analytical instrumentation. To overcome this practical limitation, various researchers have devoted their efforts at developing "sotf sensors" capable to infer concentrations from easily available measurements, such as temperatures. Various kind of approaches based on deterministic[1,2], stochastic[3,4], and artificial neural network[5,6] models have been reported in the literature.

Only recently, renewed interest has been paid to ANN-based models because of their simple structure, fast and effective computation performance, and flexibility, which make them suitable for a wide class of applications, such as system identification, closed-loop control, and fault diagnosis. The most attractive property of ANN is their capability to represent any arbitrary nonlinear functional mapping between input and output data when incorporated in black-box-like process models.

Artificial neural network technology draws its analogy from human brain. Humans often learn by trial and error; neural networks operate in a similar fashion. This means that a neural net must be trained by repeatedly showing to it the input data along with the corresponding target outcomes. After a sufficient number of training iterations, the network creates an internal approximate process model by learning to recognize the pattern (map) relating the outputs to the corresponding inputs. It is important to note that this internal model is not based on any specification of the actual process mechanism: the ANN itself generates this approximate model. This means that the neural network does not require any *a priori* fundamental understanding of the process under examination and this makes the ANN-based models suitable to describe a wide class of complex nonlinear processes.

The capability of ANN's to handle, in principle, any complex nonlinear process opens a wide range of opportunities in advanced nonlinear process control and inversion of model process operators[7]. The principal advantage of ANN models over the ones based on first principles is their capability to provide reliable and cost effective process models. In this regard, we can take advantage of ANN models in developing internal model control structures (IMC).

In this communication we show preliminary results on the application of ANN as observers for real-time estimates of product compositions of distillation columns. The idea to use an estimator to infer the outlet composition from a distillation column is not a new one and various classes of estimators have been developed and tested[8]. The results obtained for a multilayer ANN topology trained by the back-propagation algorithm are discussed. We also show that for ANN structures with multiple inputs of various nature (for instance, compositions, temperatures, flowrates, and so on) a rescaling of the inputs makes the training process more robust avoiding the numerical difficulties associated with the ill-conditioned nature of the parameter estimation task. The inferential capabilities of the developed neural net-based models are also assessed by comparing the predictions of the model with a new set of data which weren't used during the learning stage.

2. Generalities on Process Modelling with Artificial Neural Networks

The research in the area of ANN has been mainly concerned with two classes of networks: i) multilayer feed-forward ANN's (MFFANN) and ii) dynamic networks of which the recurrent nets are a particular family. The first class has been extensively and successfully used in representing static nonlinear maps, as it is the case for pattern recognition tasks. On the contrary, the second class has proved to be useful for dealing with highly nonlinear dynamic processes. Most of the available algorithms and results on neural networks have been obtained for MFFANN.

In MFFANN structures the dynamics is not inherently included because the learning process of the net is carried out in a static fashion: the net is trained by showing to it a certain number of inputs and requiring that the net recovers the corresponding output outcomes. This training procedure is static in nature, the net is taught only to recognize the outcomes corresponding to the given inputs; in other words, the net is trained to perceive nonlinear static maps. In actuality, for most of the practical situations of interest, the processes always exhibit dynamic characteristics which result in dynamic functional relationships between the inputs and the outputs. Therefore, the statically trained feed-forward nets may fail to recover primary features of the real process and this may lead to serious pitfalls when the net is used in system identification and control applications.

There are various approaches to account for the process dynamics which has been shown to be critical for neural function processing[9]. In this work, we have adopted the approach which consists in developing a neural network-based inferential estimator. In such as neural net, the natural inputs (i.e., feed temperature, feed composition, and so forth) are augmented with available experimental measurements (i.e., tray temperatures). Care must be paid in choosing the net inputs to avoid duplication of the fed information which may lead to an ill-conditioned parameter estimation task.

2.1 Feed-Forward Neural Network

As mentioned above, we have considered an estimator based on a three-layers feed-forward neural architecture as depicted in Figure 1.

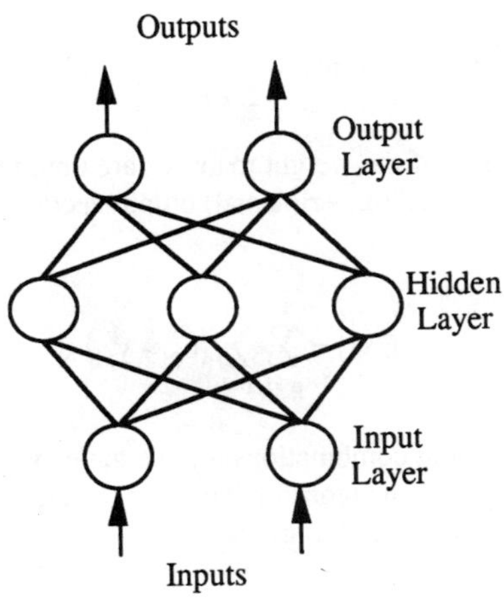

Figure 1 Three layers feed-forward artificial neural network.

A detailed discussion of the algorithm and computational issues associated to ANN can be found elsewhere[10], and thus, only a brief summary of the relevant results will be given here. As shown in Figure 1, the first (input) layer contains n_1 nodes, the second (hidden) layer contains n_2 nodes, and the third (output) layer contains n_3 nodes. The feed and hidden layers are also augmented with a bias unit which represents the threshold value. The input value of the bias node is held constant and equal to 1. The output of the jth neuron in the first layer, which depends on the input vector **I** applied to the ANN, is given by a nonlinear transformation of the activation value, S'_j, obtained as a weighted linear combination of its inputs:

$$y'_j = f\left(S'_j\right) = f\left(w'_{0j} + \sum_{i=1}^{n_1} w'_{ij}\, I_i\right), \tag{1}$$

analogously, the output of the jth neuron in the second layer is given by:

$$y''_j = f\left(S''_j\right) = f\left(w''_{0j} + \sum_{i=1}^{n_2} w''_{ij}\, y'_i\right) \tag{2}$$

where w'_{ij} e w''_{ij} are the components of the weight matrices from the input to the hidden layer neurons and from the hidden to the output layer neurons, respectively, while w'_{0j} and w''_{0j} are the components of the weight vectors related to the bias node. Here, we have adopted the conventional neuron activation function given by the sigmoid, defined as follows:

$$f(s_j) = \frac{1}{1 + e^{-S_j}} \tag{3}$$

The value of the elements of the weight matrices are obtained by minimization of the square error between the expected (experimental) output vector **O** and the ANN (predicted) output **y**":

$$E(\mathbf{w}) = \sum_{p=1}^{P} \sum_{i=1}^{n_3} \left(o_{pi} - y''_{pi}\right) \tag{4}$$

for all the available input/output combinations, P. We have evaluated the weight matrices by the traditional back-propagation algorithm based on the gradient method of the steepest descent[11] using random weights as initial guess.

3. Case Study

In this work we have developed an ANN-based estimator to infer the output stream concentrations of a distillation column. In particular, we have chosen the butane splitter tower of the SARAS alkylation plant. This is a 56 trays multicomponent distillation column, where the separation of the key components, *iso* and *n*-butane, takes place. The feed may also contain a fraction of butylenes coming from the FCC plant.

Typically, the available input measurements are temperatures, flowrates, levels, pression, reboiler duty, and feed composition, while the output measurements are the output composition of the distillate and bottom streams. Because of the set-up of the column, at each sampling time only the distillate or residue composition was available.

The selection of the available measurements to be used as inputs and how to feed them to the ANN-observer requires a bit of caution. As far as the first point is concerned, we may make use of the results obtained for conventional observers. It is well known that the goodness of the estimate generally improves when new measurements are added, however, this does not mean that one should use all the available measurements. In fact, because of the outputs correlation, additional measurements cannot provide but only a minor improvement to the estimate. Furthermore, for static estimators, the presence of dynamic lag between inputs and outputs will provide only little information, and thus, the inputs exhibiting delay with respect to the estimated variables should be disregarded. The presence of an input with a high level of noise would prevent the full recovery of the actual process dynamics, and thus, it should be neglected.

Because of this, we have chosen to use only the following measurements: temperatures (top, reflux, pilot tray, bottom and feed), reflux flowrate, reboiler duty, *iso* and *n*-butane composition in the feed, and flowrate of the stream coming from the FCC plant. The evaluation of the ANN parameters boils down to estimate the weight matrices which make minimum the square error defined by Eq.(4).

It is worth to point out that, any procedure which allows to well discriminate between two output outcomes corresponding to two different inputs will improve the numerical robustness of the parameter estimation. This is the reason why the rescaling of the input vector, which contains components of different magnitude, improves the parameter estimation.

In order to feed the maximum available information, using the mentioned inputs, all the measurements were rescaled using the following relation:

$$I_k = \frac{I'_k - I'_{med,k}}{I'_{med,k} - I'_{min,k}} \quad \text{for} \quad I'_k \le I'_{med,k}$$
$$I_k = \frac{I'_k - I'_{med,k}}{I'_{max,k} - I'_{med,k}} \quad \text{for} \quad I'_k > I'_{med,k} \tag{5}$$

where I'_k is the kth inputs as available from the experimental measurements, $I'_{med,k}$ is the average value, $I'_{min,k}$ is the minimum and $I'_{max,k}$ the maximum value of the kth inputs.

4. Discussion of the Results

The preliminary results obtained for the 56-trays butane splitter tower are discussed and few issues associated with neural net process modeling are also addressed. We have trained the various neural network architectures using a data set obtained over a period of ca four days, which corresponds to 478 sampling actions (input/output relations). The sampling was carried out at a rate of one data set every 12 minutes. The data set available was deficient in the compositions of distillate, nevertheless, the net learning process was enough robust to recover the scarce measurements available for the top of the column.

We have observed that without any rescaling of the inputs, the nets performed very poorly, and very often, the attainment of convergence was prevented.

The predictions of a three layers MFFANN (10-2-4) with rescaled inputs are shown in Figure 2. The learning process attained convergence after ca 30000 iterations. The figure shows a good agreement between the model predictions and the experimental data set used for the learning stage. It is worth to point out that, in the distillate and bottom composition measurements, the available data sets are unbalanced, because for the set-up of the column, which does not allow the simultaneous acquisition of both the concentrations.

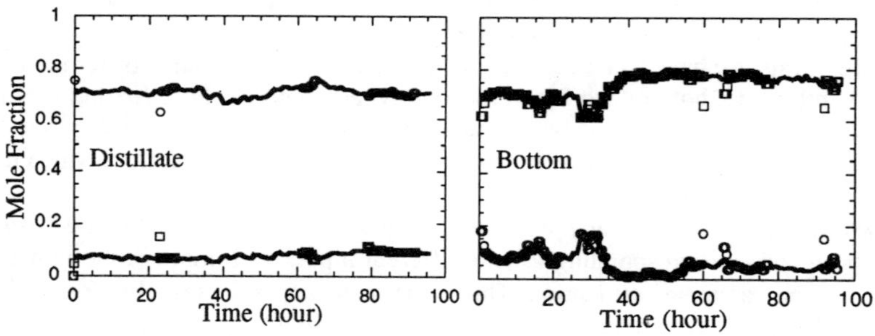

Figure 2: Comparison between the estimated (continuous line) and experimental data (*n*-butane, o *iso*-butane). MFFANN (10-2-4).

As pointed out above, the net is able to recover the top compositions, even though, the training data set was deficient. This is an indication of robustness of the learning process.

The results of a three layers MFFANN (10-4-4) are shown in Figure 3. It can be observed that changing the number of neurons of the hidden layer from 2 to 4 slightly affects the net performance, as it may be concluded from comparing Figure 4 with 3. The detected differences doesn't allow to discriminate between the two nets. Work is underway to investigate this point using nets with higher numbers of neurons and hidden layers.

We have also assessed the inferential capabilities of the developed MFFANN models by using a new set of experimental data which was not showed to the nets during the learning stage. The results are shown in Figures 4 and 5 where the predicted distillate and residue compositions of the nets (10-2-4) and (10-4-4) are compared with the experimental measurements. Both the nets exhibit good inferential capabilities (having both an average error of 4 %), and again, we cannot discriminate between the two architectures. The detrimental performance of the two nets detected at the initial data may be due to the fact that these data corresponded to feed charges containing a large amount of butylenes (about twice as much the one present in the data used for training).

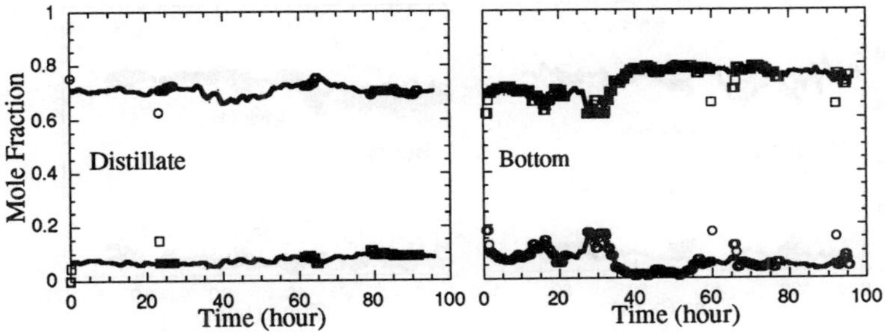

Figure 3: Comparison between the estimated (continuous line) and experimental data (*n*-butane, o *iso*-butane). MFFANN (10-4-4).

It is remarkable that the nets here considered have exhibited a good performance in predicting the dynamic behavior of the process under investigation. That is, the ANN-based estimators have shown a good capability in recovering the essential dynamic features of the process.

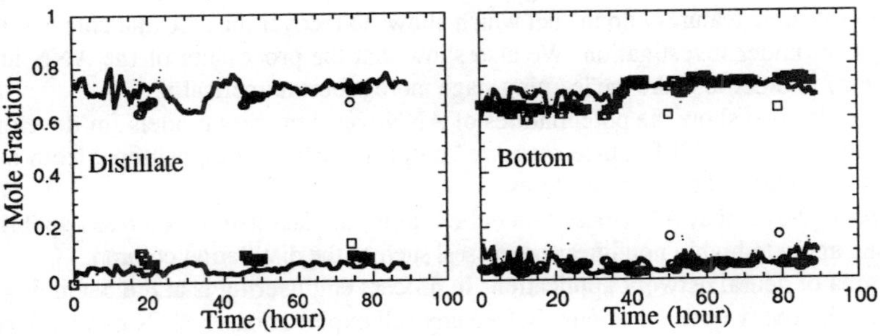

Figure 4: Comparison between the estimated (continuous line) and experimental data (*n*-butane, o *iso*-butane). MFFANN (10-2-4).

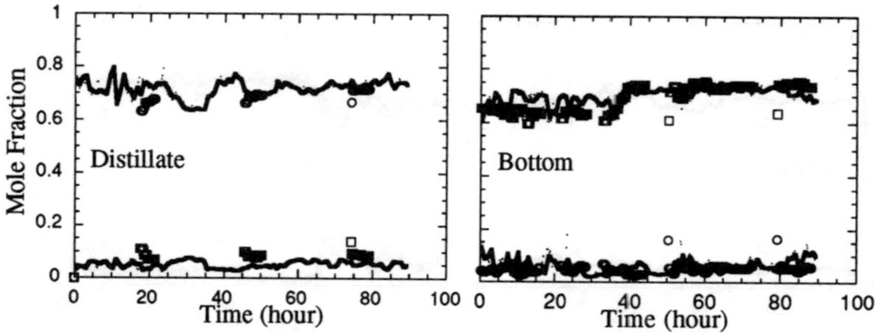

Figure 5: Comparison between the estimated (continuous line) and experimental data (*n*-butane, o *iso*-butane). MFFANN (10-4-4).

5. Conclusions

The preliminary results on the dynamic performance of ANN-based inferential estimators are rather encouraging. They also show that process modelling and intelligent control are promising areas for an extensive use of artificial neural network modelling.

The results indicate that embedding an ANN architecture within an estimator structure introduces dynamic features into the net which allows to recover the essential characteristics of the system under investigation. We also show that the processing of the ANN inputs requires care in order to make the learning stage more robust numerically.

The results also show the potentialities of ANN-based process models for developing "sotf sensors" to infer "difficult-to-measure" outputs, such as compositions, from easily measured variables, such as temperatures.

We have shown that ANN-based model exhibits satisfactorily predictive capabilities even when apply to highly non linear processes, such as the distillation column.

The area of neural network applications in process engineering is at the beginning and there are, still, many open questions. Before any full exploitation of ANN can be pursued some essential issues have to be addressed, such as: i) assessment of ANN stability and dynamics, ii) methodologies to generate the optimum net topology for the given system, iii) optimal parameter estimation.

References

1. P.H. Wallman, *Ind. Eng. Chem. Fundam.* **18** (1979) 327.
2. M. Van Dootingh, F. Viel, D. Rakotopara, J.P. Gauthier and P. Hobbes, *Computers Chem. Engng.* **16** (1992) 777.
3. E.B. Lynch and W.F. Ramirez, *AIChE J.* **21** (1975) 799.
4. R. Baratti, J. Alvarez and M. Morbidelli, *Chem. Eng. Sci.* **48** (1993) 2573.

5. J.T. Bialasiewicz, J.C. Proano and E.T. Wall, *IEEE Prodeedings Int. Symp. on Intelligent Control* (Albany, New York) (1989) 413.
6. J. Thibault, V. Van Breusegem and A. Cheruy, *Biotech. Bioeng.* **36** (1990) 1041.
7. K.J. Hunt and D. Sbarbaro, *IEEE Proceedings*, **D-138** (1991) 431.
8. T. Mejdell and S. Skogestad, *AIChE J.* **39** (1993) 1641.
9. M.J. Willis, C. Di Massimo, G.A. Montauge, M.T. Tham and A.J. Morris, *IEEE Proceedings*, **D-138** (1991) 256.
10. J. Hertz, A. Keogh and R.G. Paler, *Introduction to the Theory of Neural Computation* (Addison-Wesley, Redwood City, 1991).
11. J. Leonard and M.A. Kramer, *Computers Chem. Engng.* **14** (1990) 337.

A NEURAL NETWORK INFERENTIAL MODEL FOR CRYSTALLIZER DYNAMICS

MARCO BRAVI, ANGELO CHIANESE, CARLO MUSTACCHI, MASSIMILIANO SED
Dipartimento di Ingegneria Chimica, Università di Roma "La Sapienza"
Via Eudossiana 18, 00184 Roma, Italy

and

JIE ZHANG
Chemical Engineering Department
University of Newcastle upon Tyne
Merz Court, Newcastle upon Tyne NE1 7RU, United Kingdom

ABSTRACT

This work reports the application of an artificial neural network-based model (NNM) to estimate the average size of the crystal product from a crystallizer operating at dynamic state. The crystallization of potassium sulphate from aqueous solutions was used as case study. The input data sets to the NNM were generated by means of a phenomenological model consisting of solute mass and crystal population balance dynamics. The response of the system to pseudo random disturbances in the flow rate and concentration of the feed stream and in the rate of fines dissolution was considered. A fair estimate of the average crystal size was obtained from the adopted NNM when the time history of the magma density of fines was used as the input variable. The deviation between calculated and estimated values of the average crystal size was 2±1.5%, with a maximum error of 8%.

1. Introduction

In the operation of continuous crystallizers the main objectives are to maintain a high purity and a suitable size distribution of the product, often lumped into an *average size*. In practice, only the latter variable can be controlled on-line. The determination of the crystal size distribution is a hard task, due to the difficulties of locating a measuring device in a nonuniform suspension and of obtaining on-line measurements. Therefore, the control of the crystal size distribution from a crystallizer is usually accomplished indirectly either by pre-setting the input variables or by controlling an inferential variable such as the magma density of fines.

Aim of this work was to investigate using a neural network model as an estimator of the average size of a crystal product from a crystallizer.

This required, as a first step, the development and validation of a model for the dynamics of the crystallization unit operation and the generation of a large set of data covering the operating conditions of practical interest. Then a neural network model was selected, trained and validated. The sampling rate and the input time history length were suitably chosen.

2. Mathematical model

The crystallizer examined here is a CMSMPR (Continuous Mixed-Suspension Mixed-Product Removal) unit operating in cooling mode to produce potassium sulphate crystals from aqueous solutions. A dissolution device for the fines was included as the manipulated device to control both the amount of fines and the average size of the crystal suspension.

If we assume that the temperature in the crystallizer is held constant, the dynamic behaviour of the examined unit can be described by two balance equations relevant to the mass of the solute and the crystal population, i. e.[1],

$$V \rho_s \frac{\partial c}{\partial t} + V \frac{\partial M_T}{\partial t} = W c_{in} - W c - \frac{M_T V}{\tau} \tag{1}$$

$$\frac{\partial n(L,t)}{\partial t} + \frac{\partial [G(L,t)\, n(L,t)]}{\partial L} + \frac{n(L,t)}{\tau} = 0 \tag{2}$$

where $G(L,t)$ is the linear growth rate, dependent on the crystal size and τ is the mean residence time based on the inlet stream flow rate. Eq. (2) is based on the usual assumption that crystal breakage phenomena in the crystallizer can be neglected.

Fines dissolution, which concerns crystals smaller than a certain size (250 μm here) entails a difference in residence times between coarse crystals and fines.

Specifically, the residence time of fines, τ_f, decreases as the flow rate of fines dissolved W_f is increased, since:

$$\tau_f = \frac{V}{W + W_f} \tag{3}$$

A growth rate kinetic expression, reported by Chianese *et al.*[2], 1987), was used:

$$G(L) = 247.8\,\sigma^2 (1 + 5.87 \cdot 10^{-3}\, L) \tag{4}$$

Initial conditions for Eqs. (1) and (2) are the steady state conditions, with a further initial condition for Eq. (2) for crystal size, i. e., for $L \to 0$:

$$n_o = \frac{B_o}{G_o} \tag{5}$$

where G_o is the linear growth rate of nuclei and B_o is the nucleation rate, expressed as follows:

$$B_o = 5.6 \cdot 10^{19} (1 + 40\, M_t)\, \sigma^{9.1} \tag{6}$$

The order of heterogeneous nucleation in the solution bulk was assumed to be 9.1, practically equal to that proposed by Nývlt[3] for heterogeneous nucleation in the solution bulk.

Results for the steady-state runs have confirmed those reported in the literature[2]; the dynamic response, on the other hand, could not be confirmed because of the scarcity of data but its shape was qualitatively well reproduced.

The dynamic model described was implemented in the gPROMS package[4], a system designed for the simultaneos solution differential and algebraic equations (DAEs) in their physical unchanged form, without reducing these to ordinary differential equations (ODEs), through sparse matrix techniques and stiff integration methods.

3. The neural network. Structure and algorithm for parameter optimization

Artificial neural networks have been successfully applied to chemical process modelling recently[5 6]. The most widely used neural networks in nonlinear system modelling are multi-layer feed forward networks and radial basis function networks. Multi-layer feed forward networks generally use neurons with sigmoidal activation functions while RBF networks utilize RBF activation functions. Different types of neurons possess different information process capabilities and a neural network's information processing capability can be enhanced by incorporating different types of neurons into a single network. Real neural systems contain various types of neuron activation functions[7]. However, most of the current neural networks use only a single type of hidden neuron. It is observed that sigmodal activation function network have problems in approximating linear mappings unless a linear element is included[8]. Another issue in neural work modelling is how to decide the topology of a network, i.e. the number of hidden layers and hidden neurons in each hidden layer. This is often done through trial and error.

To tackle the above two issues, a novel type of neural network using mixed types of hidden neurons and its training algorithm are proposed by Zhang *et al.*[9]. The mixed node network is used in this paper to develop dynamic models for the estimation of average crystal size. In the following, the mixed nodes neural network and the sequential orthogonal training algorithm are briefly introduced. Details about these can be found in the referenced paper.

The network studied in this paper has one hidden layer. Three types of hidden neurons, linear neurons, sigmoidal neurons, and RBF (Gaussian) neurons, are incorporated in a single network. The output layer neuron is a linear neuron.

The network can be trained sequentially using a sequential orthogonal training algorithm. At the first training stage, the first hidden neuron is used to model the relationships between the inputs and outputs. At the second training stage, the second hidden neuron is introduced to model the relationships between the inputs and model residuals from the previous stage. Hidden neurons are sequentially added until the trade-off between model accuracy and complexity, expressed by Akaike's Information Criterion (AIC)[10], attained its minimum. When adding a hidden neuron, the new information it introduces is due to that part of its output vector which is orthogonal to the space spanned by the output vectors of the previously obtained hidden neurons. The classical Gram-Schmidt orthogonalisation method is used at each training stage to form a set of orthogonal bases for the space spanned by the output vectors of the hidden neurons. The hidden layer weights associated with each hidden neuron are found through optimization method such as the gradient descent method or the conjugate gradient method. The output layer weights associated with each hidden neuron are obtained using the least squares regression. At each stage of network training, all the three types of hidden neurons are tried and the one resulting in the minimum sum of squared model errors is selected. Through this sequential orthogonal training, different types of hidden neurons are gradually incorporated into a single network and the appropriate network size is discovered when AIC attains its minimum value.

4. Selection and validation of the neural network

The network selection and validation has been performed in successive steps. At first, we attempted to model the input-output relationship: $L_m = f(W, W_f, c_{in})$. Validation exhibited significant relative errors for such a model.

Successively, an attempt was made to model the relationship between an internal variable (M_{T_f}) and the average size of the crystal product (L_m) (see Fig. 1), since a high cross-correlation existed and M_{T_f} is a variable easily measured on-line.

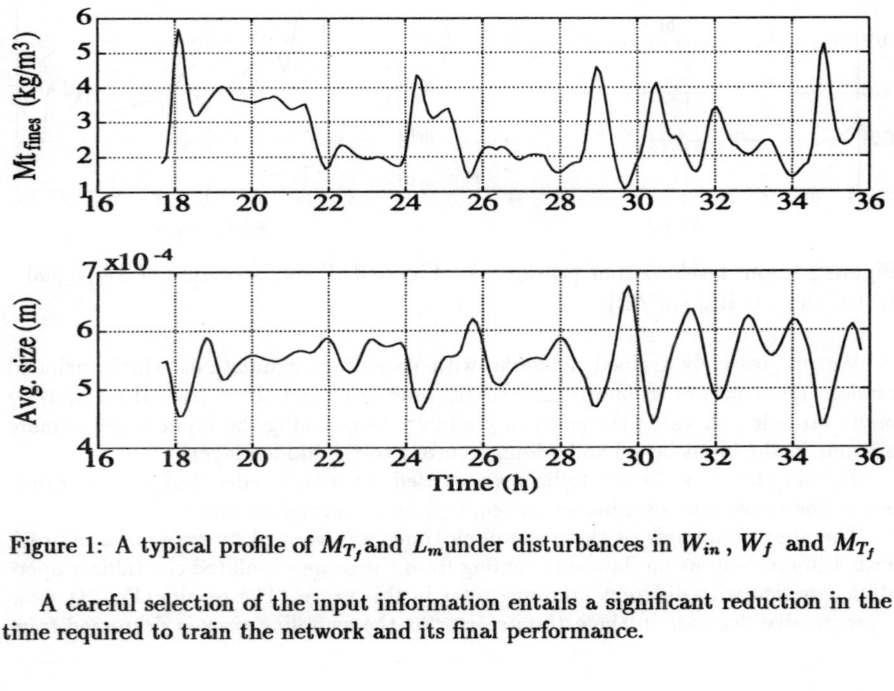

Figure 1: A typical profile of M_{T_f} and L_m under disturbances in W_{in}, W_f and M_{T_f}

A careful selection of the input information entails a significant reduction in the time required to train the network and its final performance.

A number of criteria can be applied in selecting the input information carriers. Here we have used a frequency analysis-based method. The frequency behaviour of the simulated crystallizer was analyzed by observing the power spectrum of the output produced by several types of disturbances: in particular, the output from a step disturbance and the output from a pseudo-random sequence (Fig. 2) were used.

The results show that a dominant frequency of about 5 rad/h characterizes the behaviour of the simulated crystallizer. The frequency spectrum then decays progressively to a plateau (Fig. 3). The highest frequency of interest was selected as that corresponding to a decay to 1% of the signal power of the dominant frequency, that is 10 rad/h. The sample rate utilized was 20 rad/h, i.e. twice the Nyquist frequency.

The choice of time history length for the input variables utilized was based on the criterion that, for a given waveform, a minimum of three samples is necessary to calculate three parameters (i.e., amplitude, phase and frequency). A total number of samples of at least three has therefore been used.

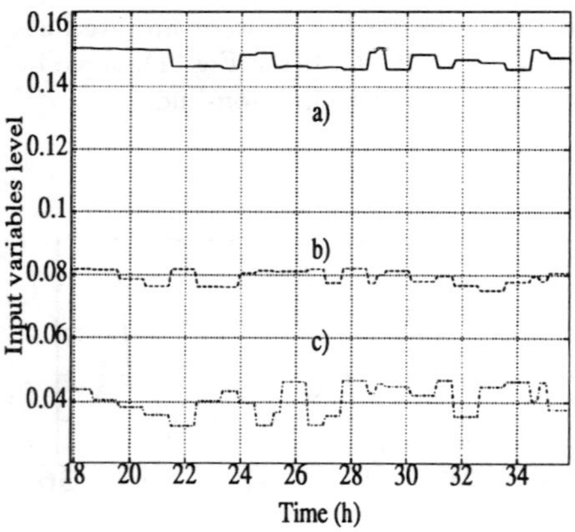

Figure 2: Input levels vs time [a): c_{in}; b): W_f(m^3/h); c): W_{in} (m^3/h)]

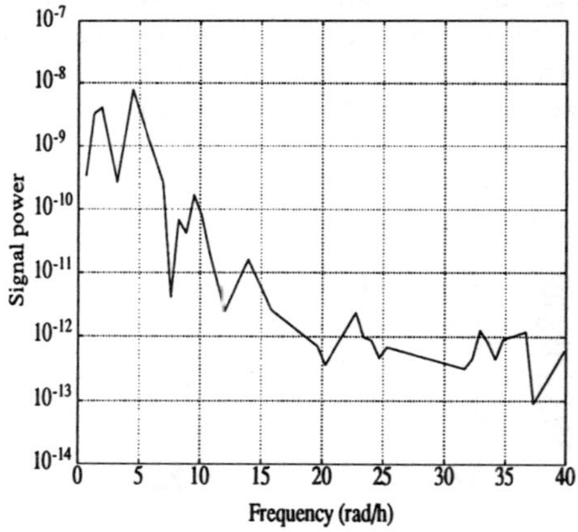

Figure 3: Power spectrum of M_{T_f}signal

We progressively trained networks with data from simulations which included random disturbances in one (c_{in}), two (c_{in} and W_f) and three (c_{in}, W_f and W_{in}) input variables, checking the effect of gradually lengthening the input history (more neurons in the input layer) and adding neurons to the hidden layer.

The simulation of the crystallizer was carried out with a cadence of variable printout in the range 5 to 48 minutes, as required by the sampling rate.

The appropriateness of these sampling times was verified by training a network with 4 input neurons on data sets coming from the same simulated crystallizer operation sampled with different sampling rates in this range. The results (Fig. 4) show a progressive decrease in the estimate error as the sampling time is decreased from

5' to 18.6'. Then the error is almost constant in the range 18.6' to 36' and increases again beyond this limit. It appears therefore that sampling times of 18.6' to 36' entail no difference in the quality of the estimate. It also appears that the frequency component given by the free damped oscillation of the system under a ramp disturbance remains the main component. However, the sampling frequency formerly selected was retained to improve the estimate in the neighbourhood of spikes.

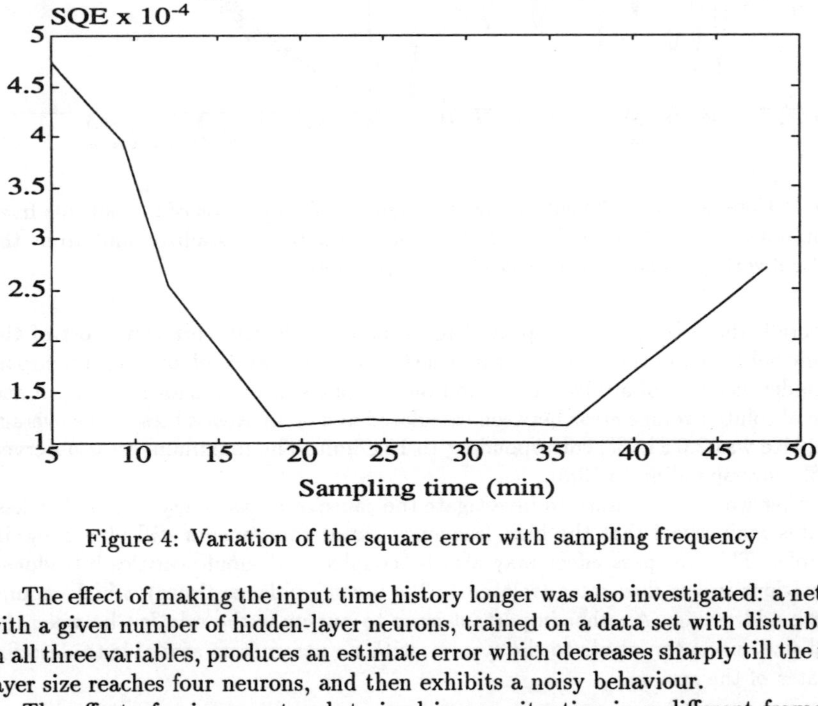

Figure 4: Variation of the square error with sampling frequency

The effect of making the input time history longer was also investigated: a network with a given number of hidden-layer neurons, trained on a data set with disturbances in all three variables, produces an estimate error which decreases sharply till the input layer size reaches four neurons, and then exhibits a noisy behaviour.

The effect of using a network trained in one situation in a different framework examined by carrying out forecasts with disturbances singled out from the complete set. This test was performed to analyze possible problems arising from disturbances of a different class from those used for the training and original validation, and to ascertain whether a central limit effect had simplified the classification task of the network by smoothing out singularities.

The network trained with disturbances in three and two variables was therefore validated with data sets with disturbances in two and one variables respectively. The estimate error (*reduced* data set origin) is smaller than the validation error (*complete* data set origin) for each network. As expected, a reverse test (network originally validated for simplex or duplex disturbances, tested on a triplex one) yielded, in most cases, poorer results.

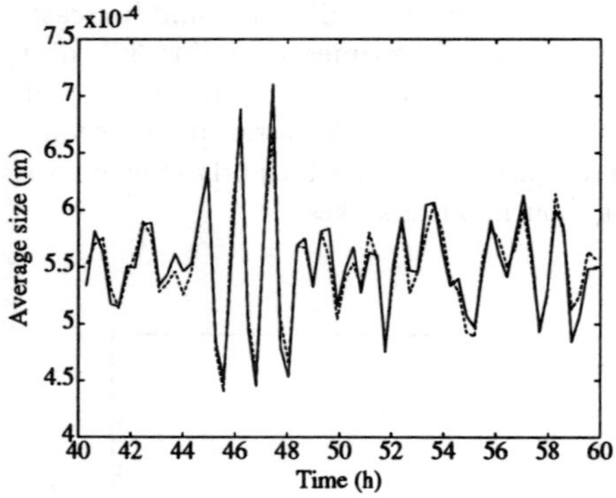

Figure 5: Comparison of the outputs from one simulated crystallizer and from the neural network (—: calculated value)

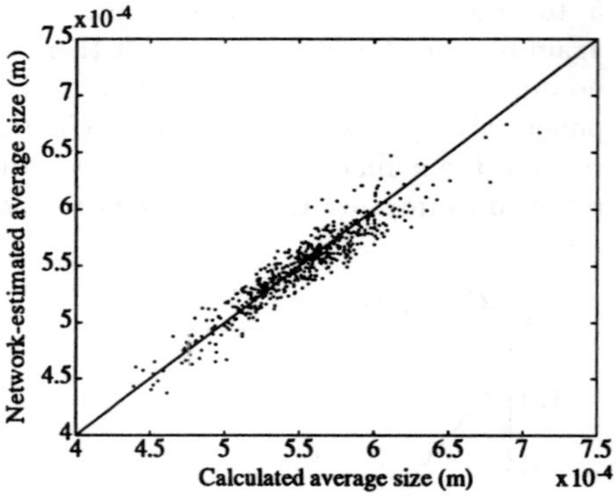

Figure 6: Comparison of the outputs from one simulated crystallizer and from the neural network

Overall, the neural network proved to be able to identify correctly most of the dynamic behaviour of the crystallizer with a structure composed of four input neurons, four hidden neurons of a mixed type, and one output neuron as shown in Fig. 5 and 6. The absolute average error between calculated and estimated values of the average crystal size was 2.0±1.5%, corresponding to 10±8μm. The maximum error observed was 8%, corresponding to 53μm.

Further work is necessary to investigate the cause of occasionally rounded spikes, since this may reveal that the high frequency signal components still play a significant role. This low-pass effect may also be coupled with nonlinearities introduced by variations in the flow rates (related to the mean residence times), which require separate addressing. Another important issue to pursue is the relation between the crystal size and outer inputs to the system, with no parameters available to describe the states of the system.

5. Conclusions

The output of a dynamic crystallizer model was used as input to a neural network in order to model the relationship between some easily measured variable and its output, i.e. the average size which is to be controlled.

The sampling rate and time history length were optimized and their effects on the final network performance were assessed; network structures with satisfactorily low estimate error were obtained for the case of simplex, duplex and triplex disturbance and the effect of training with a given set of disturbances carrying out estimates in environments with different classes of disturbances was evaluated.

6. Notations

B_o	Nucleation rate	$m^{-3}\,s^{-1}$
c	Concentration of the solute inside the crystallizer	$kg_{K_2SO_4}\,kg^{-1}_{solut.}$
c_{in}	Concentration of the inlet solution	$kg_{K_2SO_4}\,kg^{-1}_{solut.}$
G	Linear growth rate of crystals	$m\,s^{-1}$
L	Crystal size	m
$L_a v$	Average size of the crystal product	m
M_T	Magma density	$kg_{K_2SO_4}\,m^{-3}_{solut.}$
M_{T_f}	Magma density of fines	$kg_{K_2SO_4}\,m^{-3}_{solut.}$
n	Crystal population density	m^{-4}
T	Temperature of the crystallizer solution bulk	$^\circ C$
V	Crystallizer liquid holdup	m^3
W_{in}	Flow rate of the inlet stream	$kg_{H_2O}\,s^{-1}$
W_f	Flow rate of the stream to dissolution	$kg_{H_2O}\,s^{-1}$
ρ_s	Solution density	$kg_{solut.}\,m^{-3}_{solut.}$
σ	Relative supersaturation	
τ	Mean residence time in the crystallizer	s
τ_f	Residence time of the fines in the crystallizer	s

Acknowledgements

Financial support for the bilateral research project (University of Rome – University of Newcastle upon Tyne) by the Italian Research Council (CNR) is gratefully acknowledged.

We are also grateful to the Imperial College of London and Dr. C. C. Pantelides for permitting the use of the gPROMS package.

References

1. Sed, M., *Simulazione dinamica di un cristallizzatore continuo.* Graduation thesis, University of Rome "La Sapienza" (1993).
2. Chianese, A., Di Cave, S., Mazzarotta, B., *Quaderni dell'Ing. Chim. Ital.*, **23** (1987) p. 8.
3. Nývlt, J., Rychly, R., Gottfried, J. and Wurzelová, J. *J. of Crystal Growth* **6** (1970) pp. 151-161.
4. Barton, P. I., *The Modelling and Simulation of Combined Discrete/Continuous Processes.* PhD thesis, University of London, 1992.

5. Willis, M. J., C. Di Massimo, G. A. Montague, M. T. Tham and A. J. Morris. *Proceedings of IEE* **138 D** (1991) pp. 256-266.
6. C. Di Massimo, G. A. Montague, Willis, A. J., M. T. Tham and A. J. Morris, *Computers Chem. Engng.* **16** (1992) pp. 283-291.
7. Ballard, D. H. in *Vision, Brain and Cooperative Computation.* eds. M. Arbib and Hamson (MIT Press, 1988), pp. 563-621.
8. Holcomb, T. and Morari, M., *Computers chem. Engng.* **16** (1992) pp. 393-411.
9. J. Zhang, A. J. Morris, G. A. Montague and M. T. Tham. *IFAC ADCHEM'94, Advanced Control of Chemical Processes*, May 25-27, 1994, Kyoto Research Park, Kyoto, Japan.
10. Akaike, H. *IEEE Transactions on Automatic Control*, **19** (1974) pp. 716-723.

INTEGRALS OVER FRACTAL SETS AND SPATIAL DEFINITION OF MULTIFRACTALITY

MASSIMILIANO GIONA
Dipartimento di Ingegneria Chimica, Universitá di Cagliari
Piazza d'Armi, 09123 Cagliari, ITALY

ABSTRACT

Starting from the definition of integrals of continuous functions over fractal sets, a spatial representation of the multifractal properties of a measure is derived from the pointwise scaling of the Gauss-Fourier transform of the measure.

1. Introduction

The fractal characterization of invariant measures makes use of the multifractal formalism by defining the spectrum of generalized dimensions $D(q)$ or equivalently the $f(\alpha)$ spectrum[1,2]. For a fixed value of the exponent α, $f(\alpha)$ represents the fractal (Hausdorff) dimension of the subset of the measure support having α as Hölder exponent. If the $f(\alpha)$-spectrum collapses into a point, $\alpha = f(\alpha) = D_H$ (D_H being the Hausdorff dimension of the set), the measure has no multifractal character and is a regular Hausdorff measure.

In practical application to chaotic attractors and random fractals, estimation of multifractal spectra involves evaluation of the sums

$$\sum_i p_i(\varepsilon)^q \sim \varepsilon^{(q-1)D(q)}, \qquad \varepsilon \to \infty , \tag{1}$$

i.e. of the probability measure $\{p_i(\varepsilon)\}$ at the lengthscale ε for values of the parameter q ranging on the real line $q \in (-\infty, \infty)$. Evaluation of the sums (1) may become problematic with the increasing of the phase-space dimensionality, as discussed by many authors[3,4] in connection with box-counting methods of evaluation of the fractal dimension of chaotic attractors.

In this paper a different approach to multifractality is proposed. In particular, two points are addressed: the pointwise (spatial) characterization of the multifractal spectrum and the possibility of obtaining information about the multifractal structure of a measure without making use of box-counting techniques to evaluate $\{p_i(\varepsilon)\}$. These results are based on the definition of integrals of continuous functions over fractal sets and on the possibility of evaluating these integrals by means of temporal averages (ergodic hypothesis). The examples considered refer to onedimensional measures generated by means of Iterated Function Systems[5].

2. Integrals over fractal sets of continuous functions

The definition of integrals over fractal sets can be derived from the definition of Hausdorff sums and Hausdorff measures[6]. Let $\mathcal{C} \subset E^d$ (E^d is the Euclidean d-dimensional space) be a measurable Hausdorff set with Hausdorff dimension D_H and $f(x)$ a continuous function $f : E^d \to E^d$. The integral of f over $\mathcal{C}$, which can be called D_H-integral of f, $I^{D_H}(f,\mathcal{C})$, can be defined as

$$I^{D_H}(f,\mathcal{C}) = \lim_{\delta\to 0} \inf_{\{A_i\}_\delta} \sum_i f(\xi_i)|A_i|^{D_H} , \tag{2}$$

where $\{A_i\}_\delta$ is a δ-covering of $\mathcal{C}$, $|A_i|$ the diameter of the set A_i, and ξ_i a generic point $\xi_i \in A_i$. If $f(x) = 1$, equation (2) coincides with the Hausdorff D_H-measure of $\mathcal{C}$, $\text{mis}_{D_H}(\mathcal{C})$, i.e.

$$I^{D_H}(1,\mathcal{C}) = \text{mis}_{D_H}(\mathcal{C}) . \tag{3}$$

If $\text{mis}_{D_H}(\mathcal{C})$ is normalized to unity (probability measure), Eq. (2) represents the average $\overline{f}$ of f with respect to the uniform Hausdorff measure defined on $\mathcal{C}$, $\overline{f} = I^{D_H}(f,\mathcal{C})$. Hereafter probability measures will be always considered.

Definition (2) is formal and cannot be directly applied to practical problems deriving from the experimental measurements of chaotic attractors or random fractal structures. In the usual applications, we have at our disposal a time series $\{x_i\}$ associated with the evolution of a system having the set $\mathcal{C}$ as a limit attractor. From the knowledge of a time series, the usual algorithm of chaotic analysis can be applied in order to estimate generalized dimensions, Ljapunov exponents, entropies and all the other parameters characterizing both the dynamic and the geometric properties of the system and of its limit attractor $\mathcal{C}$. For this reason, hereafter we assume that the system generating the time series $\{x_i\}$ is ergodic. In practical terms, this means that the temporal average $< f >$ of any continuous function f

$$< f >= \lim_{N\to\infty} \frac{1}{N} \sum_{i=1}^{N} f(x_i) \tag{4}$$

coincides with the corresponding ensemble average of f, $\overline{f}_\mu$, evaluated with respect to the unique invariant ergodic measure μ associated with the time series $\{x_i\}$. It should observed that $\overline{f}_\mu$ is equal to $\overline{f}$ if μ is a uniform Hausdorff measure, i.e. if, for any $\mathcal{A} \subseteq \mathcal{C}$, $\mu(\mathcal{A}) = I^{D_H}(1,\mathcal{A})$.

3. Moment analysis of onedimensional measures

Information on the multifractal scaling of a given measure μ can be obtained from the analysis of the (ensemble or temporal) average of a sequence of continuous functions $\{f_n\}$. In order to develop this concept, let us consider some simple onedimensional measures which are invariant ergodic measures of Iterated Function

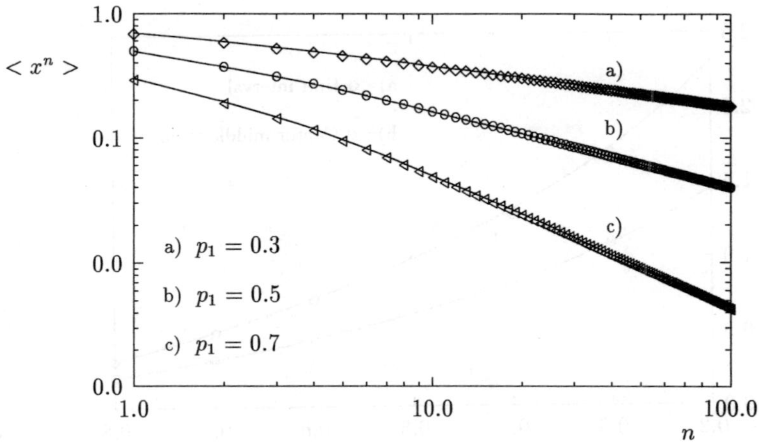

Figure 1: Behaviour of the moment hierarchy $< x^n >$ vs n associated with the invariant measure μ of the IFS (5) for different values of p_1; $l_1 = l_2 = l = 3$ and the limit set $\mathcal{C}$ is the Cantor middle third.

Systems (IFS) on the unit interval [5,7]. The class of IFS considered is represented by the random dynamical system $x_{n+1} = \phi(x_n, p_1)$

$$\phi(x, p_1) = \begin{cases} f_1(x) = x/l_1 & \text{probability}\ ,\ p_1 \\ f_2(x) = (l_2 - 1 + x)/l_2 & \text{probability}\ ,\ p_2 = 1 - p_1\ , \end{cases} \tag{5}$$

with $1/l_1 + 1/l_2 \leq 1$. The last condition ensures that the system converges towards a limit set $\mathcal{C} \subseteq [0,1]$ and that the points $x = 0$ and $x = 1$ belong to $\mathcal{C}$. The Hausdorff dimension D_H of $\mathcal{C}$ is the solution of the trascendental equation $l_1^{-D_H} + l_2^{-D_H} = 1$ and the multifractal properties can be readily obtained in closed form since the system (5) represents a binomial multiplicative process on the line.

For $l_1 = l_2 = 2$, $\mathcal{C} = [0,1]$; and for $l_1 = l_2 = 3$, $\mathcal{C}$ is equal to the Cantor middle third set. In order to simplify the analysis, let us further consider the simple case of equal scaling factors, $l_1 = l_2 = l$, which corresponds to self-similar limit sets $\mathcal{C}$. Since $x = 1$ belongs to $\mathcal{C}$, a convenient sequence of continuous functions can be represented by the powers $f_n(x) = x^n$, $n = 0, 1, ...$ The resulting averages, $\{< x^n >\}$, represent the moments associated with the measure μ (for self-affine IFS, the moment hierarchy can be obtained in closed form, see Appendix). With the increasing of n, $f_n(x)$ tends to be significant only in a neighbourhood of $x = 1$.

Figure 1 shows the result of the scaling of $< x^n >$, in the case of the Cantor middle third, for different values of the probability weight p_1. The temporal averages were performed over an asymptotic time series of $N = 10^6$ points generated by the IFS, Eq. (5).

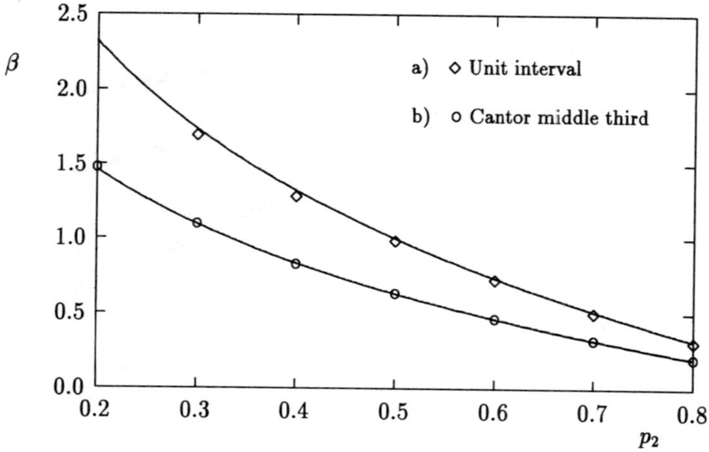

Figure 2: Behaviour of the exponent β vs p_2: a) Unit interval; b) Cantor middle third. The solid lines represent Eq. (7) and the points are the numerical results obtained from the scaling of the moment hierarchy, Eq. (6).

For $n \to \infty$, $< x^n >$ scales as

$$< x^n > \sim \frac{1}{n^\beta}, \qquad n \to \infty , \tag{6}$$

β being a function of the probability weights (p_1, p_2) and ultimately of the measure structure. In the case of $l = 2$, $p_1 = 1/2$, $< x^n > = (1+n)^{-1}$. From simple scaling arguments, it follows that

$$\beta = -\frac{\log p_2}{\log l} . \tag{7}$$

Eq. (7) derives from the observation that, with the increase of n, the interval in which x^n is less than an arbitrary positive constant ε is given by $[0, \varepsilon^{1/n}]$, and therefore for small ε,

$$< x^n > \simeq \mu([1 - \varepsilon^{1/n}, 1]) \sim \mu([1 - 1/n, 1]) . \tag{8}$$

To an interval of length $1/n$ close to $x = 1$, there corresponds the invariant measure

$$\mu([1 - 1/n, 1]) \simeq p_2^{n^*} , \quad \text{with} \quad 1/n \simeq 1/l^{n^*}. \tag{9}$$

Eq. (7) is recovered by solving the last scaling relation for n^* and inserting the result into the previous equation.

Figure 2 shows the values of the exponent β as a function of p_2 evaluated numerically from the scaling of the first $n = 100$ moments $< x^n >$, compared with Eq. (7) in the case of the unit interval ($l = 2$) and of the Cantor middle third ($l = 3$). The errors in the numerical estimate of β are within 5 percent. To conclude, if the measure is a regular Hausdorff measure, the exponent β is equal to the Hausdorff (fractal) dimension; if the invariant measure exhibits multifractal features, the exponent β is equal to the Hölder exponent in the neighbourhood of $x = 1$.

4. Gauss-Fourier transform and pointwise scaling

In the previous section we have seen that the scaling of the averages (ensemble averages with respect to the measure μ or temporal averages) of a sequence of continuous functions makes it possible to obtain information about the properties of a measure μ. The choice of the moment hierarchy $\{< x^n >\}$ is justified by the fact that in the examples considered $x = 1$ belongs to the support of the measure.

Nevertheless, moment hierachy does not enable us to obtain a complete description of the local (pointwise) behaviour of the Hölder exponent α. For a generic measure, different function sequences should be considered in order to achieve a pointwise characterization of the singular structure of the measure.

In particular, an appropriate choice could be represented by Gauss functions. Let $x_o \in \mathcal{C}$ and λ be a positive integer. The *Gauss-Fourier transform* $\mathcal{G}(x_o, \lambda)$ of the measure in the point $x = x_o$ is then defined as

$$\mathcal{G}(x_o, \lambda) =< \exp(-\lambda|x - x_o|^2) >= \int_{\mathcal{C}} \exp(-\lambda|x - x_o|^2) d\mu(x) . \tag{10}$$

For sufficiently large λ

$$\mathcal{G}(x_o, \lambda) \sim \frac{1}{\lambda^{g(x_o)}} , \quad \lambda \to \infty , \tag{11}$$

and the exponent $g(x_o)$ depends on the position x_o. The relationship between $g(x_o)$ and the pointwise Hölder exponent $\alpha(x_o)$ is given by

$$2g(x_o) = \alpha(x_o) , \tag{12}$$

which reduces to $2g(x_o) = D_H$ uniformly in the case of regular Hausdorff measures.

Figures 3-6 show the behaviour of the exponent $\alpha(x_o)$ evaluated from the scaling of the Gauss-Fourier transform, as a function of the position x_o, for the binomial multiplicative measures on the unit interval discussed in the previous section for different values of p_1. Of course, an highly heterogeneous distribution of $g(x_o)$ is an indicator of the multifractal nature of the measure. The set of points $\{x_{o,i}\}$ $(i = 1, .., N_x)$ at which $\alpha(x_o)$ is evaluated is randomly selected with a distribution induced by the invariant measure μ.

The simplest way to characterize the spatial distribution of $\alpha(x_o)$ is to analyze its mean and variance $< \alpha >_\mu$, $\sigma^2_{\alpha,\mu}$, defined as

$$< \alpha >_\mu = \frac{1}{N_x} \sum_{i=1}^{N_x} \alpha(x_{o,i}) , \quad \sigma^2_{\alpha,\mu} = \frac{1}{N_x} \sum_{i=1}^{N_x} [\alpha(x_{o,i}) - < \alpha >_\mu]^2 . \tag{13}$$

The subscript μ indicates that, according to the random selection of the points $\{x_{o,i}\}$ induced by the invariant measure, these quantities are averaged with respect to μ.

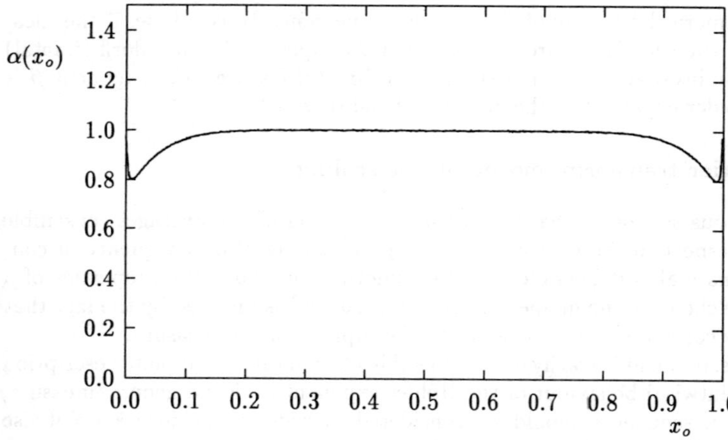

Figure 3: Pointwise distribution of $\alpha(x_o)$ vs x_o: $l = 2$, $p_1 = 0.5$.

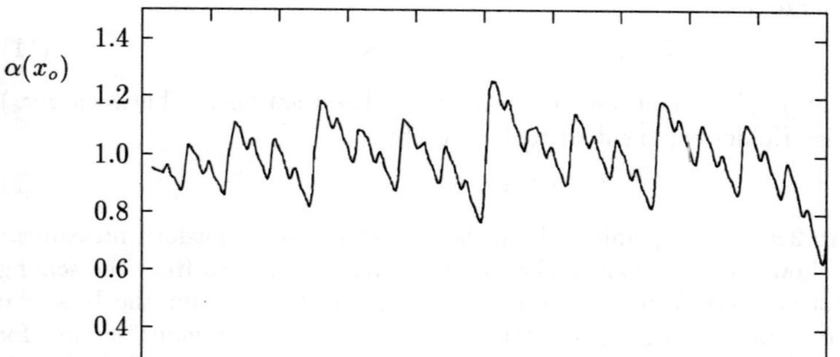

Figure 4: Pointwise distribution of $\alpha(x_o)$ vs x_o: $l = 2$, $p_1 = 0.4$.

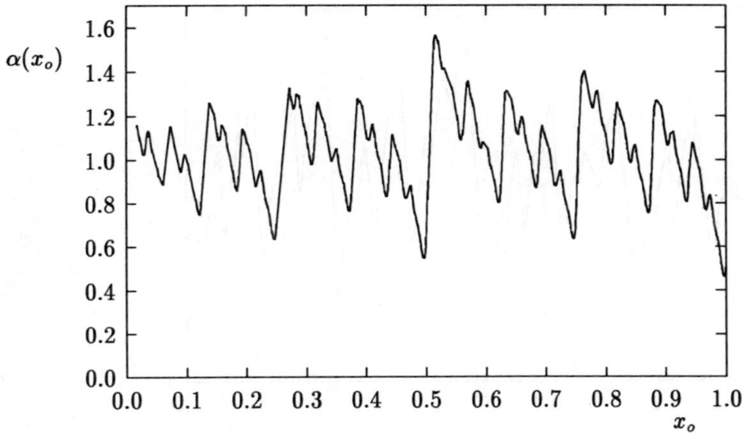

Figure 5: Pointwise distribution of $\alpha(x_o)$ vs x_o: $l = 2$, $p_1 = 0.3$.

However, it should be borne in mind that the number of points N_x employed in estimating $< \alpha >_\mu$ and $\sigma^2_{\alpha,\mu}$ is finite ($N_x = 10^3$) and that these quantities therefore represent an approximation of the ensemble average and variance of the Hölder exponent $< \alpha(n) >$, $< \sigma^2_{\alpha(n)} >$, evaluated at a finite lengthscale or equivalently at a finite order of iteration n of the multiplicative process. Therefore, in order to compare the numerical results deriving from the scaling of the Gauss-Fourier transform with the theoretical predictions, we require the order of iteration n of the multiplicative process corresponding to the lengthscale considered. The order of iteration n can be approximately obtained by considering the average distance between consecutive points point $d_o =< |x_{o,i+1} - x_{o,i}| >$ is related to n through the relation $d_o \simeq 1/l^n$. From the value of d_o deriving from numerical simulation, the following values of n were obtained: $n = 10$ for the unit interval, and $n = 6$ for the Cantor middle third. The theoretical predictions for these statistical quantities can be obtained in the case of binomial multiplicative processes by considering that the distribution of the Hölder exponent $\alpha(k, n)$ at a given order of iteration n is given by

$$\alpha(k,n) = -\frac{k}{n}\frac{\log p_1}{\log l} - \left(1 - \frac{k}{n}\right)\frac{\log p_2}{\log l}\ , \quad k = 0, 1.., n\ , \tag{14}$$

with probabilities

$$p_n(k) = \frac{n!}{k!(n-k)!}\ . \tag{15}$$

Correspondingly, the theoretical mean $< \alpha(n) >$ and variance $< \sigma^2_{\alpha(n)} >$ of α at a

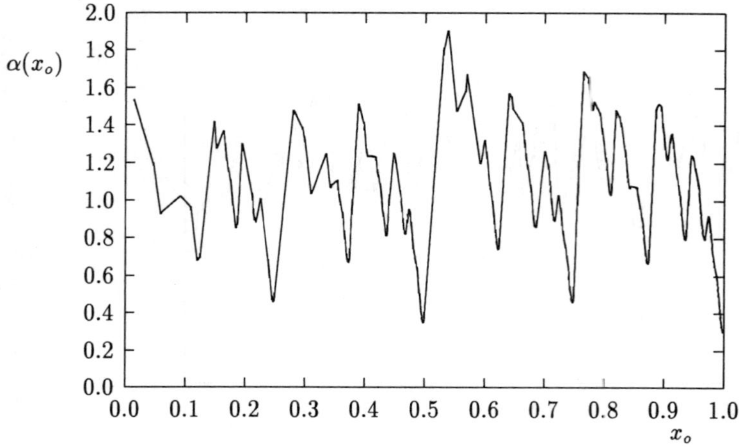

Figure 6: Pointwise distribution of $\alpha(x_o)$ vs x_o: $l = 2$, $p_1 = 0.2$.

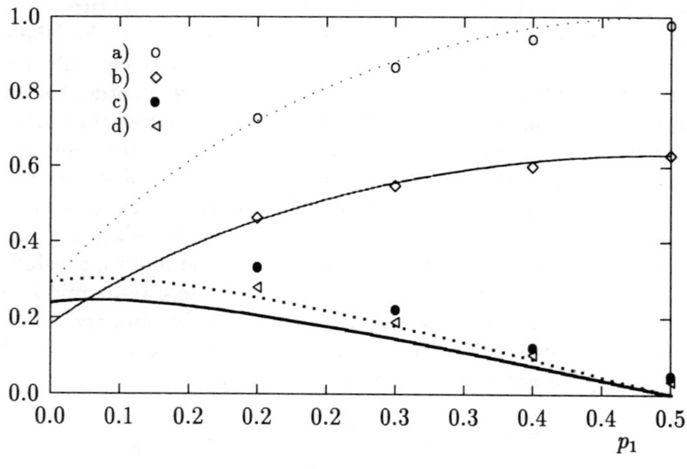

Figure 7: Mean Hölder exponent $< \alpha >_\mu$ and variance $\sigma^2_{\alpha,\mu}$ vs p_1 for the unit interval and Cantor middle third. The lines are the theoretical results, Eq. (16), obtained from the expressions for $\alpha(k,n)$ and $p_n(k)$, Eq. (14)-(15); the points the simulation values. a) $< \alpha >_\mu$ unit interval; b) $< \alpha >_\mu$ Cantor middle third; c) $\sigma^2_{\alpha,\mu}$ unit interval; d) $\sigma^2_{\alpha,\mu}$ Cantor middle third.

given order of interation n is given by

$$< \alpha(n) >= \sum_{k=0}^{n} p_n(k)\alpha(k,n) \; , < \sigma^2_{\alpha(n)} >= \sum_{k=0}^{n} p_n(k)[\alpha(k,n) - < \alpha(n) >]^2 \qquad (16)$$

and can be compared with the numerical results, figure 7. As can be observed, the agreement is fairly satisfactory, especially for the averages $< \alpha >_\mu$. Indeed, the averages $< \alpha(n) >$ do not depend on the iteration n and are an intrinsic property of the distribution whereas, on the contrary, the variance $< \sigma^2_{\alpha(n)} >$ decreases with the order of iteration n of the multiplicative process.

5. Concluding remarks

A spatial approach to multifractality has been proposed. The method outlined in this article is based on the scaling of the Gauss-Fourier transform associated with a given measure. Because multifractal properties of a measure are obtained from the scaling of a pointwise function $< \exp(-\lambda|x - x_o|^2) >$, the intrinsic problems underlying the application of box-counting methods can be overcome and the accuracy of the estimate rests exclusively on the number of points used in obtaining temporal averages. All the analysis presented is valid for ergodic measures, i.e. on condition that the temporal and ensemble averages coincide. It has been also shown that the results obtained from the scaling of the Gauss-Fourier transform can be interpreted in terms of the usual multifractal formalism.

The results obtained in this article for simple onedimensional measures can be extended to more complex measures associated with chaotic attractors and random fractals. In particular, another way to characterize multifractal measures is to look at the properties of the invariant orthonormal polynomial system associated with the measure[8,9]. Finally, it should be noted in passing that the present analysis can be also applied in connection to the multifractal characterization and filtering of images[10]. in signal processing and in fluid dynamics (analysis of turbulence).

Acknowledgments

The author wishes to thank B. Forte for pointing out the closed form result discussed in the Appendix and for other useful discussions.

Appendix

Where the IFS consists of affine transformations

$$w_i(x) = a_i x + b_i \; , \quad \text{probability } \; p_i \qquad (i = 1, .., N) \; , \qquad (17)$$

the moments hierarchy associated with the invariant measure can be obtained in closed form, as shown by Forte and Vrscay[11].

Let M be the Markov operator associated with the IFS,

$$M[\nu] = \sum_{i=1}^{N} p_i \nu \circ w_i^{-1} , \tag{18}$$

where ν is a measure defined on the compact metric space X in which the IFS is defined. By making use of the property that the invariant measure μ is a fixed point of the Markov operator

$$M[\mu] = \mu , \tag{19}$$

for the moment hierarchy,

$$< x^n >= m_n = \int x^n d\mu(x) , \tag{20}$$

the following recursive relation can be derived

$$\left[1 - \sum_{i=1}^{N} p_i a_i^n\right] m_n = \sum_{k=1}^{n-1} \binom{n}{k} \left[\sum_{i=1}^{N} p_i a_i^k b_i^{n-k}\right] m_k , \tag{21}$$

which can be solved starting from $m_0 = 1$.

References

1. T. C. Halsey, M. H. Jensen, L. P. Kadanoff, I. Procaccia and B. I. Shraiman, *Phys. Rev.* **A 33** (1986) 1141.
2. B. B. Mandelbrot, *J. Fluid. Mech.* **62m** (1974) 331.
3. H. S. Greenside, A. Wolf, J. Swift and T. Pignataro, *Phys. Rev.* **A 25** (1982) 3453.
4. Hao Bai-lin, *Elementary Symbolic Dynamics* (World Sci., Singapore, 1989).
5. M. Barnsley, *Fractals Everywhere* (Academic Press, New York, 1988).
6. K. J. Falconer, *The Geometry of Fractal Sets* (Cambridge Univ. Press, Cambridge, 1985).
7. A. Lasota and M. C. Mackey, *Chaos, Fractals and Noise* (Springer-Verlag, Berlin, 1994).
8. M. Giona, F. Lentini and V. Cimagalli, *Phys. Rev.* **A 44** (1991) 3496.
9. L. Colace and A. R. Giona, in this book.
10. J. Levy Vehel and J-P. Berroir, in *Fractals in Natural and Applied Sciences*, ed. M. M. Novak (Elsevier, Amsterdam, 1994), 261.
11. B. Forte and E. R. Vrscay, *Appl. Prob.* (1994), in press.

Author Index

Subject Index